The Philosophy of Manufactures

Documents in American Industrial History
Michael Brewster Folsom, general editor

Volume 1
The Philosophy of Manufactures: Early Debates over Industrialization in the United States, edited by Michael Brewster Folsom and Steven D. Lubar, 1982

Volume 2
The New England Mill Village, 1790–1860, edited by Gary Kulik, Roger Parks, Theodore Penn, 1982

The Philosophy of Manufactures

Early Debates over Industrialization
in the United States

Edited by Michael Brewster Folsom and Steven D. Lubar

The MIT Press
Cambridge, Massachusetts
London, England

Merrimack Valley Textile Museum
North Andover, Massachusetts

© 1982 by The Massachusetts Institute of Technology

All rights reserved. No part of this book may be reproduced in any form or by any means, electronic or mechanical, including photocopying, recording, or by any information storage and retrieval system, without permission in writing from the publisher.

This book was set in VIP Optima, Palatino, and Baskerville by DEKR Corporation and printed and bound in the United States of America.

Library of Congress Cataloging in Publication Data
Main entry under title:

The Philosophy of manufactures.

(Documents in American industrial history; v. 1)
Bibliography: p.
1. United States—Industries—History—Sources.
I. Folsom, Michael. II. Lubar, Steven D. III. Series.
HC105.P55   338.0973   81-12334
ISBN 0-262-06076-0   AACR2

Contents

General Editor's Preface
xv

Acknowledgments
xvii

Introduction
xix

Benjamin Rush

*Speech to the United Company of
Philadelphia for Promoting American
Manufactures (1775)*
3

Lord Sheffield

*Observations on the Commerce of the
American States (1783)*
11

Thomas Jefferson

The Present State of Manufactures (1785)
15

Thomas Jefferson

Letters (1785–1816)
19

Tench Coxe

Address to an Assembly Convened to Establish a Society for the Encouragement of Manufactures and the Useful Arts (1787)
33

"A Plain, but Real Friend to America"

Three Letters on Manufactures (1787)
63

John Morgan

Agriculture Preferable to the Mechanic Arts (1789)
71

Tench Coxe

A Brief Examination of Lord Sheffield's Observations on the Commerce of the United States (1791)
75

Alexander Hamilton

Report on the Subject of Manufactures (1791)
81

Alexander Hamilton

Prospectus of the Society for Establishing Useful Manufactures (1791)
95

George Logan

Letters Addressed to the Yeomanry of the United States (1792)
103

Tench Coxe

Observations on the Letters of George Logan (1792)
111

John Beal Bordley

Intimations on Manufactures (1794)
117

George Logan

*Constitution of the Lancaster County
Society for Promoting of Agriculture,
Manufactures, and the Useful Arts (1800)*
123

David Humphreys

*A Poem on the Industry of the United
States of America (1804)*
129

Robert Southey

Espriella's Letters (1807)
143

George W. P. Custis

*An Address on the Importance of
Encouraging Agriculture and Domestic
Manufactures (1808)*
149

Charles Brockden Brown

Address on the Utility and Justice of Restrictions upon Foreign Commerce (1809)
157

Henry Clay

Speech on Domestic Manufactures (1810)
167

James Mease

Address to the Cultivators, the Capitalists, and Manufacturers (1811)
173

Thomas Cooper

Prospectus of the Emporium of Arts and Sciences *(1813)*
187

Daniel Webster

Debate on Repeal of the Embargo (1814)
195

American Society for the Encouragement of
Domestic Manufactures

*Address to the People of the United States
(1817)*
199

James Swan

*Address on the Question for an Inquiry into
the State of Agriculture, Manufactures and
Commerce (1817)*
225

John Taylor of Caroline

Tyranny Unmasked (1822)
241

Henry Niles

Morality of Manufactures (1823)
251

Thomas Cooper

The Disadvantage of Machinery (1823)
255

Thomas Carlyle

Signs of the Times (1829)
259

Edward Everett

Fourth of July at Lowell (1830)
281

Timothy Walker

Defence of Mechanical Philosophy (1831)
295

Alexander Everett

*Memorial of the New York Convention of
the Friends of Domestic Industry (1831)*
305

American Quarterly Review

Review of The Results of Machinery
(1832)
317

Zachariah Allen

The Practical Tourist (1832)
331

John P. Kennedy

Address to the American Institute (1833)
345

John Adolphus Etzler

Paradise within the Reach of All Men (1833)
353

Andrew Ure

The Philosophy of Manufactures (1835)
365

Isaac Hill

*Message to Both Houses of the New
Hampshire Legislature (1836)*
385

Daniel Webster

*Lecture before the Society for the Diffusion
of Useful Knowledge (1836)*
389

Henry Hubbard

Gubernatorial Address (1842)
403

Henry David Thoreau

Paradise (to be) Regained (1843)
411

John Greenleaf Whittier

The City of a Day (1845)
421

William Gregg

Essays on Domestic Industry (1845)
427

Daniel Webster

Opening of the Northern Railroad (1847)
437

Ralph Waldo Emerson

Wealth (1856, 1860)
447

General Editor's Preface

To most students, teachers, and the general public the industrial origins of the United States remain largely unknown. Aside from a few textbook cliches about "fathers" and "birthplaces" of the "Industrial Revolution," we learn little about this central aspect of our past. Scholarly literature does exist, but primary materials are not easily available. Teaching industrial history too often resembles teaching Shakespeare with only secondary studies for texts, learning about the works from what the critics quote and scholars argue. Conventional historiography, when it does touch on the industrial past, often disappoints. Social, cultural, economic, political historians—even labor and business historians—often do not understand the technology itself. Historians of technology, with significant exceptions, traditionally have confined their study to machines abstracted from the human origins and consequences of technical innovation.

Just why this should be so—why the most highly industrialized and technology-dependent society should ignore the very source of its world power—is a cultural and intellectual problem well worth study in its own right. What concerns us more immediately is one practical consequence of this problem and one practical step toward its solution. Much of the documentary record in social, cultural, and political history is available to the nonspecialist in published collections. This is not true for industrial history. The abundant primary materials remain, for the most part, in archives. Our aim is to stimulate inquiry into these materials by publishing a series of collections which draw upon archival resources, make available the most important and representative documents, and demonstrate the ways in which greater familiarity with the industrial record can illuminate our material past.

This volume of the series provides the intellectual and ideological groundwork for study of the rise of modern industry in the New World, the early debates over "the encouragement of manufactures," which began at the time of the Revolution and continued in various guises through the middle of the nineteenth century. In one way or another, this national argument raised questions about the very fundaments of our civilization and probed a matter Americans no longer find themselves at liberty to debate: whether the United States should be an industrial nation.

The next three volumes cover the rise and demise of the first characteristic manufacturing centers in the United States, the mill villages and industrial cities of New England. Volume 2 contains the documentary history of the New England mill village. Volumes 3 and 4 cover the history of urban textile manufacture. The first of these two volumes documents the "Yankee Era" up to the Civil War, when most of the "operatives" were native born farm women; the other deals with the period of maturity and decline, from the Civil War to the Great Depression.

Anticipated steps beyond these four volumes are documentary histories of other major industries of the region: paper, glass, and especially the shoe industry, which was second only to textiles. There are also plans for volumes on the coal industry, the iron and steel industry, and the industrialization of the South.

Note: When portions of the text have been deleted from the documents, an asterisk appears in the margin. Note numbers in the margin refer to editorial comments, which appear at the end of the documents.

Acknowledgments

The editors wish to thank the following libraries for allowing us to photocopy documents for publication in this volume: Free Library of Philadelphia; Harvard University libraries, including Baker Library, Houghton Library, Kress Library, and Widener Library; Library Company of Philadelphia; MIT libraries; Old Sturbridge Village Research Department; and the University of Chicago Library.

Introduction

The achievement of political independence in America coincided with the beginnings of modern industrial capitalism. Ever since, the careers of manufacturing and democracy have intermeshed. Former colonists could look at England not only as an old oppressor but also as a new manufacturing nation, where machines produced goods with remarkable economy, speed, and profit. At the time when the American states were first trying out new democratic institutions and ideas, the new social and economic structures of the British manufacturing city were plain to see, as were the cultural, aesthetic, and moral consequences of factory life. Although the material advantages of manufacturing were obviously great, Americans were of several minds about the opportunities and dangers before them, and for decades after independence an open and fundamental debate continued as to whether or just how the factory system should come to these shores.

Strange to our ears, statesmen and patriots could argue then that manufacturing in the modern sense had no place among a free and virtuous people. Early sponsors of industry, for their part, often felt called upon to justify themselves elaborately with arguments and projects shaped to respond to strong opposition. Just how firmly rooted and persistent was the debate over manufactures may be seen in a passage from a little book called *Lowell, As it Was, and As it Is,* by the Reverend Henry A. Miles. This early retrospective account of the premier American industrial city was published in 1845, twenty years after the Lowell mills paid their first generous dividend and almost three quarters of a century after independence:

Lowell has been highly commended by some, as a model community, for its good order, industry, spirit of intelligence, and general freedom from vice. It has been strongly condemned, by others, as a hot-bed of corruption, tainting and polluting the whole land. . . . We are destined to be a great manufacturing people. The influences that go forth from Lowell, will go forth from many other manufacturing villages and cities. If these influences are pernicious, we have great calamity impending over us. Rather than endure it, we should prefer to have every factory destroyed; the character of our sons and daughters being of infinitely more importance than any considerations "wherewithal they shall be clothed."[1]

In Miles's hands, the debate had atrophied to rhetoric. His suggestion that manufacturing might be abandoned—indeed, "destroyed"—was disingenuous. As the passage suggests, Miles's book firmly endorses our manufacturing "destiny." But the fact that he still found it even rhetorically advantageous to acknowledge the anti-industrial argument suggests the tenacity of that point of view.

The debate over America's industrial "destiny" that Miles echoed had its origin in the earliest manifestations of American thought and social policy and its elaborations in every sector of the American community. The two conflicting but interwoven points of view articulated in the early debate over manufactures are still recognizable in our continuing discussions of just what the good life in the New World might be.

On one side was the argument for what proponents universally called the "encouragement of manufactures." This argument starts from, among other places, the Protestant assumption that industry, in its several senses, is good for us. Social welfare as well as individual salvation is served by hard work.[2] David Humphreys, one of the first American manufacturers, put the old notion in verse:

From Industry, the sinews strength acquire,
The frame dilates, the bosom feels new fire.
Unwearied INDUSTRY pervades the whole,
Nor lends more force to body than to soul.
Beyond all other aid, this Pow'r alone,
Gives to man's character a manlier tone,
Exalts the purpose, dignifies the mind,
And adds unconquer'd firmness to mankind.[3]

Such sentiments tallied with projects that demanded labor from the idle for the purposes of social control, material progress, and private gain. Attitudes and arrangements that might people a wilderness or keep village widows, orphans, and the landless off the public dole and out of trouble might also flourish in a cotton mill.[4] Thus the first industrialists, in the modern sense of the term, found approval outside their modest ranks for the enforced hierarchies of factory production and the long hours of factory labor.

The encouragement of manufactures in America was also the result of the periodic scarcity of imported goods and the failure of household production ever to supply the needs of the entire community. As early as 1640, the General Court of the Massachusetts Bay Colony was taking measures to make sure children learned how to spin and weave.[5] Other colonial legislatures and town councils did likewise from time to time, sometimes granting bounties for the production of cloth. The proprietors of many new towns offered special encouragement to persons who would contract to build the sawmills and gristmills necessary to the expansion of agrarian society. There was well-founded suspicion that such governmental sponsorship of manufactures favored private economic interests, especially as a result of experience with the manipulative mercantilist policies of the British government. Nevertheless, the idea that the state had a legitimate and necessary role in guiding productive economic activity was well established, and it survived the Revolution intact. It did not have to be invented to serve the new manufacturing interest.[6]

By the end of the eighteenth century, when large-scale industrial concentrations became a possibility, Americans were not strangers to the character of city life or to the motives of private economic interest that flourished there.[7] They had heard entrepreneurial and civic virtues praised often.[8] Benjamin Franklin's Poor Richard and his "Way to Wealth," for example, came out of the "pre-industrial" world of urban commerce but presented a firm ideological basis for industrial entrepreneurship.[9] At the same time, strong traditions of American thought and social behavior laid a firm basis for principled opposition to industrialization.

The arguments against manufactures varied widely. None put it more eloquently or with more blithely unexamined assumptions than Thomas Jefferson, who observed in 1785 that "Those who labor in the earth are the chosen people of God, if ever He had a chosen people, whose breasts He has made His peculiar deposit for substantial and genuine virtue."[10] Much more modestly, almost half a century later, a young English immigrant "just stepping into the world" looked forward to his American prospects in a letter to an uncle:

> . . . dont you think farming the best, and surest way of getting a living? Manufacturing is a very unsteady business, somtimes up, and somtimes down, some few gets Rich, and thousands are ruined by it. Rogues, Rascals, Knaves and vagabonds are connected with it. Some persons that you trade with will cheat you in spite of your teeth, and you must cheat others in return to make ends meet and tie. In Short no honnest man can live by it. A Factory too, is liable to be burnt down, but a Farm cannot be easily burnt up. Manufactoring breeds lords and Aristocrats, Poor men and slaves. But the Farmer the American farmer, he, and he alone can be independent, he can be industrious, Healthy and Happy. I am for Agriculture.[11]

The benefits of agrarian life often might seem enhanced, as in this letter, by the presumed faults of industrial life, not the least of which was the threat of urban disorder. Jefferson argued that manufacturing was bound to bring with it a new class of dependent urban workers that would dislocate the economy, social structure, and political assumptions of a fledgling democracy. The noxious condition of European factory centers—Manchester, England, the most notorious—was evident before the first factory was conceived in the New World. Antipathy to the prospect of such blight was felt even by Americans most deeply committed to industrial projects. Zachariah Allen, propagandist for American manufacturing and factory proprietor himself, had qualms about the consequences of industrialization that amounted to downright terror: "God forbid," he wrote, ". . . that there ever may arise a counterpart of Manchester in the New World."[12]

Well established to resist the promotion of manufactures, in addition to our broadly agrarian tradition, was that aspect of Protestant belief which found economic self-interest as great a threat to the community as idleness.[13] From the beginning, democratic opinion was suspicious of the concentrations of wealth and power private industry required and created. Men and women who had recently fought for liberty from the king could look at the new industrial capitalist and find a "purse proud aristocrat," a "Tory in disguise."[14] Americans were wedded to the self-sustaining family farm as the norm of social and economic organization and the locus of public virtue. Such a people had an understandably difficult time accommodating the system of production, labor, and rewards that came with the factory. Industrial production radically altered the accustomed link between work and value—between the quantity of work invested and the economic value realized and between the quality of work done and the value of self-esteem. In this new "factory system," initiative, autonomy, and skill began to be replaced by passive attention, external discipline, and a predetermined, uniform product. Industrial labor also disrupted traditional sex roles and family relations, among other ways, by providing opportunities for independent income to, or more efficient exploitation of, wives, children, and sisters.[15]

The established class of commercial men—bankers, merchants, and shippers—sometimes took one side of the debate, sometimes the other. These were the people who controlled the capital, economic expertise, and access to markets that made American industrialization possible. Most of the first manufacturers in the United States were drawn from this class, but significant opposition to domestic manufactures did arise among commercial people. One reason was inertia. Conservative merchants had made their way to wealth and distrusted innovation.[16] When Francis Cabot Lowell organized the Boston Manufacturing Company in 1813, he is reported to have been warned by Brahmin well-wishers against such a "visionary" scheme.

Other reasons for opposition to domestic manufactures among men of commerce had a more material base in the realities of trade. Merchants who did well importing English goods had little reason to encourage domestic production, especially if that meant increasing the price and thus shrinking the market. Especially in the matter of the protective tariff, which was often the key proposal for encouraging American industry, those with commercial interests were divided among themselves.[17]

Just how great a change industrialization represented to those most intimately affected by it is a matter of much debate.[18] Indeed, doubt has been cast on whether the "industrial revolution" seemed at all revolutionary when it was happening. What looks to us like a major break in historic periods, a fundamental change in the means of production, as well as in the forms of social and economic organization, could look to many of its initiators, at least in the United

States, more like a shift in emphasis, a modest improvement, a reform nestled into the traditional forms of life and work. This was especially true during the first several decades of the debate over manufactures, when resistance led proponents of industry to adopt forms of rhetoric, architecture, and social organization that would minimize the sense that they were working some drastic change upon the nation.[19]

An important aspect of the response to industrialization is the terms in which people stated their perceptions. In the documents reprinted in this volume, the words "manufacture" and "industry" were still tied to their root meanings, "hand work" and "hard work," and could be used without reference to power machine production. A "manufacturer" was someone who made things by hand, not one who employed others to work machines. When Benjamin Rush, in the address reprinted in this collection, called for inviting "manufacturers" and establishing "manufactories" in Philadelphia in 1775, he had in mind arrangements under which people who worked spinning jennies and looms by hand for a living were gathered in a central workplace. Such production arrangements were similar to those of modern industry in capitalist societies. Workers used equipment owned by the proprietor of the place. But this pattern was at least as old as the sixteenth century and distinctly "pre-industrial," in that hand craft skill, rather than mechanical power, was the basis of production.

When, on the other hand, the name of the workplace came to be abbreviated to "factory," the term seems often to have connoted something quite new—distinctly a place of powered machine production. Well into the nineteenth century, people spoke of the "factory system" to distinguish it from other systems of production. The Englishman Andrew Ure, whose *Philosophy of Manufactures* (1835) was a vigorous and influential apology for young industrial capitalism, insisted precisely that a "factory" was a distinct innovation:

> The term *Factory,* in technology, designates the combined operation of many orders of work-people, adult and young, in tending with assiduous skill a system of productive machines continuously impelled by a central force. . . . Some authors, indeed, have comprehended under the title *factory,* all extensive establishments wherein a number of people co-operate towards a common purpose of art. . . . But I conceive that this title, in its strictest sense, involves the idea of a vast automaton, . . . acting in uninterrupted concert for the production of a common object, all of them being subordinated to a self-regulated moving force.[20]

Ure here employs the term "technology" in an early sense: the study of practical and industrial techniques. During the period of these debates, what we call "technology" would have usually been called the "mechanic arts," or simply the "arts." These terms suggest the tendency (or desire) to conceive new industrial realities as though they were but variations on an older world of artisan labor.

The role of manufactures in the development of the United States was discussed not only as a social and moral issue but also as an economic one, of course. The debate took place against a background of the economic beliefs of the time. Throughout the documents reprinted in this volume, one finds appeal to the three major schools of economic thought that were influential in the later eighteenth and early nineteenth century: the Physiocrats, the followers of Adam Smith, and the neomercantilists or Nationalists.

The Physiocrats were an important influence on Jefferson, John Taylor of Caroline, and others who believed in the primacy of agriculture. The Physiocratic school, founded by François Quesnay in France about 1750, argued that agriculture was the only source of wealth and that manufactures merely changed the form, not the value, of things.

In his *Wealth of Nations* (1776), Adam Smith emphasized free trade and described how "the invisible hand" of self-interest could bring about a harmonious world. Smith believed commerce and agriculture to be of equal importance. His follower and popularizer J. B. Say (1767–1832) elevated manufactures to the same level. The *Wealth of Nations* was the bible of writers who favored free trade and believed that the government should not interfere in economic matters.

Opposed to the followers of Adam Smith were the neomercantilists or Nationalists, who emphasized the importance of the state in the economy. Alexander Hamilton and Mathew Carey were the most prominent of this school in the United States. They both favored government protection and sponsorship of industry.

Each of the documents in this collection reflects to some degree the economic beliefs of its author. We exclude, however, theoretical economic writings, which are not pointedly concerned with the "philosophy of manufactures." American economic theory has been much discussed and reprinted elsewhere.[21]

The materials collected in this volume have been well considered by cultural and intellectual historians but are widely scattered and not easily accessible.[22] This collection means to encourage further independent inquiry by reprinting key documents in the debate over industrialization. From the vast literature on the topic, we have chosen those documents that discuss *whether* to encourage manufactures. *How* to do so is another debate that cannot be ignored, for practical measures were often a central concern in theoretical discussions. Documents that treat only specific measures without reference to the larger issues are excluded from this collection, but, in their outlines, these measures need to be kept in mind.

The protective tariff (or simply "protection") was the most important practical measure. It was proposed, among other reasons, because capital and labor were more expensive in the United States than in England because they were more scarce. American goods could only compete if the price of imported goods were forced up to an equal (or greater) level. Just how high a tariff should be was almost more vexing an issue than whether it should be imposed at all. The appeal of the tariff was much enhanced by the fact that it added to the federal treasury at the same time that it encouraged manufactures.[23] Opposition to the tariff was not always identical with opposition to the factory system itself. In addition to merchants, who wanted to keep their costs low and their markets wide, artisans and small farmers also opposed measures that would increase the price of manufactured goods. Southern planters opposed tariffs for their own purposes. The encouragement of an American cotton textile industry might seem to have been to the cotton growers' advantage, but they already had an established market overseas. And they had no desire to promote the economic and political power of the North, where nascent industry was centered.[24]

Accumulation of capital was another area in which legislation could encourage industry. In the immediate post-Revolution decades, the United States was "capital poor." Commercial fortunes were but reluctantly enticed into highly speculative industrial investments.[25] On occasion state legislatures authorized special lotteries, with the proceeds going to the capitalization of a manufacturing experiment. Legislatures could also grant land, water rights, and the right of eminent domain, by which private parties could gain property and other resources desired for industrial development.[26] This was often an area of direct confrontation between agrarian and industrial interests. Farmers who found their fields flooded for mill ponds and their pastures transected by railroads tended to oppose such developments.[27]

One of the most important innovations that came with manufacturing in the United States was the invention of the private industrial corporation, which was another means of legislative encouragement of manufactures. The capital requirement of the new industries quickly became so large and the sources of capital so diverse, including many investors not directly involved in the management of the concern, that the simple partnership would no longer suffice as the form of private economic organization. The corporation, in which a group is constituted an independent body, a virtual person at law, had long existed, but only as a public body gathered for public purposes, such as a town. After the Revolution, entrepreneurs increasingly turned the corporate form of organization to their private advantage. They persuaded legislatures to grant them corporate privileges for their projects, which they argued were quasi public in nature, a benefit to the prosperity and security of the state and the nation. Corporations then pressed for further advantage, gaining (as in Massachusetts in 1829) the privilege of "limited liability," so that no individual member of a corporation could be held responsible for the debts of the corporation. This measure greatly increased the ability of corporations to attract investors.[28]

Another practical measure of importance to industry was the patent. Originally a patent was any privilege granted by government document, as in *letters patent,* "patent" meaning open, public. A government might grant arbitrary patents amounting to monopolies, as the English government had done for centuries, allowing trading companies absolute right to do business in certain goods or at certain places. Much suspicion of monarchic abuse lingered as the idea of patenting inventions matured. Under just what conditions a person might be granted sole right to exploit a machine process was not easy to determine. Given the fact that the primary inducement and reward for industrial development was money, new industrial interests exerted pressure to establish safeguards for the investment of time and capital in technological innovation.[29] Innovation itself then became an industrial product, which could only be marketed if the law reserved the rights to private parties. The first great American industrial corporation, the Boston Manufacturing Company, made a significant early profit selling other textile companies rights to the machine patents it held.[30]

Legislation to ease the difficulty of finding workers for early manufacturing enterprises was not common. Generally manufacturers devised their own means of getting and keeping labor. On several early occasions, however, exemptions from taxes or militia duty were offered industrial workers. There was also some effort, both public and private, to encourage skilled industrial workers and proprietors to immigrate from England.[31]

The debate over whether the United States should industrialize shades into and involves a number of other matters that are not easy to ignore. Early labor protest, for instance, often raised fundamental questions about the propriety of manufacturing. When the working women of Lowell called their first recorded strike in 1834 and denounced management as "Tories in disguise," they therewith excluded factory owners, rhetorically at least, from the American political consensus as firmly as Jefferson would originally have excluded factories from American shores. But we have not included documents of labor protest in this collection. They are more appropriate among documents on the early history of industrial life in America, which will appear in subsequent volumes in this series.

One particularly intriguing line of dissent that we also leave for later volumes is the matter of who should make the decisions and hold the power of early modern industry in democratic America. This consideration, recently opened anew by Anthony F. C. Wallace, cannot easily be ignored in the debate our volume does record. As Wallace puts it:

By hindsight, it may appear that because large-scale manufacturing required great capital . . . existing sources of private capital would necessarily take control. But in the late eighteenth and early nineteenth centuries, before industrial capitalism became the dominant economic institution, there were many thoughtful people who still saw alternative ways of using the new machinery. Much of the social conflict in England and America in the early nineteenth century can be regarded as a struggle—a brief struggle, to be sure—for control of the machinery of the Industrial Revolution.[32]

Another source of anti-industrial argument appears later in the nineteenth century in the form of Southern attacks on Northern capitalism. As the moral and political pressure on chattel slavery grew in the 1840s and 1850s, numerous Southern spokesmen observed the inhumanity of "wage slavery" and railed against the peculiar institutions of the factory city. Apologists for slavery could sound like Marxists in their attack on capitalism. In Northern industry, one Southern propagandist argued, "the profits, made from free labor, are the amounts of the products of such labor, which the employer, by means of the command which capital or skill gives him, takes away, exacts, or 'exploitates' from the free laborer."[33] But this was not really an argument against manufactures. Its author meant only to insist that industrial labor was, no less than slave labor, a "trade in flesh." His aim was to undermine the moral basis of Northern abolitionism, not to abolish industrial capitalism. Though such tracts can be adduced in opposition to manufacturing, they do not properly belong among the early debates over its development in the United States.

One diverging strand of argument that takes rise from the debates we document is the broadly literary critique of industrialism, starting with Thomas Carlyle's influential essay "Signs of the Times" (1829) and located largely, but not entirely, in the American Transcendentalist movement. Carlyle moved into the mainstream of American debate when his essay was answered at length by Timothy Walker, writing in the *North American Review* on "The Defence of Mechanical Philosophy."[34] Carlyle's influence on mid-nineteenth-century American writing can hardly be underestimated. His critique of modern mechanism is reflected especially in the work of Thoreau. Melville's story, "Tartarus of Maids," is its most explicit manifestation in fiction. Thoreau and John Greenleaf Whittier bring John Adolphus Etzler's machine utopianism into the debate. Orestes Brownson's attack on industrial capitalism in 1840 (reprinted in volume 3 of this series) is as probing a critique as any before the rise of American Marxism. And, among Transcendentalist literati, Emerson went furthest to accommodate his philosophy to the powers of the factory system. Too often literary materials—even ones bearing so heavily on the central facts of our civilization—are studied in isolation from the less graceful but more practically influential realms of opinion and argument from which most of our documents arise. We believe that the power of the literary sensibility and imagination to grasp untoward portents and latent grandeur in the industrial machine can be better appreciated in tight proximity to the political, economic, ideological, and ethical discourse that was its immediate contemporary public context.

Carlyle, Etzler, and the American Transcendentalists tend to consider the machine, often figuratively, abstracted from industrial work sites and from the social, cultural, and aesthetic character of such sites. In fact, the literary materials illustrate especially well how the process of industrialization was sometimes distinguished from the process of mechanization and the progress of technological innovation per se. Some of those who opposed the promotion of manufactures responded with pleasure to the ingenuities of machinery. George Logan and John Taylor of Caroline, whose attacks on industrial interests are included in this volume, were participants in the widespread movement to promote technological innovation in agriculture. Jefferson's delight in the ingenious practicalities of nonindustrial mechanism is well known. John Calhoun, strong opponent of Northern manufacturing interests in Congress, leavened that opposition with "feelings little short of enthusiasm" for the machinery of industrial progress. What liberated his rhetorical fancy was abstraction of industrial processes and methods from the work site and "mere pecuniary interests." What he could talk about with pleasure were the "mechanic arts" and their "progress" in conquering the "great agents of nature":

> I now am, and ever have been decidedly friendly to them, the manufacturing interests, though I cannot concur in all of the measures which have been adopted to advance them. I believe, considerations higher than any question of mere pecuniary interest, forbids [sic] their use. But subordinated to these higher views of policy, I regard the advancement of mechanical and chemical improvements in the arts with feelings little short of enthusiasm; not only, as the prolific source of national and individual wealth, but as the great means of enlarging the domain of man over the material world. . . .[35]

Proponents of "labor-saving" machinery, for their part, could quite ignore the industrial origins and consequences of such mechanism. Mark Twain's *Connecticut Yankee* is a particularly good example of the possibility of imagining an entire machine civilization without acknowledging its industrial base and urban character. Most of the American industrial utopians from Etzler to Bellamy did likewise. Such abstract discussions of mechanization, however, often touch on important aspects of the larger historical process—especially on the means, both literal and figurative, of the shift from craft work to industrial labor in the experience of the individual. When Carlyle spoke of "the living artisan" being "driven from his workshop to make room for a speedier, inanimate one," he was talking about a "Mechanical Age," not an "industrial" one, and he focused on the point of contact between person and machine: "The shuttle drops from the fingers of the weaver, and falls into iron fingers that ply it faster." Still, that is some part of what the coming of urban industrial life meant.

The documents that record the American debate over the encouragement of manufactures do not nicely organize themselves by theme, date, place, or author. We begin with early calls for American industrial independence and move forward historically as evenly as possible, breaking chronological sequence when necessary to group closely related materials. For instance, the Thomas Jefferson documents spanning more than thirty years are clustered to reveal his changing opinions.

Most of the documents assume that the promotion or discouragement of manufactures is a national concern, vital to the prosperity or dangerous to the character of the whole people. But the provenance of these documents shifts perceptibly. The earlier materials in this collection, from the Revolution to the time of the War of 1812, are largely works of the Mid-Atlantic region. Most of the documents after 1815 or so derive from New England. This is representative of what happened in the economy and culture of the nation during these years. Between the War of 1812 and the Civil War, New England became the industrial center of the country. It was easy to assume that when one celebrated (or denounced) New England manufactures, one was talking about the nation. The locus of debate became all the more tightly provincial because New England also assumed the role of American cultural center. During the middle decades of the nineteenth century, when the United States gained a clear literary voice, most important literary intellectuals lived within a very few miles of the Boston counting houses whence American industrial power largely emanated. Any gathering of contributions to the national debate over manufactures inevitably draws heavily from this local source.

We are aware that, read serially, this collection of documents may seem not to constitute a debate at all but rather the record of overwhelming opinion in favor of the inevitable. The arguments against the promotion of manufactures, mechanization, the private corporation, "wage slavery," and the industrial city lack intellectual, programmatic, and political coherence. No credible alternative emerged. Opponents often foundered on illusions about the viability of agrarian values and did not astutely judge the vast powers of fledgling industrial institutions. They often based their arguments on narrow interests and petty anxieties. Read sympathetically, however, these anti-industrial documents may lead to an appreciation of the extent to which industrial capitalism did arouse opposition. In the face of the growing economic and political power behind manufactures, this opposition stressed ethical, humanitarian, and democratic concerns that have not lost their appeal. Perhaps the most persuasive aspect of the argument against manufactures is implicit in the extent to which proponents of industry had to mitigate their proposals and hew their arguments in response to their opposition. The rhetorical stratagems of Tench Coxe, the "pastoral" vision of industrialist Zachariah Allen, and the much advertised benevolence of the Lowell corporations are testimony to the conditions America exacted for its early accommodation to the industrial machine.

What our volume asks in resurrecting these opposing documents is this: How might things have been done differently? How might the social benefits of modern manufacturing have come to the United States without the record of recklessness and brutality the industrial corporation has stacked up to balance its vast material accomplishments? How might it yet be done otherwise? In the conclusion of her 1931 history of the early New England cotton manufacture, Caroline Ware asked the ultimate questions posed by the introduction of modern industry as private enterprise:

Could political democracy encompass industrial autocracy, could it harbor a working class and a moneyed power and survive? Hamilton, who had cared most for industry, had cared least for democracy. These problems which New England faced before 1860 have confronted other American communities as one by one they have experienced the process of industrialization. Their solution still lies in the future.[36]

Their solution *still* lies in the future, and we hope that this book will make some contribution toward it.

Notes

1. Reverend Henry A. Miles, *Lowell, As it Was, and As it Is* (Lowell, Mass., 1845), p. 2.

2. The seminal examination of the Protestant work ethic is, of course, Max Weber's *The Protestant Ethic and the Spirit of Capitalism,* trans. Talcott Parsons (New York: Scribner's, 1958). For a discussion of Weber's ideas in an American context, see Stuart Bruchey, *The Roots of American Economic Growth, 1607–1861: An Essay in Social Causation* (New York: Harper & Row, 1965), pp. 42–48. On the vicissitudes of the work ethic in a later period, see Daniel T. Rodgers, *The Work Ethic in Industrial America, 1850–1920* (Chicago: University of Chicago Press, 1978).

3. David Humphreys, *A Poem on Industry. Addressed to the Citizens of the United States of America* (Philadelphia, 1794), pp. 17–18.

4. See, for example, Gary B. Nash, "The Failure of Female Factory Labor in Colonial Boston," *Labor History,* volume 20, number 2 (Spring 1979), pp. 165–188.

5. William R. Bagnall, *The Textile Industries of the U.S., Including Sketches of Cotton, Woolen, Silk and Linen Manufactures in the Colonial Period* (Cambridge, Mass., 1893), volume 1, pp. 4–8.

6. For a discussion of the extent to which mercantilist policies survived the Revolution, see William Appleman Williams, "The Age of Mercantilism; An Interpretation of the American Political Economy, 1763 to 1828," *William and Mary Quarterly,* 3rd Series, volume 15 (1958), pp. 419–487.

7. Of course the United States did not become a predominantly urban society quickly or without anxiety. The point to stress is only that the city was not simply a creature and consequence of industrialization, even in the United States. For a good study of the development of civic values in America, see Thomas Bender's *Towards an Urban Vision* (Lexington, Ky.: University Press of Kentucky, 1975).

8. For the intellectual traditions out of which the entrepreneurial ethic grew, see Albert O. Hirschman, *The Passions and the Interests: Political Arguments for Capitalism Before Its Triumph* (Princeton, N.J.: Princeton University Press, 1977).

9. See Franklin's compendium of entrepreneurial aphorisms in "The Way to Wealth," *Papers of Benjamin Franklin,* ed. William Wilcox (New Haven: Yale University Press, 1963), volume 7, pp. 326–350. Among Franklin's contribu-

tions to American arguments for the promotion of manufactures is "On the Laboring Poor," *Papers,* volume 15, pp. 103–107.

10. Thomas Jefferson, "The Present State of Manufactures," Query 19, *Notes on the State of Virginia* (Philadelphia, 1788), first American ed. The entire statement on manufactures in Jefferson's *Notes* is reprinted in this volume. For a comprehensive treatment of the agrarian social ideal, see Rex Burns, *Success in America: The Yeoman Dream and the Industrial Revolution* (Amherst: University of Massachusetts Press, 1976). The indispensable literary studies of American pastoralism are Henry Nash Smith, *The Virgin Land* (New York: Vintage, 1972), and Leo Marx, *The Machine in the Garden* (New York: Oxford University Press, 1964), especially chapter 3. On the history of the idea that "farming as a family enterprise is the backbone of democracy," see A. Whitney Griswold, *Farming and Democracy* (New York: Harcourt, Brace, 1948).

11. Jabez Hollingworth, letter to William Rawcliff, 8 November 1830, from *The Hollingworth Letters: Technical Change in the Textile Industry, 1826–1837,* ed. Thomas W. Leavitt (Cambridge: MIT Press and the Society for the History of Technology, 1969), p. 93. The Hollingworths were a family of English mechanics, and young Jabez was thoroughly familiar with factory life when he wrote this letter.

12. The passage in which Allen's plea appears is reprinted in this volume. It concludes a lengthy account of his visit to Manchester, in which Allen's admiration of industrial ingenuity is mixed with his observations of depravity and suffering among the working population. The excerpt from Robert Southey's "Espriella's Letters" (1807), also reprinted in this volume, suggests the kind of negative perception of industrial England prevalent in the United States; Southey's work went through a number of American editions between 1807 and 1822. For the most vivid later account of Manchester, see Frederick Engels, *The Condition of the Working Class in England* (1845), trans. and ed. by W. O. Henderson and W. H. Chaloner (Stanford, Calif.: Stanford University Press, 1968) pp. 50–87. A number of scholars argue that American manufacturers who established the first factory cities were conscientious about the welfare of their new employees and eager not to duplicate English urban conditions. Thomas Bender discusses Francis Cabot Lowell's reaction to British industrial life in *Towards an Urban Vision,* pp. 32–36. Also see Charles L. Sanford, "The Intellectual Origins and New Worldliness of American Industry," *Journal of Economic History,* volume 18 (March 1938), pp. 1–16. Bender, however, stretches a few hints in Lowell's correspondence rather far, and Sanford relies heavily on the writing of Nathan Appleton, who was one of the most pious of the early Boston manufacturers.

13. J. E. Crowley's *This Sheba, Self: The Conceptualization of Economic Life in Eighteenth-Century America* (Johns Hopkins University Studies in Historical and Political Sciences, Ninety-Second Series, Number 2, 1974) documents

the resistance of Protestant religious opinion in the New World to the fundamental goals and psychology of economic self-interest.

14. John Kasson's *Civilizing the Machine: Technology and Republican Values in America, 1776–1900* (New York: Grossman, 1976) explores some of the problems Americans faced in squaring industrialism and "republican" values. Hugo Meier's "Technology and Democracy, 1800–1860," *Mississippi Valley Historical Review,* volume 43 (March 1957), pp. 618–640, is valuable, though Meier finds that democratic America apprehended the danger of technology only in "exaggerated materialism." He does not make much of democratic opposition to corporate capitalism.

15. The effect of industrialization on the lives of women in America is considered at length in numerous scholarly and popular studies. See Percy Bidwell, "The Agricultural Revolution in New England," *American Historical Review,* volume 25 (July 1921), pp. 683–702, for an analysis of the effects of industrialization on women in the farm family. Hannah Josephson's *The Golden Threads: New England's Mill Girls and Magnates* (New York: Duel, Sloan and Pearse, 1949) remains the best of the popular works. See also Thomas Dublin, *Woman at Work: Female Operatives in the Lowell Mills, 1830–1860* (New York: Columbia University Press, 1979).

16. The conservative response to industrial innovation is traced through nineteenth-century New England in Frederic Cople Jaher's "The Boston Brahmin in the Age of Industrial Capitalism," *The Age of Industrialism in America* (New York: The Free Press, 1968), pp. 188–262. See also Paul Goodman, "Ethics and Enterprises: The Values of a Boston Elite, 1800–1860," *American Quarterly,* volume 18, (Fall 1966), pp. 437–451; and Robert V. Spaulding, "The Boston Mercantile Community and the Promotion of the Textile Industry in New England, 1813–1860," Ph.D. dissertation, Yale University, 1963. Spalding makes clear the effects of the merely speculative interest of the most respectable Boston merchant industrialists in the fate of the manufacturing communities they created.

A comparison of the views on industry of the elite of Boston and Philadelphia can be found in E. Digby Baltzell, *Puritan Boston and Quaker Philadelphia: Two Protestant Ethics and the Spirit of Class Authority and Leadership* (New York: Free Press, 1979), pp. 220–230.

17. Those with mercantile interests tended to identify themselves increasingly with manufacturing as the latter became more and more obviously a source of great wealth. The career of Edward Everett makes this plain. In the mid-1820s, he spoke for the merchants of Boston against high tariffs ("The Tariff Question," *North American Review,* volume 19 [1824], pp. 246–249). By the end of the decade, he was a wholehearted booster of industrial progress. See his oration, "Fourth of July at Lowell," reprinted in this volume.

18. The standard older study of workers' response to industrialization is Norman Ware's *The Industrial Worker, 1840–1860: the Reaction of American*

Industrial Society to the Advance of the Industrial Revolution (Boston: Houghton Mifflin, 1924). Among more recent studies, the following are important: Herbert Gutman, *Work, Culture and Society in Industrializing America: Essays in the American Working-class and Social History* (New York: Knopf, 1976); Susan E. Hirsch, *Roots of the American Working Class: The Industrialization of Crafts in Newark, 1800–1860* (Philadelphia: University of Pennsylvania Press, 1978); Merritt Roe Smith, *The Harpers Ferry Armory* (Ithaca, N.Y.: Cornell University Press, 1977); and Anthony F. C. Wallace, *Rockdale: The Growth of An American Village in the Early Industrial Revolution* (New York: Knopf, 1978). Also see Milton Cantor, ed., *American Workingclass Culture: Explorations in American Labor and Social History* (Westport, Conn.: Greenwood, 1979). The burden of scholarship is on the side of a complex argument that mechanization and industrialization worked great hardship on American labor. One must not romanticize "pre-industrial" labor, however, or overemphasize the negative effect of manufacturing on the conditions of work. The unskilled agricultural laborer and the unpaid milkmaid-steamstress often found advantages in factory work.

19. A recent work that stresses the way public rhetoric might be molded to minimize the sense that industrialism deviated from traditional norms of earlier nineteenth-century American culture is Carl Siracusa's *A Mechanical People: Perceptions of the Industrial Order in Massachusetts, 1815–1860* (Middletown, Conn.: Wesleyan University Press, 1979). For a discussion of whether the Industrial Revolution was "revolutionary," see Stuart Bruchey, *The Roots of American Economic Growth, 1607–1861: An Essay in Social Causation* (New York: Harper & Row, 1968), pp. 88–90, and Eric Hobsbawm, *The Age of Revolution* (Cleveland: World Publishing Co., 1962), chapter 2.

20. Andrew Ure, *The Philosophy of Manufactures* (London: 1835; reprint London: Frank Cass, 1967), pp. 13–14. The section of Ure's book in which this passage appears is reprinted in this volume.

21. See Henry William Spiegel, ed., *The Rise of American Economic Thought* (Philadelphia: Chilton, 1960); Virgle Glenn Wilhite, *Founders of American Economic Thought and Policy* (New York: Bookman Associates, 1958); Joseph Dorfman, *The Economic Mind in American Civilization* (New York: Viking Press, 1946–1959); Michael Hudson, *Economics and Technology in Nineteenth-Century American Thought* (New York: Garland Publishers, 1975); and the series of reprints edited by Michael Hudson for Garland Publishers, *The Neglected American Economists*. The response of economic thinkers to the Industrial Revolution is discussed in Maxine Berg, *The Machinery Question and the Making of Political Economy, 1815–1848* (Cambridge University Press, 1980).

22. Leo Marx's *The Machine in the Garden: Technology and the Pastoral Ideal in America* (New York: Oxford University Press, 1964) is the standard study of the literary response to American industrialization. Chapters 3 and

4 are particularly helpful in discussing many of the documents in this collection. Thomas Bender's *Towards an Urban Vision* considers America's coming to terms with the industrial city, especially in chapters 2 and 3. John Kasson's *Civilizing the Machine* describes the clash of American culture and new technology. The ideological response to industrialism is considered in Samuel Reznick, "The Rise and Early Development of Industrial Consciousness in the United States, 1760–1830," *Journal of Economic and Business History*, volume 4 (August 1932), pp. 784–812.

23. The tariff was of great importance to every sector of the American economy, and spokesmen representing every group were responsible for hundreds of articles, pamphlets, and books on the subject. The best general history of the tariff is F. W. Taussig, *The Tariff History of the United States* (New York: G. P. Putnam's Sons, 1888). On the debates over the tariff, see Orin Leslie Elliott, *The Tariff Controversy in the United States, 1789–1833* (Palo Alto, Calif.: published by the University [Stanford], 1892), and Jonathan J. Pincus, *Pressure Groups and Politics in Antebellum Tariffs* (New York: Columbia University Press, 1977). Important documents of the debate on the tariff are collected in F. W. Taussig, ed., *State Papers and Speeches on the Tariff* (Cambridge: Harvard University Press, 1893).

24. The political interrelations between northern manufacturers and southern planters are discussed at length in Thomas O'Connor, *Lords of the Loom: The Cotton Whigs and the Coming of the Civil War* (New York: Scribner's, 1968). Southern views on the tariff are discussed in Elliot, *Tariff Controversy*, and Pincus, *Pressure Groups and Politics*.

25. Robert F. Dalzell, "The Rise of the Waltham-Lowell System," *Perspectives in American History*, volume 9 (1975), pp. 229–268, argues that Boston mercantile investment in the nascent textile industry was essentially a cautious and conservative matter of diversification rather than adventurous speculation or a necessity determined by the closure of other investment opportunities as many have suggested.

26. See Oscar Handlin and Mary Flug Handlin, *Commonwealth: A Study of the Role of Government in the American Economy: Massachusetts, 1774–1861*, revised ed. (Cambridge: Belknap Press of Harvard University Press, 1969); and Louis C. Hunter, *A History of Industrial Power in the United States, 1780–1893. Volume I: Waterpower in the Century of the Steam Engine*, published for the Eleutherian Mills-Hagley Foundation by University Press of Virginia, 1979, pp. 28–36.

27. The legal and political problems of waterpowered mills and farmers' dislike for them are discussed in Hunter, *A History of Industrial Power*, pp. 139–151. The controversy over industrial exploitation of eminent domain was especially sharp in New Hampshire in the 1840s. See the addresses of Daniel Webster and Governors Hill and Hubbard. Also see the documents concerning controversy over water rights in volume 2 of this series.

28. The shift of the American corporation from public to private purposes is discussed in Morton J. Horwitz, *The Transformation of American Law, 1790–1860* (Cambridge: Harvard University Press, 1977), pp. 109–114. See also Oscar Handlin and Mary F. Handlin, "Origins of the American Business Corporation," *Journal of Economic History,* volume 5 (May 1945), pp. 1–23; Ronald E. Seavoy, "The Public Service Origins of the American Business Corporations," *Business History Review,* volume 52, number 1 (Spring 1979), pp. 30–60; and Shaw Livermore, "Unlimited Liability in Early Corporations," *Journal of Political Economy,* volume 40 (1935), pp. 674–687. For examples of early corporations see Joseph Stancliffe Davis, *Essays in the Earlier History of American Corporations* (Cambridge: Harvard University Press, 1919).

29. The early history of American patent law is discussed in P. J. Federico, ed., "Outline History of the United States Patent Office," *Journal of the Patent Office Society,* volume 18, number 7 (July 1936). The first casebook of American patent law, which contains a great deal of information on both specific patents and the idea of protection for invention is Thomas Green Fessenden, *An Essay on the Law of Patents for New Inventions* (Boston, 1810).

30. The importance of the licensing of textile patents is discussed in Kenneth Mailloux, "The Boston Manufacturing Company of Waltham Massachusetts, 1813–1848: The First Modern Factory in America," Ph.D. dissertation, Boston University Press, 1957; and in George S. Gibb, *The Saco-Lowell Shops: Textile Machinery Building in New England, 1813–1849* (Cambridge: Harvard University Press, 1950).

31. See Herbert Heaton, "The Industrial Immigrant in the U.S., 1783–1812," *Proceedings of the American Philosophical Society,* volume 95 (1951), pp. 519–527; and Rowland Tappan Berthoff, *British Immigrants in Industrial America, 1790–1950* (Cambridge: Harvard University Press, 1953).

32. Wallace, *Rockdale,* p. 245.

33. The quotation is from George Fitzhugh's *Cannibals All!* (Richmond, Va., 1857; reprinted Cambridge: Harvard University Press, 1960), p. 15. Though extreme in his position, Fitzhugh was not the only Southerner to condemn the hypocrisy and inhumanity of labor relations in the free North. At times Southern controversialists engaged in direct debate with spokespersons for Northern industry. See Harriet Farley's *Operative's Reply to Hon. Jere. Clemens* (Lowell, 1850), portions of which are reprinted in volume 3 of this series. Other related materials will appear in a subsequent volume on the industrialization of the South.

34. Leo Marx discusses Carlyle's essay and Walker's response in detail in his *Machine in the Garden,* pp. 170–191. Both documents are reprinted in this volume.

35. John C. Calhoun, Letter to Frederick W. Symes, 26 July 1831, *The Papers of John C. Calhoun,* ed. Clyde N. Wilson (Columbia, S.C.: University of South Carolina Press, 1978), volume 11, p. 438.

36. Caroline Ware, *The Early New England Cotton Manufacture: A Study in Industrial Beginnings* (Boston and New York: Houghton Mifflin, 1931), p. 298.

The Philosophy of Manufactures

Reproduced courtesy of Eleutherian Mills Historical Library.

Benjamin Rush

Speech to the United Company of Philadelphia for Promoting American Manufactures (1775)

Benjamin Rush of Philadelphia (1745–1813) was a leading American physician. He was a reformer in many fields in addition to medicine—an abolitionist, founder of the American temperance movement, signatory to the Declaration of Independence. His argument for "establishing manufactories" was patriotic in substance, part and parcel of his advocacy of independence from Great Britain. The nationalistic terms of Rush's argument differ sharply from those employed two decades earlier by Benjamin Franklin in his essay on "The Increase of Mankind" (1751): ". . . Britain should not too much restrain Manufactures in her Colonies. A wise and good Mother will not do it. To distress, is to weaken, and weakening the Children, weakens the whole Family."

The immediate impetus for Rush's address was the Non-importation Agreement recently passed by the Continental Congress. Rush started out by considering the possibility of the colonies becoming self-sufficient in the production of textiles, arguing that, with encouragement, sufficient wool and cotton could be produced locally. The second half of his address, reprinted here, considers the necessity of domestic production.

This speech was given at the founding of the United Company of Philadelphia for Promoting American Manufactures. Founded to fur-

---

[Benjamin Rush], "A Speech delivered in Carpenter's Hall, March 16th [1775], before the Subscribers towards a Fund for establishing Manufactories of Woollen, Cotton, and Linen, in the City of Philadelphia. Published at the Request of the Company." *The Pennsylvania Evening Post*, 11 April 1775, p. 133, and 13 April 1775, pp. 137–138. For information about the situation in Philadelphia leading to the foundation of the United Company, see Charles S. Olton, *Artisans for Independence: Philadelphia Mechanics and the American Revolution* (Syracuse, N.Y.: Syracuse University Press, 1975).

ther American economic independence from Britain, this company was the first attempt at large-scale cotton manufacturing in the colonies. It employed some four hundred women, mostly in hand spinning, and possessed the first spinning jenny in America. The company lasted only from 1775 until the British occupation of Philadelphia in 1777.

I come now to point out the advantages we derive from establishing the woollen, cotton, and linen manufactories among us. The first advantage I shall mention is, we shall save a large sum of money annually in our province. The province of Pennsylvania is supposed to contain 400,000 inhabitants. Let us suppose that only 50,000 of these are clothed with the woollens, cottons, and linens of Great Britain, and that the price of clothing each of these persons upon an average amounts to five pounds sterling a year. If this computation be just, then the sum annually saved in our province by the manufactory of our clothes, will amount to 250,000 £ sterling. Secondly, Manufactories, next to agriculture, are the basis of the riches of every country. Cardinal Ximenes is remembered, at this day in Spain, more for the improvement he made in the breed of sheep, by importing a number of rams from Barbary, than for any other services he rendered his country. King Edward IV and Queen Elizabeth, of England, are mentioned with gratitude by historians for passing acts of Parliament to import a number of sheep from Spain; and to this mixture of Spanish with English sheep, the wool of the latter owes its peculiar excellence and reputation all over the world. Louis XIV, King of France, knew the importance of a woollen manufactory in this kingdom, and in order to encourage it allowed several exclusive privileges to the company of woollen drapers in Paris. The effects of this royal patronage of this manufactory have been too sensibly felt by the English, who have within these thirty or forty years had the mortification of seeing the trade up the Levant for woollen cloths in some measure monopolized by the French. It is remarkable that the riches, and naval power of France have encreased in proportion to this very lucrative trade.

Thirdly, By establishing these manufactories among us, we shall employ a number of poor people in our city, and that too in a way most agreeable to themselves, and least expensive to the company, for according to our plan, the principal part of the business will be carried on in their own houses. Travellers thro' Spain inform us that in the town of Segovia, which contains 60,000 inhabitants, there is not a single beggar to be seen. This is attributed entirely to the woollen manufactory, which is carried on in the most extensive manner in that place, affording constant employment to the whole of their poor people. Fourthly, By establishing the woollen, cotton and linen manufactories in this country, we shall invite manufacturers from every part of Europe, particularly from Britain and Ireland, to come and settle among us. To men who want money to purchase lands, and who from habits of manufacturing are disinclined to agriculture, the prospect of meeting with employment as soon as they arrive in this country in a way they have been accustomed to, would lessen the difficulties of emigration, and encourage thousands to come and settle in America. If they encreased our riches by encreasing the value of our property, and if they added to our strength by adding to our numbers only, they would be a great acquisition to us. But there are higher motives which should lead us to invite strangers to settle in this country. Poverty, with its other evils, has joined with it in every part of Europe, all the miseries of slavery. America is now the only assylum for liberty in the whole world. The present contest with Great Britain was perhaps intended by the Supreme Being, among other wise and benevolent purposes, to show the world this assylum, which, from its remote and unconnected situation with the rest of the globe, might have remained a secret for ages. By establishing manufactories we stretch forth a hand from the ark to invite the timid manufacturers to come in. It might afford us pleasure to trace the new sources of happiness which would immediately open to our fellow creatures from their settlement in this country. Manufactories have been accused of being unfriendly to population. I believe the charges should fall upon slavery. By bringing manufacturers into this land of liberty and plenty, we recover them from the torpid state in which they existed in their own country, and place them in circumstances which enable them to become husbands and fathers, and thus we add to the general tide of human happiness. Fifthly, The establishment of manufactories in this country, by lessening our imports from Great Britain, will deprive European luxuries and vices of those vehicles in which they have been transported to America. The wisdom of the Congress cannot be too much admired in putting a check to them both. They have in effect said to them, "Thus far shall ye go, and no farther." Sixthly, By establishing manufactories among us, we erect an additional barrier against the encroachments of tyranny.

A people who are entirely dependant upon the foreigners for food or clothes, must always be subject to them. I need not detain you in setting forth the misery of holding property, liberty and life, upon the precarious will of our fellow subjects in Britain. I beg leave to add a thought in this place, which has been but little attended to by the writers upon this subject, and that is, that poverty, confinement and death are trifling evils, when compared with that total depravity of heart which is connected with slavery. By becoming slaves, we shall lose every principle of virtue. We shall transfer unlimited obedience from our Master to a corrupted majority in the British House of Commons, and shall esteem their crimes the certificates of their divine commission to govern us. We shall cease to look with horror upon the prostitution of our wives and daughters to those civil and military harpies, who now hover-around the liberties of our country. We shall cheerfully lay them both at their feet. We shall hug our chains. We shall cease to be men. We shall be slaves.

I shall now consider the objections which have been made to the establishment of manufactories in this country.

The first and most common objection to manufactories in this country is, that they will draw off our attention from agriculture. This objection derives great weight from being made originally by the Duke of Sully against the establishment of manufactories in France. But the history of that country shows us that it is more founded in speculation than fact. France has become opulent and powerful in proportion as manufactories have flourished in her, and if agriculture has not kept pace with her manufactories, it is owing entirely to that ill judged policy which forbad the exportation of grain. I believe it will be found upon enquiry that a greater number of hands have been taken from the plough, and employed in importing, retailing and transporting British woollens, cottons and linens, than would be sufficient to manufacture as much of them as would clothe all the inhabitants of the province. There is an endless variety in the geniuses of men, and it would be to preclude the exertion of the faculties of the mind, to confine them entirely to the simple arts of agriculture. Besides, if these manufactories were conducted as they ought to be, two thirds of the labor of them will be carried on by those members of society who cannot be employed in agriculture, namely, by women and children.

A second objection is, that we cannot manufacture clothes so cheap here, as they can be imported from Britain. It has been the misfortune of most of the manufactories, which have been set up in this country, to afford labor to journeymen only for six or nine months in the year, by which means their wages have necessarily been so high as to support them in the intervals of their labor. It will be found upon inquiry that those manufactories, which occupy journeymen the whole year, are carried on at as cheap a rate as they are in Britain. The expence of manufacturing clothes will be lessened from the great share women and children will have in them; and I have the pleasure of informing you that the machine lately brought into this city for lessening the expence of time and hands in spinning, is likely to meet with encouragement from the legislature of our province. In a word, the experiments which have been already made among us, convince us that woollens and linens of all kinds may be made and bought as cheap as those imported from Britain, and I believe every one, who has tried the former, will acknowledge that they wear twice as well as the latter.

A third objection to manufactories is, that they destroy health, and are hurtful to population. The same may be said of navigation, and many other arts which are essential to the happiness and glory of a state. I believe that many of the diseases to which the manufacturers in Britain are subject are brought on not so much by the nature of their employment, but by their unwholesome diet, damp houses, and other bad accommodations, each of which may be prevented in America.

A fourth objection to establishing manufactories in this country is a political one. The liberties of America have been twice and we hope will be a third time preserved by a non-importation of British manufactures. By manufacturing our own clothes we deprive ourselves of the only weapon by which we can hereafter effectually oppose Great Britain. Before we answer this objection it becomes us to acknowledge the obligations we owe to our merchants for consenting so cheerfully to a suspension of their trade with Britain. From the benefits we have derived from their virtue, it would be unjust to insinuate that ever there will be the least danger of trusting the defence of our liberties to them; but I would wish to guard against placing one body of men only upon that forlorn hope to which a non-importation agreement must always expose them. For this purpose I would fill their shores with the manufactures of American looms, and thus establish their trade upon a foundation that cannot be shaken. Here then we derive an answer to the last objection that was mentioned; for in proportion as manufactories flourish in America, they must decline in Britain, and it is well known that nothing but her manufactories have rendered her formidable in all our contests with her—These are the foundation of all her riches and power. These have made her merchants nobles, and her nobles princes. These carried her so triumphantly through the late expensive war, and these are the support of a power more dangerous to the liberties of America than her fleets and armies, I mean the power of corruption—I am not one of those vindictive patriots who exult in the prospect of the decay of the manufactories of Britain. I can forgive her late attempts to enslave us, in the memory of our once mutual freedom and happiness. And should her liberty—her arts—her fleets and armies and her empire ever be interred in Britain, I hope they will all rise in British garments only in America.

Lord Sheffield

Observations on the Commerce of the American States (1783)

John Baker Holroyd, First Earl of Sheffield (1735–1821), was a conservative British peer of great distinction and political influence. This pamphlet attacked a bill introduced in Parliament in 1783 to relax restrictions on trade between England the United States. Sheffield's attitude was pragmatic: "Instead of exaggerating the loss suffered by the dismemberment of the empire, our thoughts may be employed to more advantage in considering what our situation really is, and the greatest advantage that can be derived from it." Sheffield believed that the British Empire should be self-sufficient and thus favored reducing trade with the United States to a minimum. The colonies, he argued, had committed economic suicide by declaring independence, and he staunchly opposed any reconciliation with them. He believed that the United States needed England much more than England needed the United States, especially in the production of manufactured goods.

Sheffield was widely read in both England and the United States, and several pamphlets appeared in answer to him. Sheffield's argument that American manufactures were doomed to failure, summarized in the brief passage and swollen footnote reprinted here, was rebutted by Tench Coxe in his *Brief Examination of Lord Sheffield's Observations* (1791), which is also reprinted in this collection.

John Baker Holroyd, First Earl of Sheffield, *Observations on Commerce of the American States with Europe and the West Indies; Including the several Articles of Import and Export; and On the Tendency of a Bill now depending in Parliament* (London, 1783), pp. 65–67.

. . . it will be a long time before the Americans can manufacture for themselves. Their progress will be stopped by the high price of labour, and the more pleasing and more profitable employment of agriculture, while fresh lands can be got; and the degree of population necessary for manufactures cannot be expected, while a spirit of emigration, especially from the New-England provinces to the interior parts of the continent, rages full as much as it has ever done from Europe to America.

*

If manufacturers should emigrate from Europe to America, at least nine-tenths will become farmers; they will not work at manufactures when they can get double the profit by farming *

---

* The emigrants from Europe to the American States will be miserably disappointed; however, having got into a scrape, they may wish to lead others after them. When the numberless difficulties of adventurers and strangers are surmounted, they will find it necessary to pay taxes, to avoid which probably they left home, and in the case of Britons, gave up great advantages. The same expence, the same industry that become absolutely necessary to save them from sinking in America, if properly employed in most parts of Europe, would give a good establishment, and without the entire sacrifice of the dearest friends and connections, whose society will be ever lamented, and whose assistance, although not to be exerted at the moment, might at other times be most important.

The Philosophy of Manufactures

No American articles are fo neceffary to us, as our manufactures, &c. are to the Americans, and almoft every article of the produce of the American States, which is brought into Europe, we may have at leaft as good and as cheap, if not better, elfewhere.

The abfolute neceffity of great exertions of induftry and toil, added to the want of opportunity of diffipation in the folitary life of new fettlers, and the difficulty and fhame of returning home, alone fupport them there. They find their golden dream ends, at moft, in the poffeffion of a tract of wild uncultivated land, fubject in many cafes to the inroads of the proper and more amiable owners, the Indians.

Emigration is the natural refource of the culprit, and of thofe who have made themfelves the object of contempt and neglect; but it is by no means neceffary to the induftrious.

Reproduced courtesy of Merrimack Valley Textile Museum.

Thomas Jefferson

The Present State of Manufactures (1785)

Thomas Jefferson (1743–1826) composed *Notes on the State of Virginia* in 1780, a bleak period in the course of the Revolution, as an answer to a series of questions posed by the secretary of the French legation at Philadelphia, François Barbe-Marbois. After the war, Jefferson revised and expanded his report for publication in book form. His answer to "Query XIX"—"The present state of manufactures, commerce, interior and exterior trade?"—remains the most eloquent and absolute early argument against the industrialization of the New World. It was, however, a *theoretical* argument, as Jefferson soon acknowledged, and one he would gradually back off from under the pressure of events.

Thomas Jefferson, *Notes on the State of Virginia* (Philadelphia, 1788; first American edition).

QUERY XIX.

THE present state of manufactures, commerce, interior and exterior trade?

We never had an interior trade of any importance. Our exterior commerce has suffered very much from the beginning of the present contest. During this time we have manufactured within our families the most necessary articles of cloathing. Those of cotton will bear some comparison with the same kinds of manufacture in Europe; but those of wool, flax and hemp are very coarse, unsightly, and unpleasant: and such is our attachment to agriculture, and such our preference for foreign manufactures, that be it wise or unwise, our people will certainly return as soon as they can, to the raising raw materials, and exchanging them for finer manufactures than they are able to execute themselves.

The political œconomists of Europe have established it as a principle that every state should endeavour to manufacture for itself: and this principle, like many others, we transfer to America, without calculating the difference of circumstance which should often produce a difference of result. In Europe the lands are either cultivated, or locked up against the cultivator. Manufacture must therefore be resorted to of necessity not of choice, to support the surplus of their people. But we have an immensity of land courting the industry of the husbandman. It is best then that all our citizens should be employed in its improvement, or that one half should be called off from that to exercise manufactures and handicraft arts for the other? Those who labour in the earth are the chosen people of God, if ever he had a chosen people, whose breasts he has made his peculiar deposit for substantial and genuine virtue. It is the focus in which he keeps alive that sacred fire, which otherwise might escape from the face of the earth. Corruption of morals in the mass of cultivators is a phænomenon of which no age nor nation has furnished an example. It is the mark set on those, who not looking up to heaven, to their own soil and induſtry, as does the huſbandman, for their ſubſiſtance, depend for it on the caſualties and caprice of cuſtomers. Dependance begets ſubſervience and venality, ſuffocates the germ of virtue, and prepares fit tools for the deſigns of ambition. This, the natural progreſs and conſequence of the arts, has ſometimes perhaps been retarded by accidental circumſtances: but, generally ſpeaking, the proportion which the aggregate of the other claſſes of citizens bears in any ſtate to that of its huſbandmen, is the proportion of its unſound to its healthy parts, and is a good-enough barometer whereby to meaſure its degree of corruption. While we have land to labour then, let us never wiſh to ſee our citizens occupied at a work-bench, or twirling a diſtaff. Carpenters, maſons, ſmiths, are wanting in huſbandry: but, for the general operations of manufacture, let our work-ſhops remain in Europe. It is better to carry proviſions and materials to workmen there, than bring them to the proviſions and materials, and with them their manners and principles. The loſs by the tranſportation of commodities acroſs the Atlantic will be made up in happineſs and permanence of government. The mobs of great cities add juſt ſo much to the ſupport of pure government, as ſores do to the ſtrength of the human body. It is the manners and ſpirit of a people which preſerve a republic in vigour. A degeneracy in theſe is a canker which ſoon eats to the heart of its laws and conſtitution.

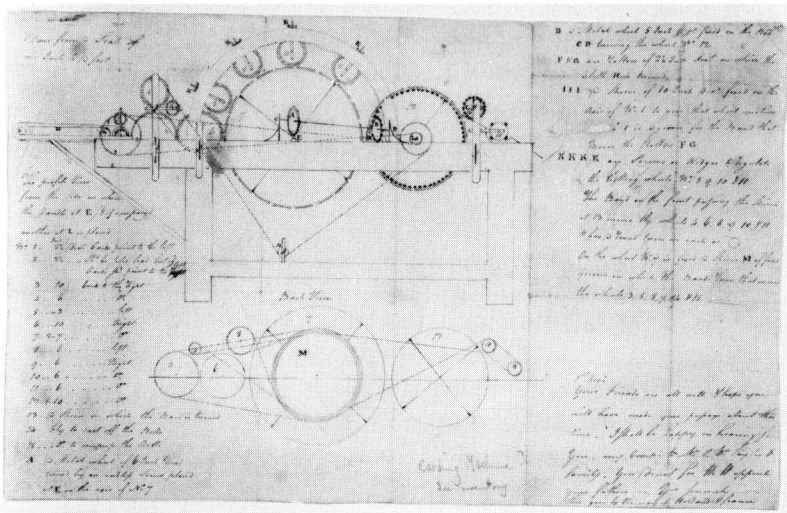

Carding engine drawn by Thomas Jefferson, ca. 1787. Reproduced courtesy of the Massachusetts Historical Society.

Thomas Jefferson

Letters (1785–1816)

This series of letters to friends and associates documents Jefferson's changing attitudes toward the encouragement of domestic industry after the publication of his *Notes on the State of Virginia*. They also reveal something of his own private practical involvement in manufacturing during these years.

Thomas Jefferson, to John Jay, August 23, 1785 (*The Papers of Thomas Jefferson*, ed. Julian P. Boyd. [Princeton, N.J.: Princeton University Press, 1953], volume 8, pp. 426–428). To G. K. van Hogendorp, October 13, 1785 (*The Works of Thomas Jefferson*, ed. P. L. Ford. [New York, 1904], volume 8, pp. 631–634). To Thomas Digges, June 19, 1788 (*The Writings of Thomas Jefferson*, collected and ed. Paul Leicester Ford [New York, 1898], volume 5, pp. 408–410). To M. de Meusnier, April 29, 1795 (*Writings*, volume 8, pp. 173–176). To Archibald Stuart, January 3, 1796 (*Writings*, volume 8, pp. 212–214). To Mr. Lithson, January 4, 1805 (*Works*, volume 11, pp. 55–56). To Colonel David Humphreys, January 20, 1809 (*Works*, volume 12, pp. 235–236). To John Adams, January 21, 1812 (*Writings*, volume 9, pp. 332–333). To Thaddeus Kosciusko, June 28, 1812 (*Works*, volume 13, pp. 168–172). To James Ronaldson, January 12, 1813 (*Works*, volume 13, p. 205). To John Melish, January 13, 1813 (*Works*, volume 13, pp. 206–207). To Benjamin Austin, January 9, 1816 (*Writings*, volume 11, pp. 500–505).

To John Jay

DEAR SIR   Paris Aug. 23, 1785.

    I shall sometimes ask your permission to write you letters, not official but private. The present is of this kind, and is occasioned by the question proposed in yours of June 14 'Whether it would be useful to us to carry all our own productions, or none?' Were we perfectly free to decide this question, I should reason as follows. We have now lands enough to employ an infinite number of people in their cultivation. Cultivators of the earth are the most valuable citizens. They are the most vigorous, the most independent, the most virtuous, and they are tied to their country and wedded to it's liberty and interests by the most lasting bands. As long therefore as they can find emploiment in this line, I would not convert them into mariners, artisans, or any thing else. But our citizens will find emploiment in this line till their numbers, and of course their productions, become too great for the demand both internal and foreign. This is not the case as yet, and probably will not be for a considerable time. As soon as it is, the surplus of hands must be turned to something else. I should then perhaps wish to turn them to the sea in preference to manufactures, because comparing the characters of the two classes I find the former the most valuable citizens. I consider the class of artificers as the panders of vice and the instruments by which the liberties of a country are generally overturned. However we are not free to decide this question on principles of theory only. Our people are decided in the opinion that it is necessary for us to take a share in the occupation of the ocean, and their established habits induce them to require that the sea be kept open to them, and that that line of policy be pursued which will render the use of that element as great as possible to them.

*

    If a war with England should take place it see[ms] to me that the first thing necessary would be a resolution to abandon the carrying trade because we cannot protect it. Foreign nations must in that case be invited to bring us what we want and to take our productions in their own bottoms. This alone could prevent the

The Philosophy of Manufactures

loss of those productions to us and the acquisition of them to our enemy. Our seamen might be emploied in depredations on their trade. But how dreadfully we shall suffer on our coasts, if we have no force on the water, former experience has taught us. Indeed I look forward with horror to the very possible case of war with an European power, and think there is no protection against them but from the possession of some force on the sea. Our vicinity to their West India possessions and to the fisheries is a bridle which a small naval force on our part would hold in the mouths of the most powerful of these countries. I hope our land office will rid us of our debts, and that our first attention then will be to the beginning a naval force of some sort. This alone can countenance our people as carriers on the water, and I suppose them to be determined to continue such.

*

To G. K. van Hogendorp

DEAR SIR   Paris Oct. 13, 1785.

Having been much engaged lately, I have been unable sooner to acknolege the receipt of your favor of Sep. 8. What you are pleased to say on the subject of my Notes [on the State of Virginia] is more than they deserve. The condition in which you first saw them would prove to you how hastily they had been originally written: as you may remember the numerous insertions I had made in them from time to time, when I could find a moment for turning to them from other occupations. I have never yet seen Monsr. de Buffon. He has been in the country all the summer. I sent him a copy of the book, and I have only heard his sentiments on one particular of it, that of the identity of the Mammoth and Elephant.

*

You ask what I think on the expediency of encouraging our states to be commercial? Were I to indulge my own theory, I should wish them to practice neither commerce nor navigation, but to stand with respect to Europe precisely on the footing of

China. We should thus avoid wars, and all our citizens would be husbandmen. Whenever indeed our numbers should so increase as that our produce would overstock the markets of those nations who should come to seek it, the farmers must either employ the surplus of their time in manufactures, or the surplus of our hands must be employed in manufactures, or in navigation. But that day would, I think be distant, and we should long keep our workmen in Europe, while Europe should be drawing rough materials and even subsistence from America. But this is theory only, and a theory which the servants of America are not at liberty to follow. Our people have a decided taste for navigation and commerce. They take this from their mother country: and their servants are in duty bound to calculate all their measures on this datum: we wish to do it by throwing open all the doors of commerce and knocking off it's shackles. But as this cannot be done for others, unless they will do it to us, and there is no great probability that Europe will do this, I suppose we shall be obliged to adopt a system which may shackle them in our ports as they do us in theirs.

*

To Mr. Thomas Digges

SIR,   Paris, June 19, 1788.

I have duly received your favor of May 12, as well as that of the person who desires information on the state of cotton manufactures in America, and for his interest & safety I beg leave to address to you the answers to his queries without naming him.

In general it is impossible that manufactures should succeed in America from the high price of labour. This is occasioned by the great demand of labour for agriculture. A manufacturer going from Europe will turn to labour of other kind if he find more to be got by it, & he finds some emploiment so profitable that he can soon lay up money enough to buy fifty acres of land, to the culture of which he is irresistibly tempted by the independence in which that places him, & the desire of having a wife & family around him. If any manufactures can succeed there, it will be that of cotton. I must observe for his information that this plant grows

The Philosophy of Manufactures

nowhere in the United States Northward of the Potowmack, and not in quantity till you get Southward as far as York & James rivers. I know nothing of the manufacture which is said to be set up at Richmond. It must have taken place since 1783, when I left Virginia. In that state (for it is the only one I am enabled to speak of with certainty) there is no manufacture of wire or of cotton cards: of if any, it is not worth notice. No manufacture of stocking-weaving, consequently none for making the machine: none of cotton cloths of any kind whatever for sale; tho in almost every family some is manufactured for the use of the family, which is always good in quality, & often tolerably fine. In the same way they make excellent knit stockings of cotton, weaving it in like manner carried on principally in the family way: among the poor, the wife weaves generally, & the rich either have a weaver among their servants or employ their poor neighbors. Cotton cost in Virginia from 12*d*. to 18*d*. sterling the pound before the war, probably it is a little raised since. Richmond is as good a place for a manufactory as any in that State, & perhaps the best as to it's resources for this business. Cotton clothing is very much the taste of the country. A manufacturer on his landing should apply to the well informed farmers and gentlemen of the country. Their information will be more disinterested than that of merchants, and they can better put him into the way of disposing of his workmen in the cheapest manner till he has time to look about him & decide how & where he will establish himself. Such is the hospitality in that country, & their disposition to assist strangers, that he may boldly go to any good house he sees, and make the inquiry he needs. He will be sure to be kindly received, honestly informed, and accommodated in a hospitable way, without any other introduction than an information who he is & what are his views. It is not the policy of the government in that country to give any aid to works of any kind. They let things take their natural course without help or impediment, which is generally the best policy. More particularly as to myself I must add that I have not the authority nor the means of assisting any persons in their passage to that country.

To M. de Meusnier

Monticello, Virginia, Apr. 29, '95.

In our private pursuits it is a great advantage that every honest employment is deemed honorable. I am myself a nail-maker. On returning home after an absence of ten years, I found my farms so much deranged that I saw evidently they would be a burden to me instead of a support till I could regenerate them; & consequently that it was necessary for me to find some other resource in the meantime. I thought for awhile of taking up the manufacture of pot-ash, which requires but small advances of money. I concluded at length however to begin a manufacture of nails, which needs little or no capital, & I now employ a dozen little boys from 10. to 16. years of age, overlooking all the details of their business myself & drawing from it a profit on which I can get along till I can put my farms into a course of yielding profit. My new trade of nail-making is to me in this country what an additional title of nobility or the ensigns of a new order are in Europe.

To Archibald Stuart

DEAR SIR, Monticello, Jan. 3, '96.

I troubled you once before on the subject of my nails, and must trouble you once more, but hope my present plan will protect you from all further embarrasment with it. I set out with refusing to retail, expecting the merchants of my neighborhood and the upper country would have given a preference to my supplies, because delivered *here* at the *Richmond whole prices,* and at hand to be called for in small parcels, so that they need not to keep large sums invested in that article & lying dead on their hands. The importing merchants however decline taking them from a principle of suppressing every effort towards domestic manufacture, & the merchants who purchase here being much under the influence of the importers, take their nails from them with their other goods. I have determined therefore to establish deposits of my nails to be retailed at Milton, Charlottesville, Staunton, Wormester, & Warren, but first at the three first places, because I presume my present works, which turn out a ton a month, will fully furnish them, and two additional fires which will be at work in a short time, will raise it to a ton and a half a month, and enable me to extend my supplies to Wormester & Warren. I shall retail at Richmond *wholesale* prices, laying on 5 percent at Milton & Charlottesville for commission to the retailers, and 10. percent at the other places for commission & transportation. My present retailing prices at Staunton would be

Sixes	$12\frac{1}{2}^d$ per lb.	equal to	$7/3\frac{1}{2}$ per M
Eights	12^d	"	equal to 10/
Tens	$11\frac{1}{2}^d$	"	equal to $12/5\frac{1}{2}$
Twelves	11^d	"	equal to 14/8
Sixteens	$10\frac{1}{2}^d$	"	equal to 17/6
Twenties	10^d	"	equal to 20/10

It is tolerably certain that the moment my deposit opens there will be an entire stoppage to the sale of all imported nails, for a body can *retail* them in the upper country at the Richmond *wholesale* prices, advanced only 5 or 10 percent. and as I mean to employ only one person in each place to retail, it will be of some advantage to the merchant who will undertake it, to have the entire monopoly of the nail business, & so draw to his store every one who wants nails, besides the commission of 5 percent, which in an article to be sold for ready money only, and where he does not employ a farthing of his own capital, I am advised is a sufficient allowance for commission. I should expect them to send me a copy of their sales once a month, and to hold the proceeds ready for my draughts at stated periods, say monthly. I trouble you to engage some person whom you can recommend for punctuality, to retail for me.

*

To Mr. Lithson

DEAR SIR,   Washington, January 4, 1805.

    Your favor of December 4th has been duly received. Mr. Duane informed me that he meant to publish a new edition of the Notes on Virginia, and I had in contemplation some particular alterations which would require little time to make. My occupations by no means permit me at this time to revise the text, and make those changes in it which I should now do. I should in that case certainly qualify several expressions in the nineteenth chapter, which have been construed differently from what they were intended. I had under my eye, when writing, the manufacturers of the great cities in the old countries, at the time present, with whom the want of food and clothing necessary to sustain life, has begotten a depravity of morals, a dependence and corruption, which renders them an undesirable accession to a country whose morals are sound. My expressions looked forward to the time when our own great cities would get into the same state. But they have been quoted as if meant for the present time here. As yet our manufacturers are as much at their ease, as independent and moral as our agricultural inhabitants, and they will continue so as long as there are vacant lands for them to resort to; because whenever it shall be attempted by the other classes to reduce them to the minimum of subsistence, they will quit their trades and go to laboring the earth. A first question is, whether it is desirable for us to receive at present the dissolute and demoralized handicraftsmen of the old cities of Europe? A second and more difficult one is, when even good handicraftsmen arrive here, is it better for them to set up their trade, or go to the culture of the earth? Whether their labor in their trade is worth more than their labor on the soil, increased by the creative energies of the earth? Had I time to revise that chapter, this question should be discussed, and other views of the subject taken, which are presented by the wonderful changes which have taken place here since 1781, when the Notes on Virginia were written. Perhaps when I retire, I may amuse myself with a serious review of this work; at present it is out of the question. Accept my salutations and good wishes.

The Philosophy of Manufactures

To Colonel David Humphreys

SIR, Washington, January 20, 1809.

I have to acknowledge the receipt of your favor of December 12th, and to return you my thanks for the cloth furnished me. It came in good time, and does honor to your manufactory, being as good as any one would wish to wear in any country. Amidst the pressure of evils with which the belligerent edicts have afflicted us, some permanent good will arise; the spring given to manufactures will have durable effects. Knowing most of my own State, I can affirm with confidence that were free intercourse opened again to-morrow, she would never again import one-half of the coarse goods which she has done down to the date of the edicts. These will be made in our families. For finer goods we must resort to the larger manufactories established in the towns. Some jealousy of this spirit of manufacture seems excited among commercial men. It would have been as just when we first began to make our own ploughs and hoes. They have certainly lost the profit of bringing these from a foreign country. My idea is that we should encourage home manufactures to the extent of our own consumption of everything of which we raise the raw material. I do not think it fair in the ship-owners to say we ought not to make our own axes, nails, etc., here, that they may have the benefit of carrying the iron to Europe, and bringing back the axes, nails, etc. Our agriculture will still afford surplus produce enough to employ a due proportion of navigation. Wishing every possible success to your undertaking, as well for your personal as the public benefit, I salute you with assurance of great esteem and respect.

To John Adams

DEAR SIR, Monticello, January 21, 1812.

I thank you before hand (for they are not yet arrived) for the specimens of homespun you have been so kind as to forward me by post. I doubt not their excellence, knowing how far you are

The Philosophy of Manufactures

advanced in these things in your quarter. Here we do little in the fine way, but in coarse and middling goods a great deal. Every family in the country is a manufactory within itself, and is very generally able to make within itself all the stouter and middling stuffs for its own clothing and household use. We consider a sheep for every person in the family as sufficient to clothe it, in addition to the cotton, hemp and flax which we raise ourselves. For fine stuff we shall depend on your northern manufactories. Of these, that is to say, of company establishments, we have none. We use little machinery. The spinning jenny, and loom with the flying shuttle, can be managed in a family; but nothing more complicated. The economy and thriftiness resulting from our household manufactures are such that they will never again be laid aside; and nothing more salutary for us has ever happened than the British obstructions to our demands for their manufactures. Restore free intercourse when they will, their commerce with us will have totally changed its form, and the articles we shall in future want from them will not exceed their own consumption of our produce.

*

To Thaddeus Kosciusko

*
June 28, 1812.

Our manufacturers are now very nearly on a footing with those of England. She has not a single improvement which we do not possess, and many of them better adapted by ourselves to our ordinary use. We have reduced the large and expensive machinery for most things to the compass of a private family, and every family of any size is now getting machines on a small scale for their household purposes. Quoting myself as an example, and I am much behind many others in this business, my household manufactures are just getting into operation on the scale of a carding machine costing $60 only, which may be worked by a girl of twelve years old, a spinning machine, which may be made for $10, carrying 6 spindles for wool, to be worked by a girl also, another which can be made for $25, carrying 12 spindles for cotton, and a loom,

The Philosophy of Manufactures

with a flying shuttle, weaving its twenty yards a day. I need 2,000 yards of linen, cotton and woolen yearly, to clothe my family, which this machinery, costing $150 only, and worked by two women and two girls, will more than furnish. For fine goods there are numerous establishments at work in the large cities, and many more daily growing up; and of merinos we have some thousands, and these multiplying fast. We consider a sheep for every person as sufficient for their woolen clothing, and this State and all to the north have fully that, and those to the south and west will soon be up to it. In other articles we are equally advanced, so that nothing is more certain than that, come peace when it will, we shall never again go to England for a shilling where we have gone for a dollar's worth. Instead of applying to her manufacturers there, they must starve or come here to be employed. I give you these details of peaceable operations, because they are within my present sphere. Those of war are in better hands, who know how to keep their own secrets. Because, too, although a soldier yourself, I am sure you contemplate the peaceable employment of man in the improvement of his condition, with more pleasure than his murders, rapine and devastations.

To James Ronaldson

Monticello, January 12, 1813.

Household manufacture is taking deep root with us. I have a carding machine, two spinning machines, and looms with the flying shuttle in full operation for clothing my own family; and I verily believe that by the next winter this State will not need a yard of imported coarse or middling clothing. I think we have already a sheep for every inhabitant, which will suffice for clothing, and one-third more, which a single year will add, will furnish blanketing.

*

*

To John Melish

Monticello, January 13, 1813.

I had no conception that manufactures had made such progress [in the Western states] and particularly of the number of carding and spinning machines dispersed through the whole country. We are but beginning here to have them in our private families. Small spinning jennies of from half a dozen to twenty spindles, will soon, however, make their way into the humblest cottages, as well as the richest houses; and nothing is more certain, than that the coarse and middling clothing for our families, will forever hereafter continue to be made within ourselves. I have hitherto myself depended entirely on foreign manufactures; but I have now thirty-five spindles agoing, a hand carding machine, and looms with the flying shuttle, for the supply of my own farms, which will never be relinquished in my time. The continuance of the war will fix the habit generally, and out of the evils of impressment and of the orders of council a great blessing for us will grow. I have not formerly been an advocate for great manufactories. I doubted whether our labor, employed in agriculture, and aided by the spontaneous energies of the earth, would not procure us more than we could make ourselves of other necessaries. But other considerations entering into the question, have settled my doubts.

To Benjamin Austin

January 9, 1816.

You tell me I am quoted by those who wish to continue our dependence on England for manufactures. There was a time when I might have been so quoted with more candor, but within the thirty years which have since elapsed, how are circumstances changed! We were then in peace. Our independent place among nations was acknowledged. A commerce which offered the raw material in exchange for the same material after receiving the last touch of industry, was worthy of welcome to all nations. It was expected that those especially to whom manufacturing industry was important, would cherish the friendship of such customers by every favor, by every inducement, and particularly cultivate their peace by every act of justice and friendship. Under this prospect the question seemed legitimate, whether, with such an immensity of unimproved land, courting the hand of husbandry, the industry of agriculture, or that of manufactures, would add most to the national wealth? And the doubt was entertained on this consideration chiefly, that to the labor of the husbandman a vast addition is made by the spontaneous energies of the earth on which it is employed: for one grain of wheat committed to the earth, she renders twenty, thirty, and even fifty fold, whereas to the labor of the manufacturer nothing is added. Pounds of flax, in his hands, yield, on the contrary, but pennyweights of lace. This exchange, too, laborious as it might seem, what a field did it promise for the occupations of the ocean; what a nursery for that class of citizens who were to exercise and maintain our equal rights on that element? This was the state of things in 1785, when the *Notes on Virginia* were first printed; when, the ocean being open to all nations, and their common right in it acknowledged and exercised under regulations sanctioned by the assent and usage of all, it was thought that the doubt might claim some consideration. But who in 1785 could foresee the rapid depravity which was to render the close of that century the disgrace of the history of man? Who could have imagined that the two most distinguished in the rank of nations, for science and civilization, would have suddenly descended from that honorable eminence, and setting at defiance all those moral laws established by the Author of nature between nation and nation, as between man and man, would cover earth and sea with robberies and piracies, merely because strong enough to do it with temporal impunity; and that under this disbandment of nations from social order, we should have been despoiled of a thousand ships, and have thousands of our citizens reduced to Algerine slavery. Yet all this has taken place. One of these nations interdicted to our vessels all harbors of the globe without having first proceeded to some one of hers, there paid a tribute proportioned to the cargo, and obtained her license to proceed to the port of destination. The other declared them to be lawful prize if they had touched at the port or been visited by a ship of the enemy nation. Thus were we completely excluded from the ocean. Compare this state of things with that of '85, and say whether an opinion founded in the circumstances of that day can be fairly applied to those of the present. We have experienced what we did not then believe, that there exists both profligacy and power enough to exclude us from the field of interchange with other nations: that to be independent for the comforts of life we must fabricate them ourselves. We must now place the manufacturer by the side of the agriculturist. The former question is suppressed, or rather assumes a new form. Shall we make our own comforts, or go without them, at the will of a foreign nation? He, therefore, who is now against domestic manufacture, must be for reducing us either to dependence on that foreign nation, or to be clothed in skins, and to live like wild beasts in dens and caverns. I am not one of these; experience has taught me that manufactures are now as necessary to our independence as to our comfort; and if those who quote me as of a different opinion, will keep pace with me in purchasing nothing foreign where an equivalent of domestic fabric can be obtained, without regard to difference of price, it will not be our fault if we do not soon have a supply at home equal to our demand, and wrest that weapon of distress from the hand which has wielded it. If it shall be proposed to go beyond our own supply, the question of '85 will then recur, will our *surplus* labor be then most beneficially employed in the culture of the earth, or in the fabrications of art? We have time yet for consideration, before that question will press upon us; and the maxim to be applied will depend on the circumstances which shall then exist; for in so complicated a science as political economy, no one axiom can be laid down as wise and expedient for all times and circumstances, and for their contraries. Inattention to this is what has called for this explanation, which reflection would have rendered unnecessary with the candid, while nothing will do it with those who use the former opinion only as a stalking horse, to cover their disloyal propensities to keep us in eternal vassalage to a foreign and unfriendly people.

   I salute you with assurances of great respect and esteem.

Tench Coxe

Address to an Assembly Convened to Establish a Society for the Encouragement of Manufactures and the Useful Arts (1787)

Tench Coxe (1755–1824) is best remembered as assistant secretary of the treasury under Alexander Hamilton. In this capacity he was instrumental in drafting Hamilton's *Report on the Subject of Manufactures*. Coxe was a successful Philadelphia businessman with a startlingly varied political career. Early in the Revolution he appears to have been pro-British in sympathies, sufficiently so that he was arrested when the British evacuated Philadelphia. He subsequently became an active patriot, and he based much of his argument for industrialization on anti-British grounds. He was both a Federalist and a Republican at various times, and he was given preferment in government appointments both by Hamilton and by Hamilton's most vigorous opponent in matters of commerce and industry, Thomas Jefferson. Throughout his political shifts, Coxe remained staunch in his primary interest, furthering the prosperity of the American entre-

---

Tench Coxe, *An Address to an Assembly of the Friends of American Manufactures, Convened for the Purpose of Establishing a Society for the Encouragement of Manufactures and the Useful Arts, Read in the University of Pennsylvania, on Thursday the 9th of August 1787* (Philadelphia, 1787). Other important works by Coxe are "An enquiry into the principles, on which a commercial system for the United States of America should be founded; to which are added some political observations connected with the subject. Read before the society for political enquiries . . . ," *American Museum*, volume 1, number 4 (1787), pp. 432–445; and "A Plan for Encouraging Agriculture, and Increasing the Value of Farms in the Midland and More Western Counties of Pennsylvania, Applicable to Several Other Parts of that State, and to many Parts of the United States," in Coxe's *A View of the United States of America, in a Series of Papers, written at various times, between the years 1787 and 1794* (Philadelphia, 1794), pp. 384–404. For more on Coxe, see Jacob E. Cooke, *Tench Coxe and the Early Republic* (Chapel Hill: University of North Carolina Press, 1978; published for the Institute of Early American History and Culture).

preneur. Coxe was the most outspoken advocate of American manufactures of his time.

This patriotic address was read before a group of Philadelphia's commercial gentlemen gathered at the University of Pennsylvania to found the Society for the Encouragement of Manufactures and the Useful Arts on August 9, 1787. Especially interesting is Coxe's emphasis on the usefulness of power machinery in manufacturing.

AN ADDRESS

TO AN

Aſſembly of the Friends of American Manufactures.

GENTLEMEN,

WHILE I obey with ſincere pleaſure the commands of the reſpectable aſſembly whom I have now the honor to addreſs, I feel the moſt trying emotions of anxiety and apprehenſion in attempting to perform ſo difficult and ſerious a duty, as that preſcribed to me at our laſt meeting. The importance and novelty of the ſubject, the injurious conſequences of miſtaken opinions on it and your preſence neceſſarily excite feelings ſuch as theſe. They are leſſened however, by the hope of ſome benefit to that part of my fellow citizens, who depend for comfort on our native manufactures and by an ardent wifh to promote every meafure, that will give to our newborn ftates the ftrength of manhood. Supported by thefe confiderations and relying on the kind indulgence, which is ever fhewn to well meant endeavours, however unfuccefsful, I fhall venture to proceed.

PROVIDENCE has beftowed upon the United States of America means of happinefs, as great and numerous, as are enjoyed by any country in the world. A foil fruitful and diverfified—a healthful climate—mighty rivers and adjacent feas abounding with fifh are the great advantages for which we are indebted to a beneficent creator. Agriculture, manufactures and commerce, naturally 'arifing from thefe fources, afford to our induftrious citizens certain fubfiftence and innumerable opportunities of acquiring wealth. To arrange our affairs in falutary and well digested system, by which the fruits of industry, in every line, may be most easily attained, and the possession of property and the blessings of liberty may be completely secured—these are the important objects, that should engross our present attention. The interests of commerce and the establishment of a just and effective government are already committed to the care of THE AUGUST BODY now sitting in our capital. —The importance of agriculture has long since recommended it to the patronage of numerous associations, and the attention of all the legislatures—but manufactures, at least in Pennsylvania, have had but a few unconnected friends, till sound policy and public spirit gave a late, but auspicious birth, to this Society.

THE situation of America before the revolution was very unfavourable to the objects of this institution. The prohibition of most foreign raw materials—considerable bounties in England for carrying away the unwrought productions of this country to that, as well as on exporting British goods from their markets—the preference for those goods, which habit carried much beyond what their excellence would justify, and many other circumstances created artificial impediments, that appeared almost insuperable. Several branches however were carried on to good advantage. But as long as we remained in our colonial situation, our progress was very slow, and indeed the necessity of attention to manufactures was not so urgent, as it has become since our assuming an independent station. The employment of those, whom the decline of navigation has deprived of their usual occupations—the consumption of the encreasing produce of our lands and fisheries, and the certainty of supplies in the time of war are weighty reasons for establishing new manufactories now, which existed but in a small degree, or not at all, before the revolution.

WHILE we readily admit, that in taking meafures to promote the objects of this Society, nothing fhould be attempted, which may injure our agricultural interefts, they being undoubtedly the moft important, we muft obferve in juftice to ourfelves, that very many of our citizens, who are expert at manufactures and the ufeful arts, are entirely unacquainted with rural affairs, or unequal to the expences of a new fettlement; and many we may believe, will come among us invited to our fhores from foreign countries, by the bleffings of liberty, civil and religious. We may venture to affert too, that more profit to the individual and riches to the nation will be derived from fome manufactures, which promote agriculture, than from any fpecies of cultivations whatever. The truth of this remark however, will be better determined, when the fubject fhall be further confidered.

LET us endeavor firſt to diſencumber manufactures of the objections, that appear againſt them, the principal of which are, the high rate of labor, which involves the price of proviſions—the want of a ſufficient number of hands on any terms,—the ſcarcity and dearneſs of raw materials—want of ſkill in the buſineſs itſelf and its unfavorable effects on the health of the people.

FACTORIES, which can be carried on by watermills, windmills, fire, horſes and machines ingeniouſly contrived, are not burdened with any heavy expence of boarding, lodging, cloathing and paying workmen, and they multiply the force of hands to a great extent without taking our people from agriculture. By wind and water machines we can make pig and bar iron, nail rods, tire, ſheet-iron, ſheet-copper and ſheet-braſs, anchors, meal of all kinds, gunpowder, writing, printing and hanging paper, ſnuff, linſeed oil, boards, plank and ſcantling ; and they affift us in finifhing fcythes, fickles and woolen cloths. Strange as it may appear they alfo card, fpin and weave by water in the European factories. Bleaching and tanning muft not be omitted, while we are fpeaking of the ufefulnefs of water.

By FIRE we conduct our breweries, diftilleries, falt and potafh works, fugar houfes, potteries, cafting and fteel furnaces, works for animal and vegetable oils and refining drugs. Steam mills have not yet been adopted in America, but we fhall probably fee them after a fhort time in New-England and other places, where there are few mill feats and in this and other great towns of the United States. The city of Philadelphia, by adopting the ufe of them, might make a faving of above five *per cent.* on all the grain brought hither by water, which is afterwards manufactured into meal, and they might be ufefully applied to many other valuable purpofes.

Horses give us, in some instances, a relief from the difficulties we are endeavouring to obviate. They grind the tanners bark and potters clay; they work the brewers and distillers pumps, and might be applied, by an inventive mind, as the moving principle of many kinds of mills.

Machines ingeniously constructed, will give us immense assistance.—The cotton and silk manufacturers in Europe are possessed of some, that are invaluable to them. One instance I have had precisely ascertained, which employs a few hundreds of women and children, and performs the work of 12000 carders, spinners and winders. They have been so curiously improved of late years, as to weave the most complicated manufactures. In short, combinations of machines with fire and water have already performed much more than was formerly expected from them by the most visionary enthusiast on the subject. Perhaps I may be too sanguine, but they appear to me fraught with immenſe advantages to us, and full of danger to the manufacturing nations of Europe; for ſhould they continue to uſe and improve them, as they have heretofore done, their people muſt be driven to us for want of employment, and if, on the other hand, they ſhould return to manual labor, we ſhall underwork them by theſe invaluable engines. We may certainly borrow ſome of their inventions and others of the ſame nature we may ſtrike out ourſelves; for on the ſubject of mechaniſm America may juſtly pride herſelf. Every combination of machinery may be expected from a country, A NATIVE SON of which, reaching this ineſtimable object at its higheſt point, has epitomized the motions of the ſpheres, that roll throughout the univerſe*.

THE lovers of mankind, ſupported by experienced phyſicians, and the opinions of enlightened politicians, have objected to ma-

---

* David Rittenhouſe, Eſq. of Pennſylvania.

nufactures as unfavorable to the health of the people. Giving to this humane and important confideration its full weight, it furnifhes an equal argument againft feveral other occupations, by which we obtain our comforts and promote our agriculture. The painting bufinefs for inftance—reclaiming marfhes—clearing fwamps—the culture of rice and indigo and fome other employments, are even more fatal to thofe, who are engaged in them. But this objection is urged principally againft carding, fpinning and weaving, which were formerly manual and fedentary occupations. Our plan, as we have already fhewn, is not to purfue thofe modes, unlefs in cafes particularly circumftanced, for we are fenfible, that our people muft not be diverted from their farms. *Horfes, and the potent elements of fire and water, aided by the faculties of the human mind (except in a few healthful inftances) are to be our daily labourers.* After giving immediate relief to the induftrious poor, thefe unhurtful means will be purfued and will procure us private wealth and national profperity.

EMIGRATION from Europe will alfo relieve and affift us. The bleffings of civil and religious liberty in America, and the oppreffions of moft foreign governments, the want of employment at home and the expectations of profit here, curiofity, domeftic unhappinefs, civil wars and various other circumftances will bring many manufacturers to this afylum for mankind. Ours will be their induftry, and what is of ftill more confequence ours will be their fkill. Intereft and neceffity, with fuch inftructors, will teach us quickly. In the laft century the manufactures of France were next to none; they are now worth millions to her yearly. Thofe of England have been more improved within the laft twelve years, than in the preceding fifty. At the peace of 1762, the ufeful arts and manufactures were scarcely known in America. How great has been their progress since, unaided, undirected and discouraged. Countenanced by your patronage and promoted by your assistance, what may they not be 'ere such another space of time shall elapse.

WONDERFUL as it must appear, the manufacturers of beer, that best of all our commodities, have lately been obliged to import malt from England. Here must be inexcusable neglect, or a strange blindness to our most obvious interests. The cultivation of barley should certainly be more attended to, and if I mistake not exceedingly, the present abundant crop of wheat will so fill our markets, that the farmer, who shall reap barley the ensuing year, will find it the most profitable of all grains. We cannot, however, have any permanent difficulty on this article.

OF flax and hemp little need be said, but that we can encrease them as we please, which we shall do according to the demand.

WOOL must become much more abundant as our country populates. Mutton is the best meat for cities, manufactories, seminaries of learning, and poor houses, and should be given by rule as in England. The settlement of our new lands, remote from water carriage, must introduce much more pasturage and grazing, than has been heretofore necessary, as sheep, horses and horned cattle will carry themselves to market through roads impassable by waggons. The restrictions of our trade will also tend to encrease the number of sheep horses and horned cattle, used to form a great part of the New-England cargoes for the English West-India islands. These animals are exported to those places now in very small numbers, as our vessels are excluded from their ports.—The farms, capital and men, which were formerly employed in raising them, will want a market for their usual quantity, and the nature of that country being unfit for grain, sheep must occupy a great proportion of their lands.

COTTON thrives as well in the southern states, as in any part of the world. The West-India islands and those states raised it formerly, when the price was not half what it has been for years past in Europe. It is also worth double the money in America, which it sold for before the revolution, all the European nations having prohibited the exportation of it from their respective colonies to any foreign country. It is much to be desired, that the southern planters would adopt the cultivation of an article from which the best informed manufacturers calculate the greatest profits, and on which some established factories depend.

SILK has long been a profitable production of Georgia and other parts of the United States, and may be encreased, I presume, as fast as the demand will rise. This is the ſtrongeſt of all raw materials and the great empire of China, though abounding with cotton, finds it the cheapeſt cloathing for her people.

IRON we have in great abundance, and a ſufficiency of lead and copper, were labor low enough to extract them from the bowels of the earth.

MADDER has ſcarcely been attempted, but this and many other dye ſtuffs may be cultivated to advantage, or found in America.

UNDER all the diſadvantages which have attended manufactures and the uſeful arts, it muſt afford the moſt comfortable reflection to every patriotic mind to obſerve their progreſs in the United States and particularly in Pennſylvania. For a long time after our forefathers ſought an eſtabliſhment in this place, then a dreary wilderneſs, every thing neceſſary for their ſimple wants was the work of European hands. How great

—how happy is the change. The lift of articles we now make ourfelves, if particularly enumerated would fatigue the ear, and wafte your valuable time. Permit me however to mention them under their general heads : meal of all kinds, fhips and boats, malt and diftilled liquors, potafh, gunpowder, cordage, loaf-fugar, pafteboard, cards and paper of every kind, books in various languages, fnuff, tobacco, ftarch, cannon, mufquets, anchors, nails, and very many other articles of iron, bricks, tiles, potters ware, mill-ftones, and other ftone work, cabinet work, trunks and Windfor chairs, carriages and harnefs of all kinds, corn-fans, ploughs and many other implements of hufbandry, fadlery and whips, fhoes and boots, leather of various kinds, hofiery, hats and gloves, wearing apparel, coarfe linens, and woolens, and fome cotton goods, linfeed and fifh-oil, wares of gold, filver, tin, pewter, lead, brafs and copper, clocks and watches, wool and cotton cards, printing types, glafs and ftone ware, candles, foap and feveral other valuable articles with which the memory cannot furnifh us at once.

If the nations of Europe poffefs fome great advantages over us in manufacturing for the reft of the world, it is however clear, that there are fome capital circumftances in our favor, when they meet us in our own markets. The expences of importing raw materials, which in fome inftances they labor under, while we do not—the fame charges in bringing their commodities hither—the duties we muft lay on their goods for the purpofes of revenue—the additional duties, though fmall, which we may venture to impofe without rifquing the corruption of morals or the lofs of the revenue by fmuggling —the prompt payment our workmen receive—the long credits they give on their goods—the fale of our articles by the piece to the confumer, while they fell theirs by the invoice to an intermediate purchaser—the durable nature of some American manufactures, especially of linens—the injuries theirs often sustain from their mode of bleaching—these things taken together will give us an advantage of twenty-five to fifty *per cent.* on many articles, and must work the total exclusion of several others.

Besides the difference in the qualities of American and European linens, arising from the mode of bleaching, there is a very considerable saving of expence from the same cause. So much and so powerful a sunshine saves a great loss of time and expence of bleaching-drugs and preparations, and this will be sensibly felt in our factories of linen and cotton.

We must carefully examine the conduct of other countries in order to possess ourselves of their methods of encouraging manufactories and pursue such of them, as apply to our own situation, so far as it may be in our power—Exempting raw materials, dye-stuffs, and certain implements for manufacturing from duty on importation is a very proper measure. Premiums for useful inventions and improvements, whether Foreign or American, for the best experiments in any unknown matter, and for the largest quantity of any valuable raw material must have an excellent effect. They would assist the efforts of industry, and hold out the noble incentive of honorable distinction to merit and genius. The state might with great convenience enable an enlightened Society, established for the purpose, to offer liberal rewards in land for a number of objects of this nature. Our funds of that kind are considerable and almost dormant. An unsettled tract of a thousand acres, as it may be paid for at this time, yields very little money to the state. By offering these premiums for useful inventions to any citizen of the Union, or to any foreigner, who would become a citizen, we might often acquire in the man a compenſation for the land, independent of the merit which gave it to him. If he ſhould be induced to ſettle among us with a family and property, it would be of more conſequence to the ſtate than all the purchaſe money.

It might anſwer an uſeful purpoſe, if a committee of this ſociety ſhould have it in charge to viſit every ſhip arriving with paſſengers from any foreign country, in order to enquire what perſons they may have on board capable of conſtructing uſeful machines, qualified to carry on manufactures, or coming among us with a view to that kind of employment. It would be a great relief and encouragement to thoſe friendleſs people in a land of ſtrangers, and would fix many among us whom little difficulties might incline to return.

Extreme poverty and idleness in the citizens of a free government will ever produce vicious habits and disobedience to the laws, and must render the people fit instruments for the dangerous purposes of ambitious men. In this light the employment of our poor in manufactures, who cannot find other honest means of a subsistence, is of the utmost consequence. A man oppressed by extreme want is prepared for all evil, and the idler is ever prone to wickedness, while the habits of industry, filling the mind with honest thoughts, and requiring the time for better purposes, do not leave leisure for meditating or executing mischief.

An extravagant and wasteful use of foreign manufactures, has been too just a charge against the people of America, since the close of the war. They have been so cheap, so plenty and so easily obtained on credit, that the consumption of them has been absolutely wanton. To such an excess has it been carried, that the importation of the finer kinds of coat, vest and sleeve buttons, buckles, broaches, breastpins, and other trinkets into this port only, is supposed to have amounted in a single year to ten thousand pounds sterling, which cost the wearers above 60,000 dollars. This lamentable evil has suggested to many enlightened minds a wish for sumptuary regulations, and even for an unchanging national dress suitable to the climate, and the other circumstances of the country. A more general use of such manufactures as we can make ourselves, would wean us from the folly we have just now spoken of and would produce, in a safe way, some of the best effects of sumptuary laws. Our dresses, furniture and carriages would be fashionable, because they were American and proper in our situation, not because they were foreign, shewy or expensive. Our farmers, to their great honor and advantage, have been long in the excellent œconomical practice of domestic manufactures for their own use, at least in many parts of the union. It is chiefly in the towns that this madness for foreign finery rages and destroys—There unfortunately the disorder is epidemical. It behoves us to consider our untimely passion for European luxuries as a malignant and alarming symptom, threatening convulsions and dissolution to the political body. Let us hasten then to apply the most effectual remedies, ere the disease becomes inveterate, lest unhappily we should find it incurable.

I cannot conclude this address, gentlemen, without taking notice of *the very favorable and prodigious effects upon the landed interest*, which may result from manufactures. The breweries of Philadelphia in their present infant state require forty thousand bushels of barley annually, and when the stock on hand of English beer shall be consumed, will call for a much larger quantity. Could the use of malt liquors be more generally introduced, it would be, for many reasons, a most fortunate circumstance. Without insisting on the pernicious effects of distilled liquors, it is sufficient for our present purpose to observe, that a thousand hogsheads of rum and brandy, mixt with water for common use, will make as much strong drink as will require 120,000 bushels of grain to make an equivalent quantity of beer, besides the horses, fuel, hops, and other articles of the country, which a brewery employs. The fruits of the earth and the productions of nature in America are also required by various other manufacturers, whom you will remember without enumeration. But it is not in their occupations only, that these valuable citizens call for our native commodities. They and their brethren who work in foreign articles, with their wives, children and servants necessarily consume in food and raiment a prodigious quantity of our produce, and the buildings for the accommodation of their families and bufinefs are principally drawn from our lands. Their effects upon agriculture are of more confequence than has ever been fuppofed by thofe, who have not made the neceffary eftimates. So great are the benefits to the landed intereft, which are derived from them, that we may venture to affert without apprehenfion of miftake, that the value of American productions annually applied to their various ufes, as juft now ftated, without including the manufacturers of flour, lumber and bar-iron, is double the aggregate amount of all our exports in the moft plentiful year with which Providence has ever bleffed this fruitful country.—How valuable is this market for our encreafing produce—How clearly does it evince the importance of our prefent plan. But we may venture to proceed a ftep further—Without manufactures the progrefs of agriculture muft be arrefted on the frontiers of Pennsylvania. Though we have a country practicable for roads, our western counties are yet unable to support them, and too remote perhaps to use land carriage of the most easy kind. Providence has given them, in certain prospect, a passage by water; but the natural impediments, though very inconsiderable, and the more cruel obstructions arising from political circumstances, are yet to be removed. The inhabitants of the fertile tracts adjacent to the waters of the Ohio, Potowmack and Susquehannah, besides the cultivation of grain must extend their views immediately to pasturage and grazing and even to manufactures. Foreign trade will never take off the fruits of their labor in their native state. They must manufacture first for their own consumption, and when the advantages of their mighty waters shall be no longer suspended, they must become the great factory of American raw materials for the United States. Their resources in wood and water are very great, their treasures in coal are almost peculiar. As they cannot sell their grain but for home consumption and must propagate sheep and cattle for the reasons above stated, their country will in a short time be the cheapest upon earth. Let us observe the reduction of provisions and raw materials, which even the present year will produce among them, and thence judge, with the necessary considerations, of the time to come.

How numerous and important then, do the benefits appear, which may be expected from this salutary design! It will consume our native productions now encreasing to superabundance—it will improve our agriculture and teach us to explore the fossil and vegetable kingdoms, into which few researches have heretofore been made—it will accelerate the improvement of our internal navigation and bring into action the dormant powers of nature and the elements—it will lead us once more into the paths of virtue by restoring frugality and industry, those potent antidotes to the vices of mankind and will give us real independence by rescuing us from the tyranny of foreign fashions, and the destructive torrent of luxury.

Should these blessed consequences ensue, those severe restrictions of the European nations, which have already impelled us to visit the distant regions of the eastern hemisphere, defeating the schemes of short-sighted politicians will prove, through the wisdom and goodness of Providence, the means of our POLITICAL SALVATION.

THE END.

"A Plain, but Real Friend to America"

Three Letters on Manufactures
(1787)

"Three Letters on Manufactures" were among the many articles encouraging manufacturing that appeared in Mathew Carey's *American Museum* (Philadelphia, 1787–1792). Carey reprinted news articles and pamphlets on topics of national concern and particularly on manufactures, his own special interest. The *American Museum* was received and read by some of the most influential men in America. It recorded, during the six years of its existence, the debate on the economic future of the United States.

The first two of the three letters by "A Plain, but Real Friend" are included in this collection. The third deals with more general questions of trade and answers objections to American manufacturing. The arguments given here are moral and patriotic as well as economic. The "Plain, but Real Friend" believes both that the new country needs manufactures and that all the necessary ingredients for industry are present in the country.

"A Plain, but Real Friend to America," "Three Letters on Manufactures," *American Museum*, volume 1, number 1 (1787), pp. 16–19; volume 1, number 2 (1787), pp. 116–119; volume 1, number 3 (1787), pp. 190–194.

On American manufactures.

LETTER I.

GREAT stress has been laid by several political writers, on the necessity the united states will be under, for a considerable time to come, of purchasing and importing British goods; that they will not be able to procure them any other way, or from any other quarter, on such cheap and advantageous terms; that it will be, at present, more to our interest to be almost wholly occupied in the cultivation of land, excepting that the inhabitants of our cities and towns may find it necessary to follow mercantile employments; that for America to think of manufacturing to any degree of perfection (a country so young, and in many parts but thinly peopled) is an absurd attempt, too premature to be carried into effect; that we must not think of this for many years to come; that our want of resources, the high price of labour, and a number of other things, render us utterly incapable of such undertakings.

What then is to be done?—Why we must go on in the same manner we have done since the peace. Our farmers and planters must raise as much tobacco* as possible: our merchants must buy it dear from them, and send it to England, and sell it cheap, in order to pay their debts. We must avoid manufacturing any article whatever, that can possibly be made here, because in England it can be made somewhat better, and a little cheaper, and be transported to us from thence in *British bottoms.* We must continue to import all the goods we can, of every kind, till all the houses in our towns are stores; and till there is not in the whole country a mill, a tavern, or a cross-road

NOTE.

* This essay was written in Maryland.

without a store. And what then? Why this is the way for the united states to become great and wealthy; nay—but if I mistake not, this is the direct way to learn the poor debtor's soliloquy—" To break, or not to break, that is the question."

From a strong attachment and love to my native country, I am induced to take up my pen, after having seriously considered the subject. I therefore now beg leave to obviate the foregoing propositions by the subsequent considerations.

We cannot all be cultivators of land—neither can we be all merchants.

We cannot be all cultivators of land, because every father has it not in his power to give every one of his sons a tract of land: but every father may have it in his power to have his son taught a trade. If, then, trades and manufactures were encouraged in this country, every father might place his son, some how or other, in an eligible way of procuring a decent living, by which he would become a useful member of the community. The present mode of cultivating land, requires that a planter or farmer be possessed of a considerable quantity of land: and hence it is, that many a farmer, having children to provide for, but not having land sufficient of his own where he lives —seeing no mechanical business worth following—and being desirous of settling his children in the best manner he possibly can—reasons thus: Land is too dear hereabouts for me to purchase: but if I go back, I may, with what money I can raise, purchase as much as will do for me and all my children. He sells his present plantation, and all that he possesses, takes his family, and, after many a weary step, settles himself on the frontiers. Now his troubles commence: after a world of labour and toil, his plantation is in some order: but he is now living at such a great distance from places of trade and commerce, that he would spend almost the whole produce of his plantation, were he to bring it to market. He therefore raises little more than what the wants of his family call for: and thus the state derives little or no advantage from him and his family.

But if trade and manufactures were encouraged, this need not be the case, as there would be employment for all, and our country would be more thickly settled, and increase in strength. As it is disagreeable to some to move back, in order to provide for their children, many a farmer's son is put apprentice in towns to mercantile business, with a design to learn the art of book-keeping, and to be a merchant (for there is no alternative—you must either be a farmer or merchant) though, perhaps, when the young man is out of his apprenticeship, his father may not have a single hogshead of tobacco to spare for his assistance to begin trade. Well, what is to be done? He gets a clerk's place—he dresses, powders, and waits upon the ladies—he makes friends as soon as he can—he procures letters of credit—goes or sends to England—and, after a while, sets up a store with a cargo of goods, which, in all likelihood, will never be paid for—and, in that case, becomes a bankrupt, and an injury, instead of a benefit, to society.

And what must the poor mechanic do? Why, he is told, that, for many years to come, trades and manufactures will not be encouraged: and finding daily that mechanics are very little thought of, and that nothing is to be made by his trade—that every article, almost, that ought to be manufactured in this country, is imported from abroad—in short, finding by his trade he can scarcely procure bread for his family—what must he do? Why he quits his trade, and turns merchant too; he has as good right as another. We live in a land of equal liberty, and what will not a man do rather than feel the bitterness of poverty! Here then we discover the motive that induces so many to turn merchants; the true reason of the dulness of times; necessity makes us turn merchants. We are all like to become merchants. Hence this branch is wholly overdone, and in a most ruinous state; the country full of goods, and our money all gone. What alternative? is it not clear and evident that we cannot all be merchants? If, then, American manufactures were countenanced and encouraged, this need not be the case. The time may now be come, when, if we will but open our eyes, we may see the necessity of encouraging every kind of manufactures that we can make; for supposing that we cannot manufacture every thing, it does not follow that we must do nothing in that way; let us do what we can; let us encourage, to the utmost of our abilities, the manufacturing those articles for which nature has given us resources and materials. Let us consider the benefit this will be to our country, the propriety and good policy of the measure. The sooner we begin, the sooner we shall bring them to some degree of perfection. Remember Rome itself had a beginning, and so likewise must all affairs that depend on human art and industry. And though at the first we may meet with some difficulties, what then? We have been in the practice of self denial already: a patched thread-bare coat was thought no disgrace in the time of war, and to be clothed in clean homespun was then a great honour; nor were the first women in the country ashamed of being employed in making linen for their families; none then but tories were clothed in purple and fine linen, and fared sumptuously every day.

I am happy to find that at this period, the state of Massachusetts has set us a noble example. She has lately passed a law, laying heavy duties on a variety of articles imported from Europe—several she has prohibited, under penalty of seizure.

Here let it be observed, that as it is the wisdom of America to discourage the importation from Europe of those things she is able to make—so it is equally her wisdom to encourage emigrations from Europe. The emigrants, however, that America will be chiefly visited by, and whom she most wants, are the industrious poor. The wealthy and indolent can find no temptation to cross the Atlantic, unless to make a short stay, and then carry away all the money they can. How, then, are those industrious poor to be employed? They are chiefly bred to trades and mechanical business. They know nothing of cultivating corn, tobacco, &c. but they are versed in weaving, raising and dressing flax and wool—they are clothiers, fullers, hatters, saddlers, black and white smiths, cutlers, shoemakers, &c. &c. &c. Is it not, then, a melancholy consideration, that many tradesmen of these descriptions, after coming to America, are obliged, through want of work, to return home again—while many of those who stay, must commence labourers to procure a morsel of bread? I know this to be a truth. Could such emigrants be assured of employment and encouragement in coming to America, I am persuaded that some thousands would soon visit these shores. We should soon have a competent number of workmen to carry on most kinds of manufactures. There would, in a short time, be no necessity to import a single article of cutlery, saddlery, ironmongery, tin, pewter, and copper ware, and many others. Cash, of which at present we feel such a pressing want, need not then be sent abroad in such amazing quantities, as must ever be the case upon the present plan. Numbers in our towns, who are now in a great measure idle, and who at every return of the severe season of the year, are in a perishing condition, and must be relieved by the hand of charity, might then be employed some way or other. Many of our citizens, instead of being necessitated to move to the back country, to seek land for their children, and those who emigrate to us from Europe, instead of being obliged to return back again, for want of such employment as suits them, would settle new towns and villages throughout the country; and on the banks of those fine rivers and streams that now wear a horrid, deserted aspect. We should not then see such sparse scattered settlements, but appear like the countries in some parts of Europe, full of people, and as industrious as a bee-hive.

When the minds of the people of America were really virtuous, at the beginning of the late contest, every man was convinced of the necessity of our encouraging manufactures, and employing our own people, that we might be truly independent; but now it seems we must employ manufacturers that live three thousand miles distant from us, and leave our own poor to wander in the woods and wilds of the back country, to live like Indians, and to be useless to our government. He must be a shallow politician indeed, who does not see that obliging people to go back, is weakening and prejudicial to the populous parts of the country.

A plain but real friend to America.

LETTER II.

IN my first letter I dropt a few hints on the necessity and utility of encouraging our own manufactures, together with the impropriety of our still pursuing the same ruinous system of commerce we have hitherto done, as it has a direct tendency to make us all cultivators of land, or merchants. I delivered my sentiments with some diffidence, being conscious what I advanced would not be universally relished, and knowing that the most salutary proposals are often rejected and treated with contempt, and indeed sometimes violently opposed from interested views in some, misapprehension in others, and an affectation to shine in those who conceive themselves to be men of abilities and genius. From a conviction, however, of being influenced by no other motive than the love of my native country, I endeavoured to search after truth, and to adduce none but the plainest and most natural arguments.—Considering the subject of the greatest importance, as involving the happiness or misery of these rising states, a large field opened before me, and finding it difficult to take in the whole at one view, I contented myself for that time with a partial survey, and just fluttered a little over part of its surface.

I now venture a second time to take up the subject, and with the same simplicity of language, being persuaded it is the best service I can render my country.

The common sense of mankind clearly demonstrates that it is wholly inconsistent with the welfare of any state to depend altogether on the manufactures of another country— even admitting that all the articles wanted, might be imported considerably cheaper than they can be made at home. The reasons I shall urge in order to illustrate this doctrine, will, I hope, to every candid person, be clear and satisfactory, and operate with such conviction, that they will be no longer at a loss to determine whether the united states ought, or ought not, to encourage, support, and maintain their own manufactures.

Upon a close inspection, it will, perhaps, be found, that to almost every country, nature has given something peculiar to itself : but to most countries she has given at least such things, as, by proper management and art, may be wrought into what are termed the necessary articles of life. To America nature has been profusely liberal. She enjoys the soil of every clime—stretching along an extensive sea-coast, she abounds not only with raw materials proper for manufactures—but many ingredients requisite for foreign luxuries, are also to be found deposited in her lap. With such a country, and such materials, and the blessing of free governments, shall we continue to employ manufacturers at several thousand miles distance, while a great part of our own people are idle, or employed to very little purpose—and another very considerable part likely to become bankrupts, if not beggars—and all this owing to our want of œconomy and the exertion of our native strength ? No : let us act like rational beings, and no longer be a reproach to ourselves, and to the common sense of mankind.

If, then, it be true, that nature has given those invaluable materials necessary for manufactures, I ask, why are they given to us ? Not, I conceive, that we should export them to Great Britain, or to any other part of the globe, (as, unhappily for ourselves, we seem to have hitherto supposed), there to be manufactured and worked up, and then returned back to us again. Nature does nothing in vain. Her operations are regulated by the nicest and best rules. Whatever she gives us in our own country, we may rest assured, if rightly used, will be found to be the best for us. Conduct not yourselves, therefore, my countrymen, as if you believed that nature bestowed on one country what ought

The Philosophy of Manufactures

to be given to another, which absurd idea would be chargeable on you, for your spurning at her gifts, by either wholly neglecting them, or sending them abroad to be manufactured. How contrary this to the dictates of common reason! Be wise for the future. Learn to prize the numerous blessings which God and nature have favoured you with. Know their value. Use them as you ought, and you have your remedy.

But it seems as if all the bounties of providence are to be lost upon us, and that for this reason, because nature has not formed us artists and mechanics, and given us from our birth all the wisdom and experience of old countries, and taught us the mystery of being industrious and working cheap. She would, no doubt, teach us how to attain these perfections, were we willing to follow her. But why should we throw away or lightly esteem her favours, for want of such perfections? Remember, practice alone makes perfect, to which if we never attain, the question whose fault it is, may be readily solved. Begin the work. Say no longer, we are too young. There is plenty of time on our hands. We did not think ourselves too young to enter the lists with Britain, though without arms, without money, and without every thing but resolution and courage: and the contest proved that we were of mature age. Could we then find men to compose an army? We may also now find men to work up our raw materials.—There are numbers of mechanics scattered throughout these states, who wait your call: and I now vouch for it, that whenever the attempt is made, we shall find it much easier to erect and carry on necessary manufactures, than it was to war with that powerful nation eight long years.

Go survey the continent—inspect the ruined state of commerce—anticipate the miseries ready to fall on the heads of numerous respectable merchants who are now declining and becoming bankrupts—see the poverty, which with horrid aspect threatens the industrious mechanic—and forbear to say, that the time to manufacture is not yet come. I say the time is fully come now—if ever. Before we feel the deadly symptoms of our disease—before we sicken beyond all cure, and die the death of national bankruptcy, let us apply the remedy we have in our hands, and all may yet be well. But, it may perhaps be said, though we might do something that way, yet goods cannot be made so cheap as they may be imported. What then? Is this to pass as a sufficient reason for an entire neglect of home-made articles? Examine the ground on which the matter stands. Had Great Britain totally neglected manufactures until a few years ago, is it to be presumed she could afford her goods by this time as cheap as she at present does? No, certainly. What is the inference? If we begin, and persevere, we shall in time be able to afford our goods as cheap as she can hers. What period shall we appoint for ourselves when to begin? How long will it take us until we arrive at sufficient age, and are fit to go to work?—Surely, the time of youth is a very proper period to serve our apprenticeship. This I have found in common life. We seldom see a person fond of labour in old age, who has lived an idle life when young. But if we neglect arts and manufactures much longer, we shall find it still more difficult to begin. We have many artisans and tradesmen now among us; the country every where abounds with them, and they may soon be collected: but they are turning their attention to other occupations. In some time hence, perhaps, there may not be a sufficient number found to

The Philosophy of Manufactures

carry on some of the most useful branches, for there is little encouragement to mechanic emigrants. If, therefore, it is an object with us to work well, and to work cheap, we have nothing to do but to begin. Let us then set about it without delay, and, we may depend on it, we shall, in due time, attain to every perfection in manufactures.

Whether we can import cheaper than we can manufacture, ought not so much to be the question, as, Will it be most expedient? Will it be an advantage to these states? Will it employ our own people, and keep them from idleness, which is itself a great evil? Will it save our money? Will it have a tendency to preserve us virtuous?

I have not the smallest doubt but after some years' acquaintance with manufactures, we may afford those for which we have the raw materials within ourselves, nearly, if not altogether as cheap as the same may be imported: and when we mention cheapness, let it be remembered that the saving so much money by our own industry, ought most assuredly to be taken into the account : for it is a well-known maxim, and will equally hold good in communities, as in families, that a penny saved, is a penny gained. Who does not see, that all the money spent upon such foreign articles as our own people can make, is just the same as so much treasure lost to us? For the same quantity of land could be cultivated as at present—the same quantity of wheat, corn, tobacco, &c. raised as before—and merchants sufficient found to carry on our commerce: but with this material difference, that the balance would probably be in our favour. Taking every thing then into account, we have no great cause to boast of our procuring British goods so very cheap. We pay dear enough for them, my countrymen, much more so than many of you are aware. The price, upon close inspection, will be found to be no less than the annual produce of our lands, a vast quantity of cash, and an increase to a debt already enormous : for it is my opinion, that we have contracted a debt since the peace, nearly equal to all our expences during the war.

By manufacturing ourselves, and employing our own people, we shall deliver them from the curse of idleness. We shall hold out to them a new stimulus and encouragement to industry and every useful art. We shall open an extensive field to many laudable pursuits. The speculative, the ingenious, as well as the laborious, may all employ their time and talents to valuable purposes. Idleness may be justly termed the bane of the mind, and the grand inlet to numerous vices. Nations that are remarkable for idleness and sloth, are for the most part prone to luxury, effeminacy, and extravagance. What hasty strides have we not taken since the peace to gain the summit of those refinements! O may the good genius of America now step forth, and inspire her infatuated sons, to make a solemn pause, to consider, and to amend!

What can we promise ourselves, if we still pursue the same extensive trade? What, but total destruction to our manners, and the entire loss of our virtue? Every man, in proportion as he falls into luxury, becomes more and more inclined to bribery and corruption. He finds wants and desires before unknown, and these wants and desires being merely artificial, become not easily restrained within proper bounds and limits.

It is worthy of remark, that while we were dependent on Great Britain, her policy and laws restrained us as much as possible from manufacturing. —Even the great mr. Pitt, in one of his famous speeches, was against permitting so much as a hob-nail to be made in the colonies. Why all this opposition to our working up those materials that God and nature have given us? Surely, because they were sensible, it was the only way to our real independence, and to render the habitable parts of our country truly valuable.—What countries are the most flourishing and most powerful in the world? Manufacturing countries. It is not hills, mountains, woods, and rivers, that constitute the true riches of a country. It is the number of industrious mechanic and manufacturing as well as agriculturing inhabitants. That a country, composed of agricultivators and shepherds, is not so valuable as one wherein a just proportion of the people attend to arts and manufactures, is known to every politician in Europe: and America will never feel her importance and dignity, until she alters her present system of trade, so ruinous to the interests, to the morals, and to the reputation of her citizens.

A plain, but real friend to America.

John Morgan

Agriculture Preferable to the Mechanic Arts (1789)

This speech, originally presented on March 15, 1789, to the Shandean Society, a gentleman's literary club of New Bern, North Carolina, repeats familiar arguments in favor of agriculture. The United States, says Morgan, has too few people and too little wealth for successful industry. Any attempt to encourage manufacturing would hurt agriculture and "greatly impoverish" the country. Notice that Morgan is not opposed to industry on moral grounds. Eventually, he says, some industry—wine making and cotton manufacture—may be profitable. Until such time as "our numbers and wealth" allow manufacturing, though, Morgan would have the United States depend on agriculture and trade.

John Morgan (1735–1789), a doctor, was physician-in-chief of the Continental Army from 1775–1777 and founder of the University of Pennsylvania Medical School. He was involved in several schemes for the promotion of silk and iron manufacture, none of them successful.

John Morgan, "Whether it be most beneficial to the united states, to promote agriculture, or to encourage the mechanic arts and manufactures?", *American Museum*, volume 6 (July 1789), pp. 71–74. By permission of the Houghton Library, Harvard University.

Whether it be most beneficial to the united states, to promote agriculture, or to encourage the mechanic arts and manufactures?—from a discourse, pronounced by John Morgan, M. D. F. R. S. at a meeting of the Shandean society of Newbern, North Carolina, March 15, 1789.

AGRICULTURE is the oldest employment of man, even of our first parents and primitive ancestors. It has been ever held in the highest estimation, by wise men of every nation, for the innocence that attends it, and for the health and vigour of body it produces. It has had a great number of sovereign princes, amongst it patrons and cultivators, not only for the pleasures, but also for the profits, attendant on its pursuits, as well in administering to all the most essential wants of individuals, as in producing riches to a nation. Some countries, from their high state of agriculture, becoming granaries to neighbouring nations, have abounded proportionably in wealth, population, the arts of peace and the magazines of war, as history shews to have been the case of Ægypt.

In new countries, in particular, and consequently at first but thinly inhabited, it becomes a primary object, to cultivate the earth, in preference to every other manual labour and pursuit. Wherever good lands abound, whatever can be raised from them, will be an article of worth. And whereas labour is dear from the scarcity of hands, the produce of the earth will yield greater emoluments to the husbandman, than any other species of labour. In this country especially, which is so extensive, and the number of settlers so small in proportion to the land they possess, agriculture will more abundantly supply our wants, than the manufacturing any kind of goods can do, whereof the chief value depends on the labour of many.

From the largest accounts we have, the number of inhabitants, in the united states of America, falls short of three millions; but the land, fit for tillage, pasturage and other purposes of rural life, is capable of furnishing above fifty millions of persons, without being over-crouded. Abounding with materials from the produce of the earth, the present generation can command a supply of the articles they require, in greater plenty, and of better quality, than it would be possible to manufacture ourselves. The necessaries of life are comparatively few. These are easily procured from our lands. But the articles of manufactures and commerce, which not only serve to supply our real wants, but contribute to our imaginary wants and luxury, are innumerable. In this our as yet infant state, we are therefore loudly called upon by our wants, by our interests, by the first law of nature, and good policy, to give our chief attention to agriculture: first, for the more immediate supply of our necessities, and secondly, to furnish us with the most effectual means of procuring, in the way of barter and commerce, all those things, which we cannot expect or hope to obtain by our own labour.

Mechanic arts may be justly considered, as the offspring of that plenty, which agriculture begets; but they are generally slow in their progress at first, and take a long time, before they reach to any degree of eminence. It is found policy then, and the true interest of this country, to encourage the natural disposition of the Americans to cultivate the ground, and draw from it the raw, but useful materials, of which it is so capable with little labour, and to supply the transatlantic nations of Europe, that depend upon their numbers, to manufacture for us whatever we stand in need of; which, from their skill and long experience, they can afford with greater ease and cheapness, than we can furnish ourselves.

To evince the truth of this assertion, let us reflect, with what success these states, when they were yet but colonies of Britain, pursued this plan of conduct, in adhering to their fisheries, and in clearing and cultivating the ground: thus furnishing the West Indies with lumber, iron, flour and other provisions; and Great Britain herself, and, through her, the countries subject to her dominion, and connected with her by treaties of friendship and commerce, with fish, naval stores, tobacco, pot-ash, rice, indigo, silk, hemp, flax-seed, and other materials for their different manufactures.

It requires no great extent of acquaintance with the products and exports of the different united states of America, to perceive, that our most certain and substantial riches flow from agriculture, hunting, fishing, exploring the earth, and furnishing those raw materials for commerce, which, in return, bring in the wealth and conveniences of other nations.

The plenty of codfish on the coasts of New England, as well as salmon, herring, and a variety and abundance of other species of fish, which employ a great number of their sea-faring people to catch, salt, barrel, and transport them to Portugal, Spain, Italy and the Levant, is to be considered as a rich mine, from which they derive great wealth, with comparatively little labour. The business of ship-building, the cheapness of which depends upon the quantity and convenience of timber with which the country abounds, and the interest of the husbandman to clear his ground—is another great source of power and riches. By these means, and the making of pot-ash, from the trees, they burn to clear their lands, (which is a valuable article of export) together with their lumber and naval stores, they are enabled to supply foreigners with those articles, from which they acquire ample and valuable returns. Hence, too, they are furnished with active and healthy seamen, for manning their vessels, and for carrying on their commerce with different and distant parts of the world.

The middle states, viz. New York, New Jersey, Pennsylvania. and Delaware, are, in general, fertile in their soil, and abound in all kinds of excellent grain. They also abound in mines of iron ore, from which pig and bar iron are made, and afford valuable articles of remittance to different countries, by furnishing materials for their casting and various mechanic arts. It is not my intention to enlarge upon trade, farther than to point out the raw materials, produced from agriculture and working of the earth, which may be employed to greater advantage by us, in our present state, as articles of commerce, than as mere objects of manufactures for ourselves.

I must here observe, that, where I have referred some particular products of the earth, to some states only, it is to be understood, that the same, or several of those articles, may likewise be the productions of others, or cultivated in them with advantage; although, for the sake of brevity, I have made no mention or repetition of them, as your superior knowledge of the subject will readily enable you to supply my omissions.

Tobacco has been justly considered as the great staple, and standing commodity, of Maryland and Virginia, which states are to the southward of Pennsylvania and Delaware: and it may be also raised in the three remaining states to the southward of Virginia, viz. the two Carolinas and Georgia. The tobacco, which was annually shipped to Great Britain, before the revolution, fell little short of one hundred thousand hogsheads; and the amount of the customs was above a million of pounds sterling. The three great staples of the Carolinas and Georgia, consisting of rice, indigo, and naval stores, were then computed at near half a million more. Besides which, Georgia has produced great quantities of raw silk, which, being exported to England, came into competition with, and indeed obtained the pre-eminence over, the finest silk of Piémont, for which half a million per annum had been paid. Georgia has been also engaged in making and exporting pot-ash, an article of great demand in bleaching, and in a variety of other trades and manufactures.

From this narrative it appears, of what amazing consequence it has been to North America, to confine her chief views to the improvement of her fisheries and agriculture; and to depend upon the exportation of those raw materials, which she has derived from the waters, the surface and bowels of the earth, to draw from the nations of Europe, and their dependencies, every article of commerce and manufacture, which she stood in need of, and which she could not obtain, by turning the labour of her inhabitants to manufactures and the mechanic arts. The employment of hunting, and a trade with the native Indians employed in hunting, has a connexion with this subject. Hence, we procure furs, and peltries of all sorts, which are exported, as raw materials for the manufactures of other countries, and prove a new source of wealth.

The riches not only of America, but of every other country, depend chiefly upon the product of their lands, and upon the quantity and value of the articles exported from it, above what are imported, which gives the balance of trade in favour of such country. Should we then attempt, by turning our thoughts unseasonably, and beyond what we are capable of executing with ease, to manufacture more than our necessities require, and export less of our produce, we should soon find the balance of trade against us, and ourselves greatly impoverished. Such would be the natural consequence of checking agriculture, from which our wealth immediately flows, and making it give way to mechanic arts, which cannot be carried on here with the same ease and advantage, as in older and more populous countries.

Let me repeat, that the principal articles of arts and commerce are the productions of agriculture, by means of which, after we have supplied our own demands, we are enabled to bring to us the manufactures, and productions of other countries, that we stand in need of. From a due attention to our agriculture, our fisheries and hunting, and the commerce we establish on them, the means of living become easy, early marriages are promoted, and population is increased—witness the coasts and fishing towns of New England, and the rapid encrease of the children of the industrious husbandmen. This is the consequence of the greater ease of rearing and maintaining large families. It also invites a greater number of foreigners to visit and settle in the country, who mix with us and become one people; the same in their interests, pursuits and manners.

Whenever a country is fully stocked with inhabitants, it is then in a situation to require and encourage manufactures, beyond what is practicable or prudent to attempt, in its early state. But I mean not, in denying a preference to the mechanic arts in our present circumstances, to exclude from a proper share of attention to this object, all such hands as can be well spared from agriculture and commerce, or such as may be necessary for cloathing, for building ships and houses, and for working up those materials, which can be manufactured, with more ease and profit to ourselves, than they can be imported. I even think, as grapes are the natural produce of our country, that planting vineyards, and making wines, at least for our own use and consumption, would be beneficial; and that, while the southern states give their attention to the raising of cotton, the more populous states to the northward might employ many hands and proper machines in carding, spinning and weaving it, which would be a great saving to the inhabitants of America.

I conclude, as a consequence of what I have advanced, that, whilst older and more thickly inhabited countries are employed in manufactures, the Americans ought to lay themselves out to raise all sorts of commodities, to fit them for a market, and thus to furnish other nations with the materials, of which they stand in need for carrying on their established manufactures, and so derive greater advantages from trading with them, than it is possible by following the mechanic arts and manufacturing for ourselves, till we are more capable, from our numbers and wealth, of carrying on such undertakings.

Tench Coxe

A Brief Examination of Lord Sheffield's Observations on the Commerce of the United States (1791)

Tench Coxe's lengthy answer to Lord Sheffield's *Observations* appeared in 1791, eight years after the latter was published. Coxe had three reasons for this late rebuttal. First, Sheffield's work was still read and still influential. Second, in 1791 Coxe was assistant secretary of the treasury and had just completed the first complete survey of the American economy. Thus, he had the facts with which to refute Sheffield. Finally, he was about to propose the establishment of a manufacturing corporation—the Society for the Encouragement of Useful Manufactures—and hoped that this pamphlet would serve as a springboard.

In the *Brief Examination* Coxe argues that the United States had many manufacturing establishments and was already a commercial power that England could not afford to ignore. A large part of the pamphlet was devoted to statistics of manufactures and commerce. In the section reprinted here, Coxe examines the possibilities of American manufacturing and predicts that a joint-stock company "of persons of character and judgement . . . would be sure to be a success in their manufactories." The proposal he presents here is similar to one he later suggested for the Society for Establishing Useful Manufactures.

Tench Coxe, *A Brief Examination of Lord Sheffield's Observations on the Commerce of the United States in Seven Numbers with Two Supplementary notes on American Manufactures* (Philadelphia, 1791), pp. 39–42.

The manufactures of Great Britain and Ireland are very generally good, often excellent, and almoſt always as handſome as the nature of the article will admit. Yet, there are not wanting proofs, that we ſhall take confiderable quantities of goods from other countries. Twenty-two ſhips, for example, arrived in the united ſtates from St. Peterſburg, in the year 1790, with cordage, ticking, drillings, diaper, broad linens, narrow linens, printed linens, craſh, ſheetings, ravens duck, Ruſſia duck, nail rods, and rolled iron for hoops. The remainder of their cargoes were bar iron, hemp, and flax, which were intended to be manufactured here. Nankeens, ſilks, long-cloths, porcelain and ſome ſmall articles, are imported regularly from China: and muſlins, plain, ſtriped, figured, and printed, with ſilks, and a variety of other articles, are imported from India. It being manifeſtly injurious to the manufacturing intereſt of every nation in Europe, even to import, and much more ſo to conſume theſe goods, there can be no doubt, that they will be ſupplied to us in the Eaſt Indies, with more readineſs every year; and if a few more callico printers were to eſtabliſh themſelves among us, the importation of printed callicoes and cottons might be exceedingly diminiſhed. The importation alſo of dowlas, oznaburghs, ticklenburgs, and other German linens, and of Haerlem ſtripes, and tapes, from Bremen, Hamburgh, and Amſterdam, together with the manufactory of every ton of hemp, and almoſt every ton of flax, which we raiſe or import, has very much affected the Britiſh and Iriſh linen trade. It appears from various documents, that the average exports of their manufactures to the united ſtates for ſeveral years prior to the year 1789, were near half a million of dollars leſs than the average exports of ſeveral years immediately antecedent to the war, though our population has probably doubled in the laſt twenty-five years. It is not improbable, however, that the great quantities of goods ſhipt ſince 1789, in conſequence of the jealouſy of American manufactures, the apprehenſions of a rupture with Spain, and the efforts of the Britiſh cotton manufacturers to baniſh Eaſt India goods from our markets, would ſhow a conſiderable increaſe in the laſt and preſent years. In ſhort, the united ſtates are an open market; the American merchants are men of judgment and enterpriſe; and conſequently the goods of every country in the world, which are adapted to our confumption, are found in our warehoufes. It is certainly true, that among them are very large quantities of Britifh manufactures, being much and juftly approved, and being imported on convenient credits by our merchants, and copioufly fhipped by Britifh merchants and manufacturers on their own account, to their correfpondents here. If properly conducted on both fides, it may yet be a very beneficial trade to the two countries; but it has not excluded the valuable goods of other nations, nor has it prevented a great progrefs of our own manufactures, particularly in the family way. Cordage, gunpowder, fteel, nails, paper, paper-hangings, books, ftationary, linfeed oil, carriages, hats, wool and cotton cards, ftockings, fhoes, boots, fhot, and many other articles are made in confiderable quantities, fome of them as far as fifty per centum on the demand, and others in quantities nearly equal to the confumption. Liberal wages, and cheap and excellent living, free from any excife, except a very fmall one, (compared with any in Europe) upon fpiritous liquors, operate daily to bring us manufacturers and artizans in the manual branches; and we are beginning to fee the great, and to us, the *peculiar* value of labour-faving machines. The rate of manual labour is no objection againft them, but abfolutely in their favour; for it is clear, that they yield thr greateft profit in countries where the price of labou, is the higheft. The firft judicious European capitalifts who fhall take good fituations in the united ftates, and eftablifh manufactories, by labour-faving machines, muft rapidly and certainly make fortunes. They cannot, it is prefumed, be long infenfible of this; but if they fhould continue fo, the appreciation of our public ftocks will probably bring fome of our own capitalifts into the bufinefs. The public creditors, the owners of perhaps ifteen millions fterling, of now inactive wealth, might at this moment do much towards the introduction of the cotton mills, wool mills, flax mills, and other valuable branches of machine manufacturing. It is paft a doubt, that were a company of perfons of character and judgment to fubfcribe a ftock for this purpofe, of 500,000 dollars in the public paper, they might obtain, upon a depofit of it, a loan of as much coin from fome foreign nation, at an intereft lefs than fix per cent. Was fuch a company to be incorporated, to have its ftock transferable as in a bank, to receive fubfcriptions from 400 dollars upwards, to purchafe 500 or 1000 acres of land well fituated for receiving imported materials and exporting their fabrics—were they to erect works in the centre of fuch a body of land, to lay out their grounds in a convenient town-plat, and proceed with judgment and fyftem in their plan, they would be fure of fuccefs in their manufactories; they would raife a valuable town upon their land, and would help to fupport the value of the public debt*. Were a few eftablifhments like that defcribed to take place (and there are room and funds for many of them) even the manufactories of *piece goods*, of every kind in which machinery could be applied, would foon be introduced with profit into the united ftates. It cannot, on cool reflexion, be expected, that a country remote from all the manufacturing nations, and able to produce the requifite raw materials, will continue to depend on diftant tranfmarine fources, for the mafs of her neceffary fupplies. The wonderful progrefs of other nations, which have commenced manufactures under difadvantages much greater than any we have to contend with, will powerfully incite us to exertion Until the year 1667, a piece of woolen cloth was never dyed and dreffed in England. This great manufacture was quickly after improved by the fkill of foreign emigrants, (a mean at our command); and fo rapidly has the woolen branch advanced, that it was eftimated, in 1783, at the immenfe fum of £.16,800,000 fterling (above feventy-four millions of dollars) per annum, and was equal in value to all the exports, and fuperior to all the revenues of Great Britain It may, perhaps, be afked, why manufactures were not eftablifhed in the late war? Any man, who makes a comparifon of a variety of branches as they were in 1774, and as they ftood in 1782, will perceive a great advance to have taken place, though manufactures were little encouraged, through the intermediate eight years, by reafon of the total occupation of government in the profecution of the war: their importance moreover was not duly eftimated. The Britifh manufacturers, who can now emigrate with the greateft convenience, then viewed the people of this country as enemies. Neither they, nor the people of other nations cared to rifque themfelves in an invaded country, nor would they hazard a capture in their paffages hither. Notwithftanding thefe impediments, the manufacturers of the united ftates

* This meafure, which was in contemplation at the time when thefe papers were written, has been fince digefted and commenced. The capital already engaged amounts to above 250,000 dollars.

have been found to be the moſt ſucceſsful competitors with thoſe of Great Britain in the American market. They have not made fine linens, fine cloths, ſilks, ſtuffs, and other articles requiring a great degree of ſkill, labour, or capital ; but they have made common cloths of linen, woolen, and cotton, ſteel, nails, ſheet iron, paper, gunpowder, cabinet work, carriages, ſhoes, and fabrics of the ſimple but moſt important kinds.

Alexander Hamilton

Report on the Subject of Manufactures (1791)

The *Report on the Subject of Manufactures* of Alexander Hamilton (1757–1804) was the most important document on industry written in the early years of the United States. In it Hamilton suggested that the government undertake a deliberate policy of promoting industry and proposed a system of protective duties. It is a base on which subsequent advocates of American manufactures built their arguments.

The report is not merely a defense of manufactures but a positive statement that manufacturing is necessary to the economic development of the country and that it is the duty of the government to promote it. The report consists of three parts: the reasons why manufacturing should be promoted; the methods by which manufacturing can be promoted; and a description of the present state and future possibilities of a number of industries. Portions of the first and second parts are reprinted here.

Hamilton was secretary of the treasury when he wrote the report at the request of the Congress. The federal government did not adopt the plans suggested in it, but the document was extremely influential for many years. The goals and methods of Alexander Hamilton can be seen in greater detail in his plan for the Society for Establishing Useful Manufactures, which was proposed contemporaneously with the *Report*. The S.U.M. plan is reprinted in this volume.

[Alexander Hamilton], *Report of the Secretary of the Treasury of the United States, on the subject of manufactures. Presented to the House of Representatives, December 5, 1791* (Philadelphia, 1791). By permission of the Houghton Library, Harvard University. See also the annotated text of Hamilton's *Report* and its several drafts published in *The Papers of Alexander Hamilton,* ed. Harold C. Syrett and Jacob E. Cooke, (New York: Columbia University Press, 1966), volume 10, pp. 1–340.

The SECRETARY of the TREASURY,

IN Obedience to the Order of the House *of* Representatives, *of the 15th Day of January, 1790, has applied his Attention, at as early a Period as his other Duties would permit, to the Subject of* Manufactures; *and particularly to the Means of promoting such as will tend to render the* United States *independent on foreign Nations, for Military and other essential Supplies:*

AND HE THEREUPON RESPECTFULLY SUBMITS THE FOLLOWING

REPORT.

THE expediency of encouraging manufactures in the United States, which was not long since deemed very questionable, appears at this time to be pretty generally admitted. The embarrassments which have obstructed the progress of our external trade, have led to serious reflections on the necessity of enlarging the sphere of our domestic commerce: the restrictive regulations, which in foreign markets abridge the vent of the encreasing surplus of our agricultural produce, serve to beget an earnest desire, that a more extensive demand for that surplus may be created at home: And the complete success which has rewarded manufacturing enterprise, in some valuable branches, conspiring with the promising symptoms which attend some less mature essays in others, justify a hope, that the obstacles to the growth of this species of industry are less formidable than they were apprehended to be; and that it is not difficult to find, in its further extension, a full indemnification for any external disadvantages, which are or may be experienced, as well as an accession of resources, favorable to national independence and safety.

* [Hamilton at this point anticipates the arguments of "respectable patrons of opinions, unfriendly to the encouragement of manufactures," and he disposes of each argument in turn. He then proceeds to state his case for manufactures.—Eds.]

It is now proper to proceed a step further, and to enumerate the principal circumstances, from which it may be inferred. That manufacturing establishments not only occasion a positive augmentation of the produce and revenue of the society, but that they contribute essentially to rendering them greater than they could possibly be, without such establishments. These circumstances are,

1. The division of labor.
2. An extension of the use of machinery.
3. Additional employment to classes of the community not ordinarily engaged in the business.
4. The promoting of emigration from foreign countries.
5. The furnishing greater scope for the diversity of talents and dispositions which discriminate men from each other.
6. The affording a more ample and various field for enterprize.
7. The creating in some instances a new, and securing in all, a more certain and steady demand for the surplus produce of the soil.

Each of these circumstances has a considerable influence upon the total mass of industrious effort in a community: together, they add to it a degree of energy and effect, which are not easily conceived. Some comments upon each of them, in the order in which they have been stated, may serve to explain their importance.

The Philosophy of Manufactures

1. *As to the division of labor.—*

It has juſtly been obſerved, that there is ſcarcely any thing of greater moment in the œconomy of a nation, than the proper diviſion of labor. The ſeparation of occupations cauſes each to be carried to a much greater perfection than it could poſſibly acquire, if they were blended. This ariſes principally from three circumſtances.

1ſt. The greater ſkill and dexterity naturally reſulting from a conſtant and undivided application to a ſingle object. It is evident, that theſe properties muſt increaſe, in proportion to the ſeparation and ſimplification of objects and the ſteadineſs of the attention devoted to each; and muſt be leſs, in proportion to the complication of objects, and the number among which the attention is diſtracted.

2d. The œconomy of time, by avoiding the loſs of it, incident to a frequent tranſition from one operation to another of a different nature. This depends on various circumſtances; the tranſition itſelf, the orderly diſpoſition of the implements, machines and materials employed in the operation to be relinquiſhed, the preparatory ſteps to the commencement of a new one, the interruption of the impulſe, which the mind of the workman acquires, from being engaged in a particular operation; the diſtractions, heſitations and reluctances, which attend the paſſage from one kind of buſineſs to another.

3d. An extenſion of the uſe of machinery.—A man occupied on a ſingle object will have it more in his power, and will be more naturally led to exert his imagination in deviſing methods to facilitate and abrige labor, than if he were perplexed by a variety of independent and diſſimilar operations. Beſides this, the fabrication of machines, in numerous inſtances, becoming itſelf a diſtinct trade, the artiſt who follows it, has all the advantages which have been enumerated, for improvement in his particular art; and in both ways the invention and application of machinery are extended.

And from theſe cauſes united, the mere ſeparation of the occupation of the cultivator, from that of the artificer, has the effect of augmenting the *productive powers* of labor, and with them, the total maſs of the produce or revenue of a country. In this ſingle view of the ſubject, therefore, the utility of artificers or manufacturers, towards promoting an increaſe of productive induſtry, is apparent.

II. *As to an extenſion of the uſe of machinery, a point which though partly anticipated, requires to be placed in one or two additional lights.*

The employment of machinery forms an item of great importance in the general maſs of national induſtry. 'Tis an artificial force brought in aid of the natural force of man; and, to all the purpoſes of labor, is an increaſe of hands; an acceſſion of ſtrength, *unincumbered too by the expence of maintaining the laborer.* May it not therefore be fairly inferred, that thoſe occupations, which give greateſt ſcope to the uſe of this auxiliary, contribute moſt to the general ſtock of induſtrious effort, and, in conſequence, to the general product of induſtry?

It ſhall be taken for granted, and the truth of the poſition referred to obſervation, that manufacturing purſuits are ſuſceptible in a greater degree of the application of machinery, than thoſe of agriculture. If ſo, all the difference is loft to a community, which, inftead of manufacturing for itfelf, procures the fabrics requifite to its fupply from other countries. The fubftitution of foreign for domeftic manufactures is a transfer to foreign nations of the advantages accruing from the employment of machinery, in the modes in which it is capable of being employed, with moft utility and to the greateft extent.

The cotton mill invented in England, within the laft twenty years, is a fignal illuftration of the general propofition, which has been juft advanced. In confequence of it, all the different proceffes for fpinning cotton are performed by means of machines, which are put in motion by water, and attended chiefly by women and children; and by a fmaller number of perfons, in the whole, than are requifite in the ordinary mode of fpinning. And it is an advantage of great moment that the operations of this mill continue with convenience, during the night, as well as through the day. The prodigious effect of fuch a machine is eafily conceived. To this invention is to be attributed effentially the immenfe progrefs, which has been fo fuddenly made in Great-Britain in the various fabrics of cotton.

III. *As to the additional employment of claffes of the community, not originally engaged in the particular bufinefs.*

This is not among the leaft valuable of the means, by which manufacturing inftitutions contribute to augment the general ftock of induftry and production. In places where thofe inftitutions prevail, befides the perfons regularly engaged in them, they afford occafional and extra employment to induftrious individuals and families, who are willing to devote the leifure refulting from the intermiffions of their ordinary purfuits to collateral labors, as a refource for multiplying their acquifitions or their enjoyments. The hufbandman himfelf experiences a new fource of profit and fupport from the encreafed induftry of his wife and daughters; invited and ftimulated by the demands of the neighbouring manufactories.

Befide this advantage of occafional employment to claffes having different occupations, there is another of a nature allied to it and of a fimilar tendency. This is, the employment of perfons who would otherwife be idle (and in many cafes a burthen on the community) either from the bias of temper, habit, infirmity of body, or fome other caufe, indifpofing or difqualifying them for the toils of the country. It is worthy of particular remark, that, in general, women and children are rendered more ufeful, and the latter more early ufeful, by manufacturing eftablifhments, than they would otherwife be. Of the number of perfons employed in the cotton manufactories of Great-Britain, it is computed that $\frac{4}{7}$ nearly are women and children; of whom the greateft proportion are children, and many of them of a tender age.

And thus it appears to be one of the attributes of manufactures, and one of no fmall confequence, to give occafion to the exertion of a greater quantity of induftry, even by the *fame number* of perfons, where they happen to prevail, than would exift, if there were no fuch eftablifhments.

IV. *As to the promoting of emigration from foreign countries.*

Men reluctantly quit one courfe of occupation and livelihood for another, unlefs invited to it by very apparent and proximate advantages. Many, who would go from one country to another, if they had a profpect of continuing, with more benefit, the callings to which they have been educated, will often not be tempted to change their fituation by the hope of doing better in fome other way. Manufacturers who, liftening to the powerful invitations of a better price for their fabrics, or their labor, of greater cheapnefs of provifions and raw materials, of an exemption from the chief part of the taxes, burthens and reftraints, which they endure in the old world, of greater perfonal independence and confequence, under the operation of a more equal government, and of what is far more precious than mere religious toleration, a perfect equality of religious privileges; would probably flock from Europe to the United States to purfue their own trades or profeffions, if they were once made fenfible of the advantages they would enjoy, and were infpired with an affurance of encouragement and employment, will, with difficulty, be induced to tranfplant themfelves, with a view to becoming cultivators of land.

If it be true then, that it is the intereft of the United States to open every poffible avenue to emigration from abroad, it affords a weighty argument for the encouragement of manufactures; which, for the reafons juft affigned, will have the ftrongeft tendency to multiply the inducements to it.

Here is perceived an important refource, not only for extending the population, and with it the ufeful and productive labor of the country, but likewife for the profecution of manufactures, without deducting from the number of hands, which might otherwife be drawn to tillage; and even for the indemnification of agriculture for fuch as might happen to be diverted from it. Many, whom manufacturing views would induce to emigrate, would afterwards yield to the temptations, which the particular fituation of this country holds out to agricultural purfuits. And while agriculture would in other refpects derive many fignal and unmingled advantages, from the growth of manufactures, it is a problem whether it would gain or lofe, as to the article of the number of perfons employed in carrying it on.

V. *As to the furnifhing greater fcope for the diverfity of talents and difpofitions, which difcriminate mem from each other.*

This is a much more powerful mean of augmenting the fund of national induftry than may at firft fight appear. It is a juft obfervation, that minds of the ftrongeft and moft active powers for their proper objects fall below mediocrity and labor without effect, if confined to uncongenial purfuits. And it is thence to be inferred, that the refults of human exertion may be immenfely increafed by diverfifying its objects. When all the different kinds of induftry obtain in a community, each individual can find his proper employment, and can call into activity the whole vigour of his nature. And the community is benefited by the fervices of its refpective members, in the manner, in which each can ferve it with moft effect.

If there be any thing in a remark often to be met with, namely, that there is, in the genius of the people of this country, a peculiar aptitude for mechanic improvements, it would operate as a forcible reafon for giving opportunities to the exercife of that fpecies of talent, by the propagation of manufactures.

VI. *As to the affording a more ample and various field for enterprife.*

This alfo is of greater confequence in the general fcale of national exertion, than might perhaps on a fuperficial view be fuppofed, and has effects not altogether diffimilar from thofe of the circumftance laft noticed. To cherifh and ftimulate the activity of the human mind, by multiplying the objects of enterprife, is not among the leaft confiderable of the expedients, by which the wealth of a nation may be promoted. Even things in themfelves, not pofitively advantageous, fometimes become fo, by their tendency to provoke exertion. Every new fcene which is opened to the bufy nature of man to roufe and exert itfelf, is the addition of a new energy to the general ftock of effort.

The fpirit of enterprife, ufeful and prolific as it is, muft neceffarily be contracted or expanded in proportion to the fimplicity or variety of the occupations and productions, which are to be found in a fociety. It muft be lefs in a nation of mere cultivators, than in a nation of cultivators and merchants; lefs in a nation of cultivators and merchants, than in a nation of cultivators, artificers and merchants.

VII. *As to the creating, in fome inftances, a new, and fecuring in all a more certain and fteady demand, for the furplus produce of the foil.*

This is among the moft important of the circumftances which have been indicated. It is a principal mean, by which the eftablifhment of manufactures contributes to an augmentation of the produce or revenue of a country, and has an immediate and direct relation to the profperity of agriculture.

It is evident, that the exertions of the hufbandman will be fteady or fluctuating, vigorous or feeble, in proportion to the fteadinefs or fluctuation, adequatenefs, or inadequatenefs of the markets on which he muft depend, for the vent of the furplus, which may be produced by his labor; and that fuch furplus in the ordinary courfe of things will be greater or lefs in the fame proportion.

For the purpofe of this vent, a domeftic market is greatly to be preferred to a foreign one; becaufe it is in the nature of things, far more to be relied upon.

It is a primary object of the policy of nations, to be able to fupply themfelves with fubfiftence from their own foils; and manufacturing nations, as far as circumftances permit, endeavour to procure from the fame fource, the raw materials neceffary for their own fabrics. This difpofition, urged by the fpirit of monopoly, is fometimes even carried to an injudicious extreme. It feems not always to be recollected, that nations, who have neither mines nor manufactures, can only obtain the manufactured articles, of which they ftand in need, by an exchange of the products of their foils; and that, if thofe who can beft furnifh them with fuch articles are unwilling to give a due courfe to this exchange, they muft of neceffity make every poffible effort to manufacture for themfelves; the effect of which is that the manufacturing nations abrige the natural advantages of their fituation, through an unwillingnefs to permit the agricultural countries to enjoy the advantages of theirs, and facrifice the interefts of a mutually beneficial intercourfe to the vain project of *felling every thing* and *buying nothing*.

But it is alfo a confequence of the policy, which has been noted, that the foreign demand for the products of agricultural countries, is, in a great degree, rather cafual and occafional, than certain or conftant. To what extent injurious interruptions of the demand for fome of the ftaple commodities of the United States, may have been experienced, from that caufe, muft be referred to the judgment of those who are engaged in carrying on the commerce of the country; but it may be safely affirmed, that such interruptions are at times very inconveniently felt, and that cases not unfrequently occur, in which markets are so confined and restricted, as to render the demand very unequal to the supply.

Independently likewise of the artificial impediments, which are created by the policy in question, there are natural causes tending to render the external demand for the surplus of agricultural nations a precarious reliance. The differences of seasons, in the countries which are the consumers, make immense differences in the produce of their own soils, in different years; and consequently in the degrees of their necessity for foreign supply. Plentiful harvests with them, especially if similar ones, occur at the same time in the countries which are the furnishers, occasion of course a glut in the markets of the latter.

Considering how fast and how much the progress of new settlements in the United States must encrease the surplus produce of the soil, and weighing seriously the tendency of the system, which prevails among most of the commercial nations of Europe; whatever dependence may be placed on the force of natural circumstances to counteract the effects of an artificial policy; there appear strong reasons to regard the foreign demand for that surplus as too uncertain a reliance, and to desire a substitute for it, in an extensive domestic market.

To secure such a market, there is no other expedient, than to promote manufacturing establishments. Manufacturers who constitute the most numerous class, after the cultivators of land, are for that reason the principal consumers of the surplus of their labor.

This idea of an extensive domestic market for the surplus produce of the soil is of the first consequence. It is of all things, that which most effectually conduces to a flourishing state of agriculture. If the effect of manufactories should be to detach a portion of the hands, which would otherwise be engaged in tillage, it might possibly cause a smaller quantity of lands to be under cultivation; but by their tendency to procure a more certain demand for the surplus produce of the soil, they would, at the same time, cause the lands which were in cultivation, to be better improved and more productive. And while, by their influence, the condition of each individual farmer would be meliorated, the total mass of agricultural production would probably be encreased. For this must evidently depend as much, if not more, upon the degree of improvement, than upon the number of acres under culture.

It merits particular observation, that the multiplication of manufactories not only furnishes a market for those articles which have been accustomed to be produced in abundance, in a country; but it likewise creates a demand for such as were either unknown or produced in inconsiderable quantities. The bowels as well as the surface of the earth are ransacked for articles which were before neglected. Animals, plants and minerals acquire a utility and value, which were before unexplored.

The foregoing considerations seem sufficient to establish, as general propositions, that it is the interest of nations to diversify the industrious pursuits of the individuals, who compose them—That the establishment of manufactures is calculated not only to encrease the general stock of useful and productive labor; but even to improve the state of agriculture in particular; certainly to advance the interests of those who are engaged in it.

*

The objections which are commonly made to the expediency of encouraging and to the probability of succeeding in manufacturing pursuits, in the United States, having now been discussed, the considerations which have appeared in the course of the discussion, recommending that species of industry to the patronage of the government, will be materially strengthened by a few general and some particular topics, which have been naturally reserved for subsequent notice.

I. There seems to be a moral certainty, that the trade of a country which is both manufacturing and agricultural, will be more lucrative and prosperous, than that of a country, which is merely agricultural.

One reason for this is found in that general effort of nations (which has been already mentioned) to procure from their own soils, the articles of prime necessity requisite to their own consumption and use; and which serves to render their demand for a foreign supply of such articles in a great degree occasional and contingent. Hence, while the necessities of nations exclusively devoted to agriculture, for the fabrics of manufacturing states are constant and regular, the wants of the latter for the products of the former, are liable to very considerable fluctuations and interruptions. The great inequalities, resulting from difference of seasons, have been elsewhere remarked : this uniformity of demand on one side, and unsteadiness of it, on the other, must necessarily have a tendency to cause the general course of the exchange of commodities between the parties, to turn to the disadvantage of the merely agricultural states. Peculiarity of situation, a climate and soil adapted to the production of peculiar commodities may, sometimes, contradict the rule; but there is every reason to believe that it will be found in the main, a just one.

Another circumstance which gives a superiority of commercial advantages to states, that manufacture, as well as cultivate, consists in the more numerous attractions, which a more diversified market offers to foreign customers, and in the greater scope, which it affords to mercantile enterprise. It is a position of indisputable truth in commerce, depending too on very obvious reasons, that the greatest resort will ever be to those marts, where commodities, while equally abundant, are most various. Each difference of kind holds out an additional inducement : and it is a position not less clear, that the field of enterprise must be enlarged to the merchants of a country, in proportion to the variety, as well as the abundance of commodities which they find at home for exportation to foreign markets.

A third circumstance, perhaps not inferior to either of the other two, conferring the superiority which has been stated, has relation to the stagnations of demand for certain commodities which at some time or other interfere more or less with the sale of all.—The nation which can bring to market, but few articles is likely to be more quickly and sensibly affected by such stagnations, than one, which is always possessed of a great variety of commodities : the former frequently finds too great a portion of its stock of materials, for sale or exchange, lying on hand—or is obliged to make injurious sacrifices to supply its wants of foreign articles, which are *numerous* and *urgent*, in proportion to the smallness of the number of its own. The latter commonly finds itself indemnified, by the high prices of some articles, for the low prices of others—and the prompt and advantageous sale of those articles which are in demand enables its merchants the better to wait for a favorable change, in respect to those which are not. There is ground to believe, that a difference of situation, in this particular, has immensely different effects upon the wealth and prosperity of nations.

From these circumstances collectively, two important inferences are to be drawn; one, that there is always a higher probability of a favorable balance of trade, in regard to countries, in which manufactures founded on the basis of a thriving agriculture flourish, than in regard to those, which are confined wholly or almost wholly to agriculture; the other (which is also a consequence of the first) that countries of the former description are likely to possess more pecuniary wealth, or money, than those of the latter.

Facts appear to correspond with this conclusion. The importations of manufactured supplies seem invariably to drain the merely agricultural people of their wealth. Let the situation of the manufacturing countries of Europe be compared in this particular, with that of countries which only cultivate, and the disparity will be striking. Other causes, it is true, help to account for this disparity between some of them; and among these causes, the relative state of agriculture; but between others of them, the most prominent circumstance of dissimilitude arises from the comparative state of manufactures. In corroboration of the same idea, it ought not to escape remark, that the West-India islands, the soils of which are the most fertile, and the nation, which in the greatest degree supplies the rest of the world, with the precious metals, exchange to a loss with almost every other country.

As far as experience at home may guide, it will lead to the same conclusion. Previous to the revolution, the quantity of coin, possessed by the colonies, which now compose the United States, appeared to be inadequate to their circulation; and their debt to Great-Britain was progressive. Since the revolution, the states, in which manufactures have most encreased, have recovered fastest from the injuries of the late war; and abound most in pecuniary resources.

It ought to be admitted, however, in this as in the preceding case, that causes irrelative to the state of manufactures account, in a degree, for the phenomena remarked. The continual progress of new settlements has a natural tendency to occasion an unfavorable balance of trade; though it indemnifies for the inconvenience, by that encrease of the national capital which flows from the conversion of waste into improved lands: and the different degrees of external commerce, which are carried on by the different states, may make material differences in the comparative state of their wealth. The first circumstance has reference to the deficiency of coin and the encrease of debt previous to the revolution; the last to the advantages which the most manufacturing states appear to have enjoyed, over the others, since the termination of the late war.

But the uniform appearance of an abundance of specie, as the concomitant of a flourishing state of manufactures, and of the reverse, where they do not prevail, afford a strong presumption of their favourable operation upon the wealth of a country.

Not only the wealth, but the independence and security of a country, appear to be materially connected with the prosperity of manufactures. Every nation, with a view to those great objects, ought to endeavor to possess within itself all the essentials of national supply. These comprise the means of *subsistence, habitation, clothing* and *defence.*

The possession of these is necessary to the perfection of the body politic, to the safety as well as to the welfare of the society; the want of either, is the want of an important organ of political life and motion; and in the various crises which await a state, it must severely feel the effects of any such deficiency. The extreme embarrassments of the United States during the late war, from an incapacity of supplying themselves, are still matter of keen recollection: a future war might be expected again to exemplify the mischiefs and dangers of a situation, to which that incapacity is still in too great a degree applicable, unless changed by timely and vigorous exertions. To effect this change, as fast as shall be prudent, merits all the attention and all the zeal of our public councils; 'tis the next great work to be accomplished.

The want of a navy to protect our external commerce, as long as it shall continue, must render it a peculiarly precarious reliance, for the supply of essential articles, and must serve to strengthen prodigiously the arguments in favor of manufactures.

To these general considerations are added some of a more particular nature.

Our distance from Europe, the great fountain of manufactured supply, subjects us in the existing state of things, to inconvenience and loss in two ways.

The bulkiness of those commodities which are the chief productions of the soil, necessarily imposes very heavy charges on their transportation, to distant markets. These charges, in the cases, in which the nations, to whom our products are sent, maintain a competition in the supply of their own markets, principally fall upon us, and form material deductions from the primitive value of the articles furnished. The charges on manufactured supplies, brought from Europe are greatly enhanced by the same circumstance of distance. These charges, again, in the cases in which our own industry maintains no competition, in our own markets, also principally fall upon us; and are an additional cause of extraordinary deduction from the primitive value of our own products; these being the materials of exchange for the foreign fabrics, which we consume.

The equality and moderation of individual property, and the growing settlements of new districts, occasion, in this country an unusual demand for coarse manufactures; the charges of which being greater in proportion to their greater bulk, augment the disadvantage, which has been just described.

As in most countries domestic supplies maintain a very considerable competition with such foreign productions of the soil, as are imported for sale; if the extensive establishment of manufactories in the United States does not create a similar competition in respect to manufactured articles, it appears to be clearly deducible, from the considerations which have been mentioned, that they must sustain a double loss in their exchanges with foreign nations; strongly conducive to an unfavorable balance of trade, and very prejudicial to their interests.

These disadvantages press with no small weight, on the landed interest of the country. In seasons of peace, they cause a serious deduction from the intrinsic value of the products of the soil. In the time of a war, which should either involve ourselves, or another nation, possessing a considerable share of our carrying trade, the charges on the transportation of our commodities, bulky as most of them are, could hardly fail to prove a grievous burthen to the farmer; while obliged to depend in so great degree as he now does, upon foreign markets for the vent of the surplus of his labor.

As far as the prosperity of the fisheries of the United States is impeded by the want of an adequate market, there arises another special reason for desiring the extension of manufactures. Besides the fish, which in many places, would be likely to make a part of the subsistence of the persons employed; it is known that the oils, bones and skins of marine animals, are of extensive use in various manufactures. Hence the prospect of an additional demand for the produce of the fisheries.

One more point of view only remains, in which to consider the expediency of encouraging manufactures in the United States.

It is not uncommon to meet with an opinion, that though the promoting of manufactures, may be the interest of a part of the union, it is contrary to that of another part. The northern and southern regions are sometimes represented as having adverse interests in this respect. Those are called manufacturing, these agricultural states, and a species of opposition is imagined to subsist between the manufacturing and agricultural interests.

This idea of an opposition between those two interests is the common error of the early periods of every country, but experience gradually dissipates it. Indeed they are perceived so often to succour and to befriend each other, that they come at length to be considered as one; a supposition which has been frequently abused, and is not universally true. Particular encouragements of particular manufactures may be of a nature to sacrifice the interests of landholders to those of manufacturers; but it is nevertheless a maxim well established by experience, and generally acknowledged, where there has been sufficient experience, that the *aggregate* prosperity of manufactures, and the *aggregate* prosperity of agriculture are intimately connected. In the course of the discussion which has had place, various weighty considerations have been adduced operating in support of that maxim. Perhaps the superior steadiness of the demand of a domestic market for the surplus produce of the soil, is alone a convincing argument of its truth.

Ideas of a contrariety of interests between the northern and southern regions of the Union, are in the main as unfounded as they are mischievous. The diversity of circumstances, on which such contrariety is usually predicated, authorises a directly contrary conclusion. Mutual wants constitute one of the strongest links of political connection, and the extent of these bears a natural proportion to the diversity in the means of mutual supply.

Suggestions of an opposite complexion are ever to be deplored, as unfriendly to the steady pursuit of one great common cause, and to the perfect harmony of all the parts.

In proportion as the mind is accustomed to trace the intimate connexion of interest, which subsists between all the parts of a society, united under the *same* government; the infinite variety of channels which ferve to circulate the profperity of each to and through the reft, in that proportion will it be little apt to be difturbed by folicitudes and apprehenfions which originate in local difcriminations. It is a truth as important, as it is agreeable, and one to which it is not eafy to imagine exceptions, that every thing tending to eftablifh *fubftantial* and *permanent order*, in the affairs of a country, to encreafe the total mafs of induftry and opulence, is ultimately beneficial to every part of it. On the credit of this great truth, an acquiefcence may fafely be accorded, from every quarter, to all inftitutions, and arrangements, which promife a confirmation of public order, and an augmentation of national refource.

But there are more particular confiderations which ferve to fortify the idea, that the encouragement of manufactures is the intereft of all parts of the Union. If the northern and middle ftates fhould be the principal fcenes of fuch eftablifhments, they would immediately benefit the more fouthern, by creating a demand for productions; fome of which they have in common with the other ftates, and others of which are either peculiar to them, or more abundant, or of better quality, than elfewhere. Thefe productions, principally are timber, flax, hemp, cotton, wool, raw filk, indigo, iron, lead, furs, hides, fkins and coals; of thefe articles cotton and indigo are peculiar to the fouthern ftates; as are hitherto *lead* and *coal*, flax and hemp are or may be raifed in greater abundance there, than in the more northern ftates; and the wool of Virginia is faid to be of better quality than that of any other ftate: a circumftance rendered the more probable by the reflection that Virginia embraces the fame latitudes with the fineft wool countries of Europe. The climate of the fouth is alfo better adapted to the production of filk.

The extenfive cultivation of cotton can perhaps hardly be expected, but from the previous eftablifhment of domeftic manufactories of the article; and the fureft encouragement and vent, for the others, would refult from fimilar eftablifhments in refpect to them.

If, then, it fatisfactorily appears, that it is the intereft of the United States, generally, to encourage manufactures, it merits particular attention, that there are circumftances which render the prefent a critical moment for entering with zeal upon the important bufinefs. The effort cannot fail to be materially feconded by a confiderable and encreafing influx of money, in confequence of foreign fpeculations in the funds—and by the diforders which exift in different parts of Europe.

The firft circumftance not only facilitates the execution of manufacturing enterprifes; but it indicates them as a neceffary mean to turn the thing itfelf to advantage, and to prevent its being eventually an evil. If ufeful employment be not found for the money of foreigners brought to the country to be invefted in purchafes of the public debt, it will quickly be re-exported to defray the expence of an extraordinary confumption of foreign luxuries; and diftreffing drains of our fpecie may hereafter be experienced to pay the intereft and redeem the principal of the purchafed debt.

This ufeful employment too ought to be of a nature to produce folid and permanent improvements. If the money merely ferves to give a temporary fpring to foreign commerce; as it cannot procure new and lafting outlets for the products of the country; there will be no real or durable advantage gained. As far as it fhall find its way in agricultural ameliorations, in opening canals, and in fimilar improvements, it will be productive of fubftantial utility. But there is reafon to doubt, whether in fuch channels it is likely to find fufficient employment, and ftill more whether many of thofe who poffefs it, would be as readily attracted to objects of this nature, as to manufacturing purfuits; which bear greater analogy to thofe to which they are accuftomed, and to the fpirit generated by them.

To open the one field, as well as the other, will at leaft fecure a better profpect of ufeful employment, for whatever acceffion of money there has been or may be.

There is at the prefent juncture a certain fermentation of mind, a certain activity of fpeculation and enterprife, which if properly directed, may be made fubfervient to ufeful purpofes; but which if left entirely to itfelf, may be attended with pernicious effects.

The difturbed ftate of Europe, inclining its citizens to emigration, the requifite workmen, will be more eafily acquired, than at another time; and the effect of multiplying the opportunities of employment to thofe who emigrate, may be an encreafe of the number and extent of valuable acquifitions to the population, arts and induftry of the country.

To find pleafure in the calamities of other nations would be criminal; but to benefit ourfelves, by opening an afylum to thofe who fuffer, in confequence of them, is as juftifiable as it is politic.

Reproduced courtesy of Merrimack Valley Textile Museum.

Alexander Hamilton

Prospectus of the Society for Establishing Useful Manufactures (1791)

The Society for Establishing Useful Manufactures, or, as it was commonly called, the S.U.M., was one of the largest and most important early industrial corporations in the United States. It was the creation of Alexander Hamilton and Tench Coxe. Coxe had suggested a manufacturing corporation in his *Brief Examination of Lord Sheffield's Observations* (1791) and had submitted a detailed plan of such a corporation to Thomas Jefferson in that same year. Hamilton had recently prepared his *Report on the Subject of Manufactures,* in which he urged the Congress and the public to encourage and support industry. (Hamilton submitted his report to Congress two weeks after the S.U.M. was incorporated.) In September 1791 a *Prospectus* was published in several newspapers. This document, reprinted here, was intended to interest the public in the S.U.M. and encourage investment. By November sufficient funds had been raised and the plan was submitted to the New Jersey legislature.

Before the S.U.M. could get under way, however, it was adversely affected by the Panic of 1792 and the ensuing bankruptcy of several of its officers. It was not until 1793 that production was started, and even then it did not do well. The S.U.M. failed in almost all of its attempts at manufacturing, and in 1796 its directors voted to close down all operations. Other early attempts at setting up manufacturing organizations met with fates similar to that of the S.U.M. (Documents

[Alexander Hamilton], "Prospectus of the Society for Establishing Useful Manufactures," *The Papers of Alexander Hamilton,* ed. Harold C. Sysett (New York: Columbia University Press, 1965), volume 9, pp. 144–153. For a study of the S.U.M., see Joseph S. Davis, "The 'S.U.M.': The First New Jersey Business Corporation," in *Essays in the Earlier History of American Corporations* (Cambridge: Harvard University Press, 1917), volume 1, pp. 349–504.

relating to the abortive experiments in cotton manufacture at Beverly, Massachusetts, are included in volume 2 of this series.) Bad management and labor problems contributed to their failure, which strengthened the position of those who insisted that manufactures were not only impossible in the United States but unwanted. The S.U.M. was defeated by lack of manufacturing experience and expertise. Later manufacturers learned from its mistakes.

[Philadelphia, August, 1791]

The establishment of Manufactures in the United States when maturely considered will be fo[und] to be of the highest importance to their prosperity. It [is] an almost self evident proposition that that com[muni]ty which can most completely supply its own w[ants] is in a state of the highest political perfection. [And] both theory and experience conspire to prove that a nation (unless from a very peculiar coincidence of circumstances) cannot possess much *active* wealth but as the result of extensive manufactures.

While also it is manifest that the interest of the community is deeply concerned in the progress of this species of Industry, there is [as] little room to doubt that the interest of individuals may equally be promoted by the pursuit of it. What [is] there to hinder the profitable prosecution of manufact[ures] in this Country, when it is notorious, that, independent of impositions for the benefit of the revenue and for the encouragement of domestic enterprise—the natural commercial charges of the greater part of th[ose] which are brought from Europe amount to from fiftee[n to] thirty per Cent—and when it is equally notorious that provisions and various kinds of raw materials are ev[en] cheaper here than in the Country from which our principal supplies come?

The dearness of labour and the want of Capital are the two great objections to the success of manufactures in the United States.

The first objection ceases to be formidable when it is recollected how prodigiously the proportion of manual labour in a variety of manufactures has been decreased by the late improvements in the construction and application of Machines—and when it is also considered to what an extent women and even children in the populous parts of the Country may be rendered auxiliary to undertakings of this nature. It is also to be taken into calculation that emigrants may be engaged on reasonable terms in countries where labour is cheap, and brought over to the United States.

The last objection disappears in the eye of those who are aware how much may be done by a proper application of the public Debt. Here is the resource which has been hitherto wanted. And while a direction of it to this object may be made a mean of public prosperity and an instrument of profit to adventurers in the enterprise, it, at the same time, affords a prospect of an enhancement of the value of the debt; by giving it a new and additional employment and utility.

It is evident that various fabrics, under every supposed disadvantage, are in a very promising train. And that the success has not been still more considerable may be traced to very obvious causes.

Scarcely any has been undertaken upon a scale sufficiently extensive or with a due degree of system. To insure success it is desireable to be able to enter into competition with foreign fabrics in three particulars—quality, price, term of credit. To the first, workmen of equal skill is an essential ingredient. The means employed have not generally been adequate to the purpose of procuring them from abroad and those who have been procureable at home have for the most part been of an inferior class. To cheapness of price, a capital equal to the purpose of making all necessary advances, and procuring materials on the best te[rms] is an indispensible requisite—and to the giving of [Credit] a Capital capable of affording a surplus beyond wh[at] is required for carrying on the business is not less indispensible. But most undertakings hitherto have been bottomed on very slender resources.

To remedy this defect an association of the Capitals of a number of Individuals is an obvious expedient—and the species of Capital which cons[ists of] the public Stock is susceptible of dispositions which will render it adequate to the end. There is good reason to expect that as far as shall be found necessary money on reasonable terms may be procured abroad upon an hypothecation of the Stock. It is presumeable that public Banks would not refuse their aid in the same way to a solid institution of so great public utility. The pecuniary aid even of Government though not to be counted upon, ought not wholly to be despaired of. And when the Stock shall have attained its due value so that no loss will attend the sale all such aids may be dispensed with. The Stock may then be turned into specie without disadvantage whenever specie is called for.

But it is easy to see that upon a good Capital in Stock an effective Credit may be raised in various ways which will answer every purpose in specie, independent of the direct expedient of borrowing.

To effect the desired association an incorporation of the adventurers must be contemplated as a mean necessary to their security. This can doubtless be obtained. There is scarcely a state which could be insensible to the advantage of being the scene of such an undertaking. But there are reasons which strongly recommend the state of New Jersey for the purpose. It is thickly populated—provisions are there abundant and cheap. The state having scarcely any external commerce and no waste lands to be peopled can feel the impulse of no supposed interest hostile to the advancement of manufactures. Its situation seems to insure a constant friendly disposition.

The great and preliminary desideratum then is to form a sufficient capital. This it is conceived, ought not to be less than Five hundred thousand Dollars. Towards forming this capital subscriptions ought immediately to be set on foot; upon this condition that no subscriber shall be bound to pay until an Act of Incorporation shall have been obtained—for which application may be made as soon as the sums subscribed shall amount to One hundred thousand Dollars.

As soon as it is evident that a proper Capital can be formed means ought to be taken to procure from Europe skilful workmen and such machines and implements as cannot be had here in sufficient perfection. To this the existing crisis of the affairs of certain parts of Europe appears to be particularly favourable. It will not be necessary that all the requisite workmen should be brought from thence. One in the nature of a *foreman* for each branch may in some branches suffice. In others it may be requisite to go further and have one for each subdivision. But numbers of workmen of secondary merit may be found in the United States; and others may be quickly formed.

It is conceived that there would be a moral certainty of success in manufactories of the following articles—

1st Paper and Pasteboard
2nd Paper hangings

    3rd   Sail cloth and other coarse linen cloths, such as sheetings, shirtings, diaper, oznaburgs &ca.
    4th   The printing of Cottons and linens; and as incident to this but on a smaller scale the manufacturing of the article to be printed.
    5th   Womens shoes of all kinds.
    6th   Thread, Cotton and Worsted Stockings.
    7th   Pottery and Earthen Ware.
    8th   Chip Hats
    9th   Ribbands & Tapes
    10th Carpets
    11th Blankets
    12th Brass and Iron wire.
    13th Thread and Fringes.

It will be unnecessary to enter into the det[ails] of the execution further than to observe that the employment of the labor-saving mills and machines is particularly contemplated.

In addition to the foregoing a brewery for the supply of the manufacturers, as a primary object, may be thought of.

When application shall be made for an act of Incorporation it ought to include a request that provision may be made for incorporating the Inhabitants of the district within a certain defined limit which shall be chosen by the Company as the principal seat of their factories and a further request that the Company may have permission to institute a lottery or lotteries in each year for the term of five years for a sum or sums not exceeding in one year One hundred thousand dollars. The State of Jersey if duly sensible of its interest in the measure will not refuse encouragements of this nature.

An incorporation of this sort will be of great importance to the police of the establishment. It may also be found eligible to vest a part of the funds of the Company in the purchase of ground on which to erect necessary buildings &c. A part of this ground divided into town lots may be afterwards a source of profit to the Company.

The lottery will answer two purposes. It will give a temporary command of Money and the profit arising from it will go towards indemnifying for first unproductive efforts.

The following scheme for the organisation of the Company will probably be an eligible one—

[At this point Hamilton gives details of the corporation's structure— the number of directors, frequency of meetings, and so forth.—Eds.]

We the Subscribers for ourselves respectively and not one for the other and for our respective heirs, executors and administrators do severally covenant promise and agree to and with each other and with the heirs Executors and Administrators of each other that we will respectively contribute and pay in the manner and at the times specified in the plan hereunto annexed the respective sums against our respective names hereunder set for the purpose of establishing a company for carrying on the business of manufactures in one of the States of New York New Jersey and Pennsylvania (giving a preference to New Jersey if an incorporation can be obtained from the said State on advantageous terms) according to the general principles of the plan aforesaid, but subject to such alterations as shall be agreed upon at any time previous to the obtaining an Act of Incorporation either in the principles or details thereof by the major part of us whose names are hereunto subscribed, or in the details thereof only, as shall be thought fit by the major part of the persons hereinafter named. And we do hereby jointly and severally constitute and appoint   one [1] and each of our Attorneys who or the major part of them or the major part of the survivors of them are hereby empowered as soon as the sum of One hundred thousand Dollars shall be subscribed hereto to make application on our behalf to either of the States aforesaid (giving such preference as aforesaid to the State of New Jersey) for an Act or Acts of Incorporation according to the principles of the plan aforesaid with such alterations in the details thereof as shall appear to them eligible, or with such alterations whatsoever, as shall be previously agreed upon by us; And further to take such measures at our joint expense as shall appear to them necessary and proper for engaging workmen in the several branches of manufacture mentioned in the said plan.

1. [space in original manuscript]

The Philosophy of Manufactures

Reproduced courtesy of Merrimack Valley Textile Museum.

George Logan

Letters Addressed to the Yeomanry of the United States (1792)

The Society for Encouraging Useful Manufactures aroused great controversy. It was attacked as a government-supported corporation that would drive small producers out of business. It was called a monopoly "with exclusive privileges and charters of incorporation," "a political monster" full of "great danger to the republican principles which ought to be dear to every American." Many believed it a boon to a wealthy few at the expense of the farmer and small manufacturer.

Many of these complaints are included in a pamphlet by George Logan, written under the pseudonym "A Farmer." Logan (1753–1821) was president of the Germantown Society for Promoting Domestic Manufactures and one of the most ardent defenders of Physiocratic principles in the United States. He opposed the S.U.M. on general principles. A firm believer in laissez-faire, he was opposed to any government interference with the economy. A supporter of agriculture, he opposed large-scale manufactures in any form. Of the five linked "letters" Logan published in the *American Museum*, we reprint numbers II, IV, and V.

"A Farmer" [George Logan], "Five letters addressed to the yeomanry of the united states: containing some observations on the dangerous scheme of governor Duer and mr. Secretary Hamilton, to establish national manufactories," *American Museum*, volume 12 (September 1792), pp. 159–167; volume 12 (October 1792), pp. 213–217. By permission of the Houghton Library, Harvard University.

LETTER II.

HAPPY for mankind, the present enquiry in the philosophical world is not the mechanism of the universe, or the composition of its elements, but the principles of civil society. The prodigious advantages which France has already derived from these enquiries, and the alterations which are daily taking place in other parts of Europe, in favour of the rights of man, should have some influence on the measures of the general government of the united states, which are tending, in an alarming degree, to undermine the liberties of our country—to strip the farmer, the mechanic, the manufacturer, and useful labourer, of all influence and of all importance—to consign them to contempt, or, at best, to the sad privilege of murmuring without redress.

The world has been sufficiently flooded with the blood of its inhabitants.—And free citizens, under the sanction of law, have been too often reduced to misery and wretchedness.

In this enlightened age, let American rulers beware how they proceed. Whatever may be the opinion of those characters, living upon the spoils of their fellow-citizens, or basking in the sunshine of court favour, they will find a spirit of resistance in the people, which will not submit to be oppressed, and a fund of good sense, which cannot be deceived by the arts of false reasoning or false patriotism.

Governments are tranquil, when they are adapted to the ideas and lights of the age : but whenever their regulations become unsuitable to the ideas of the times, and contrary to the opinion of the people, the rulers should look into their conduct, and remove every reason of complaint. All attempts to support unjust measures in this country, where the *people, as yet*, have so much power, are absurd, and must ultimately be unsuccessful : justice will finally take place, in spite of all efforts to suppress it.

The same spirit of arbitrary power, which, during the last year, violated the rights of the great body of yeomanry, by an *excise law*, now proposes to interfere in the *occupations* of the *mechanic* and the *manufacturer*. A government may waste the public money in erecting palaces, statues, &c. ; the evil is but temporary : but when it *assumes* principles injurious to the rights of the *people*, and, by arbitrary laws, interferes in the *occupations* of its citizens, liberty is but a name. The theory of such a government is falsehood and mockery—the practice is oppression.

Smith in his wealth of nations, vol. ii. p. 86, observes, " that to prohibit a great people from making all that they can of every part of their own produce, or of employing their stock and industry in the way that they judge most advantageous to themselves, is a manifest violation of the most sacred rights of mankind."

It may be thought improper, at this early period, to offer any observations on the justice or wisdom of the report of the secretary of the treasury on manufactures. It is true, we might have waited for the deliberations and conclusion of congress on the subject : but congress, having adopted a new method of legislating, by referring the most important business of the country to the different secretaries, and adopting their reports, experience justifies a belief, that the principles of this report will also be adopted, and will come forward under the sanction of the legislature in the form of a law. The secretary of the treasury, and his friends in New-York, have already prepared the way, by procuring one of the most unjust and arbitrary laws to be enacted by the commonwealth of New-Jersey, that ever disgraced the government of a free people.—A law granting to a few wealthy men the exclusive jurisdiction of six miles square, and a variety of unconstitutional privileges, highly injurious to the citizens of that state. This law merits your attention, not as a pattern of justice, but to convince you how dangerous it is for a free people to place their whole political safety on the conduct of any set of legislators, when surrounded by artful and designing men. Is it reasonable—is it just, that a numerous class of citizens, whose knowledge in mechanics and manufactures, not less necessary for the support of their families, than useful to their country, should be sacrificed to a wealthy few, who have no other object in view, than to add to their ill-gotten and enormous wealth?

Such being the nature of this corporation, can it be doubted, whether it violates the spirit of all just laws? Whether it subverts the principles of that equality, of which freemen ought to be so jealous? Whether it establishes a class of citizens with distinct interests from their fellow citizens? Will it not, by fostering an inequality of fortune, prove the destruction of the equality of rights, and tend strongly to an aristocracy?

There are two kinds of inequality, the one personal, that of talent and virtue, the source of whatever is excellent and admirable in society—the other, that of fortune, which must exist; because property alone can stimulate to labour; and labour, if it were not necessary to the existence, would be indispensible to the happiness of man: but though it be necessary, yet in its excess it is the great malady of civil society. The accumulation of that power which is conferred by wealth in the hands of the few, is the perpetual source of oppression and neglect of the mass of mankind. The power of the wealthy is farther concentrated by their tendency to *combination*, from which, number, dispersion, indigence, and ignorance, equally preclude the poor. The wealthy are formed into bodies by their professions, their different degrees of opulence, called ranks, their knowledge, and their small numbers:—they necessarily, in all countries, administer government; for they alone have skill and leisure for its functions. Thus circumstanced, nothing can be more evident, than their inevitable preponderance in the political scale. *The preference of partial to general interests, is, however, the greatest of all public evils*: it should, therefore, have been the object of all laws, to repress this malady; but it has been their perpetual tendency to aggravate it. Not content with the inevitable inequality of fortune, they have superadded to it honourary and political distinctions. Not content with the inevitable tendency of the wealthy to combine, they have embodied them in classes. They have fortified these conspiracies against the general interest, which they ought to have resisted, though they could not disarm. Laws, it is said, cannot equalize men,—no—but ought they, for that reason, to aggravate the inequality which they cannot cure? Laws cannot inspire unmixed patriotism: but ought they, for that reason, to foment that *corporation spirit* which is its most fatal enemy? "All professional combinations," said mr. Burke, in one of his late speeches in parliament, "are dangerous in a free state." Arguing on the same principle, the national assembly of France have proceeded further: they have conceived that " the laws ought to *create* no inequality or combination, to recognize all only in their capacities as citizens, and to offer no assistance to the natural preponderance of partial over general interests."

It is not the distinctions of titles which constitutes an aristocracy: it is the principle of partial association. The American aristocrats have failed in their attempt to establish titles of distinctions by law; yet the destructive principles of aristocracy are too prevalent among us, and ought to be watched with the most jealous eye.

LETTER IV.

FOR a government to interfere in the occupations, or in the private actions of citizens, where such actions are not injurious to the community, is not only unjust and impolitic, but is highly dangerous to the liberties of the people. All partial regulations tend to create separate interests in society, and therefore occasion jealousy and dissention among citizens, whose true interest consists in being united. On this subject the sentiments of the great body of the people in the united states, were clearly expressed during the arduous contest to establish their liberties. The constitutions of the several states, formed during that period of public animation and attention, manifest the strongest disapprobation of monopolies and exclusive privileges. Although these just and honourable sentiments have been lately suppressed by the influence of ambition and avarice, yet they are not extinct, but will re-appear with additional lustre, reflected from the glorious revolution of France. The Americans will adopt the political principles of that enlightened people, and, like them, will consider, that the *prosperity and happiness of citizens, constitute the real strength of nations*. If, under a vague, undefined idea of supporting the general welfare, congress is permitted to enact partial laws in favour of a few wealthy individuals, and to grant them exclusive privileges in any occupation in which their unbounded avarice may prompt them to engage, such regulations will inevitably destroy the infant manufactures of our country, and will consign the useful and respectable citizens, personally engaged in them, to contempt and ruin. All confidence of procuring an honourable support from any mechanic or manufacturing employ being at an end, no citizen will think of giving seven years of the prime of his life to acquire the knowledge of any profession in which he may be supplanted by a junto of monied men, under the immediate patronage and protection of government.—Whatever may be the plausible pretext of such institutions, they always promote the oppressive and injurious drudgery of the manufacturers, and the indolent luxury of the principals.

In the manufacturing towns in England, the poor appear to be in a state of the most abject servitude to their employers. The principals engaged in the various and extensive manufactories, give the poor artists and labourers six shillings sterling per week; one half of which is absorbed by government, by means of excises, duties, and stamps. To this dark system of British finance, as well as to the combination of the wealthy to keep the poor employed by them in a state of daily dependence and servitude, must be attributed the rags, the dejected eye, and squalid countenance, of a very numerous and a very useful class of citizens.

A chain does not derive its strength and utility from being composed of a few heavy links, and the remainder weak and ill conditioned—but from every link being as much as possible of equal power. The same takes place in civil society: a state is rendered more respectable and powerful by the prosperity of all its citizens, than by the overgrown wealth of the few.

Men educated in a profession which exists by the indiscriminate defence of right and wrong, will naturally support their opinions with all the art and sophistry of acute logicians. But reasoning, in proportion as it extends, and becomes complicated, does not owe its triumph always to truth; mental fatigue, or implicit faith frequently succeeds in procuring it admirers. A variety of stratagems may be made use of to deceive the people: but a minister who knows nothing more, will never lay the foundation of a great empire; the fabric raised by his duplicity and extravagance, must fall, and bury its supporters in its ruins.

The success of American manufactures will not depend on financial calculations, or legislative interference, but on the patronage and encouragement they may receive from patriotic citizens. The secretary of the treasury begins his lengthy and flimsy report, by a vague assertion, that " the expediency of encouraging manufactures in the united states, *which was, not long since, deemed very questionable*, appears at this time to be very generally admitted." It is certain, that the powerful and increasing mercantile interest has always suggested doubts respecting the propriety of giving encouragement to American artists and manufacturers. As agents, it is the immediate interest of the merchants, that every raw material, and every manufactured article, should pass through their hands: but no *real citizen* had ever a doubt of the expediency of every independent commonwealth being, as much as possible, supplied within itself, with all things necessary or useful in common life.

By giving useful employment and comfortable support to the weakest and most miserable fellow-citizen, you promote your own consequence and safety; therefore, duty, as well as interest, obliges the members of the same society to assist each other. Those, who use foreign manufactures, in preference to such as may be procured in their own country, receive protection from the government of which they are members, without complying with their duty in supporting its citizens—an injustice, that, in its increase, must be the ruin of any commonwealth. This principle, however, should be carried no farther than is confiftent with the real profperity of the ftate, as connected with the full employment, the happinefs, and independence, of all its citizens.

We ought not to defire the eftablifhment of any kind of manufacture in our country, which cannot fupport itfelf, without government granting to its agents bounties, premiums, and a variety of exclufive privileges, in violation of the rights of the people.

LETTER V.

CONGRESS have alfo been officially informed by the fecretary, that " the embarraffments of our external trade, have led to ferious reflexions on the neceffity of enlarging the fphere of our domeftic commerce; the reftrictive regulations, which in foreign markets abridge the vent of the increafing furplus of *our agricultural produce, ferve to beget an earneft defire, that a more extenfive demand for that furplus may be created at home.*"*

Commercial regulations, reftrictions, and prohibitions, may be regarded as the bane of Europe. They are univerfally fupported by the clamorous importunity of partial interefts, or by the prejudices and *contemptible cunning of financiers*, who feldom poffefs any real intention to promote the *general welfare*. Colbert, the confidential minifter of Louis XIV. did infinite injury to France, by his defire of fubjecting the commerce, the manufactures, and all the actions of the people, *to financial calculations and regulations*. That minifter, by the tariff of 1667, impofed very high duties upon a great number of foreign manufactures: upon his refufing to moderate them in favour of the Dutch, they, in 1671, prohibited the importation of the wines, brandies, and manufactures of France; and it has been thought that the war of 1672, was in part occafioned by thefe mutual injuries. The peace of Nimeguen put an end to it, in 1678, by moderating fome of thefe duties in favour of the Dutch, who, in confequence, took off their prohibitions. It was about the fame time that the French and Englifh began mutually to opprefs each other's induftry. In 1697, the Englifh prohibited the importation of bone-lace, the manufacture of Flanders; the government of that country, in return, prohibited the importation of Englifh woolens.

The motto of the French economifts, *Faire le bien c'eft le recevoir*, is as applicable to nations as to individuals; yet, in this enlightened period, do we obferve the moft polifhed nations doing every injury in their power to each other, with the unjuft and abfurd idea of deriving advantage to themfelves, by circumfcribing or deftroying the advantages of their neighbours.

Whilft we are exacting a prohibitory tonnage duty from foreigners, and annually increafing the duties on their produce and manufactures, can we expect that they will not retaliate, and in their turn obftruct the progrefs of our trade, by embarraffments particularly injurious to the agricultural intereft? If we find that fuch

---

* Secretary of the treafury on manufactures.

retaliation, on the part of foreigners, has actually taken place, and that our commercial regulations and restrictions operate to injure ourselves, by preventing a sale for the surplus produce of our agriculture—why not adopt a genuine system of policy, founded on the rights of man, and at once remove the evil, by declaring a total freedom of commerce, within the united states? Is it not more than probable that other nations would retaliate *real benefits, as well as insults and injuries?* A just and enlightened policy, of this kind, would be fully adequate to remove the complaint.

Experience justifies the assertion, that a perfect free commerce would not only take off the surplus produce of our farms, but would stimulate to industry, and would insure the prosperity of our country. The small islands of St. Thomas and Santa Cruz were under the government of an exclusive company† which had the sole right both of purchasing the surplus produce of the colonists, and of supplying them with such goods of other countries, as they wanted; and which, therefore, both in its purchases and sales, had not only the power of oppressing them, but the greatest temptation to do so. The government of an exclusive company of merchants, is, perhaps, the worst of all governments, for any country whatever. It was not, however, able to stop, altogether, the progress of these colonies, though it rendered it more slow and languid. The late king of Denmark dissolved this company; and since that time, the prosperity of these colonies has been very great. Curacoa and Eustatia, the two principal islands belonging to the Dutch, are free ports, open to the ships of all nations: and this freedom, in the midst of better colonies, whose ports are open to those of one nation only, has been the great cause of the prosperity of these two barren islands. Mr. Adam Smith, in his wealth of nations, speaking of the causes of the prosperity of new colonies, observes, " that plenty of good land, and *liberty to manage their own affairs, their own way,* seem to be the two great causes of the prosperity of all new colonies."

But, as if the measures of congress were ever to be directed by the momentary fluctuation of affairs, it is proposed, that to afford relief *to the farmer,* recourse should be had to *manufacturing establishments,* under the particular patronage and protection of government; and that congress should grant exclusive privileges, bounties, and premiums, *to a few monied men,* to encourage them to establish extensive manufactories, and to enable them to import from Europe, necessary machines and workmen. Should this scheme take place, under a *false pretext* of serving the *agricultural interest,* a valuable class of citizens, *personally engaged* in useful manufactures, will be sacrificed to the wealthy few.

While colonies of Great Britain, it was the contracted policy of that court, to prevent the progress of manufactures in this country. In a report of the lords commissioners of trade and plantations to parliament, during that period, they say, " It were to be wished, that some expedient might be fallen upon, to divert the thoughts of the colonists from manufactures, so much the rather, be-

† The tonnage duty of congress places the whole of American trade in a similar situation.

The Philosophy of Manufactures

cause these manufactures, in process of time, may be carried on to a greater degree, unless an early stop is put to their progress."

Had the court of Britain pensioned a number of men in America, to effect the ruin of the infant manufactures of our country, they could not have adopted a scheme better calculated to answer that purpose, than the scheme of Duer and Hamilton to establish *national manufactories*.—When Britain had power, she exerted it to restrain the growth of American manufactures. That country may now accomplish by finesse, what she is no longer able to effect by force. The national manufactory, however plausible its declared intention, is fully capable of answering this end.—The immense capital of that corporation will give the company an opportunity of monopolizing many raw materials already procured with difficulty, particularly in the hatting and tanning business. Their workmen, being exempt from taxes, militia duty, and enjoying other privileges, will draw off the journeymen from private manufactures, beneficially scattered through the different parts of the country. The exclusive privilege enjoyed by the company, of establishing lotteries to indemnify them for any losses, will enable them to sell the articles manufactured by them at a less price than any individual citizen, not enjoying such a privilege, can do, without certain ruin to himself.—The unjust and dangerous interference of government, in granting to a company of monied men, privileges, bounties, and favours, not enjoyed by individual citizens, honourably and usefully educated to manufacturing employ, will discourage citizens from acquiring a knowledge in professions or occupations, in which they may be at any time ruined by the arbitrary interference of government. After all, these effects are produced, and they must inevitably will be produced, should the company be successful—the act of incorporation provides for an easy dissolution of the company itself.

The declared intention of establishing national manufactories, is to carry off the surplus produce of our agriculture. The citizens of the united states, engaged in the cultivation of the ground, comprehend nine tenths of its inhabitants. This numerous, laborious, and useful class of citizens, have never come forward to government to solicit partial privileges; have never sought to be incorporated as a distinct body, forming a separate interest from the community at large; much less do they require a violation of the rights of a numerous and respectable class of citizens, *personally engaged in manufactures*.

The yeomanry of America only desire what they have a right to demand—a free unrestricted sale for the produce of their own industry; and not to have the sacred rights of mankind violated in their persons, by arbitrary laws, prohibiting them from deriving all the advantages they can from every part of the produce of their farms.

Tench Coxe

Observations on the Letters of George Logan (1792)

Tench Coxe, one of the promoters of the Society for Establishing Useful Manufactures, could not allow Logan's criticism to go unanswered. In a series of letters to the *American Museum,* using the pseudonym "A Freeman," he refuted Logan point by point. Coxe deemphasized the governmental assistance given the S.U.M., arguing at the same time that such assistance was necessary. He insisted that Logan misunderstood and misrepresented the powers given to the S.U.M. Most important, he tried to persuade his readers to believe that there was nothing evil about large manufacturing establishments. The letter reprinted here is the second of three.

"A Freeman" [Tench Coxe], "Observations on the preceding letters," *American Museum,* volume 12 (September 1792), pp. 167–170; volume 12 (October 1792), pp. 217–221; volume 12 (November 1792) pp. 272–278. By permission of the Houghton Library, Harvard University.

AN idea has been circulated, that *congress* have granted **exemptions**, privileges, and bounties to *the New-Jersey society for promoting useful manufactures*; and parts of the letters of "the Farmer," tend to confirm that mistake. The freemen of the united states will find, however, on the strictest examination, that **no vote, resolution, or act**, of the federal legislature, has been **passed** concerning that company; and of course that no bounty, **privilege, exemption, or other advantage has been given by congress to them,** their workmen, or their property, of any sort or kind. They **are** indisputably liable to the impost, tonnage, duty on di**stilled spirits,** and every other revenue of the united states, present or future, in like manner and in equal proportion with any other citizen or **owner** of property. They have no direct or indirect advantage under the acts of congress, but what every citizen has, who inclines to pursue the branches they may carry on. It is therefore *an high duty to the people*, to undeceive them in a matter, which might give them discontentment with a government, under which they are tasting a degree of prosperity, never before enjoyed by them, or any other nation.

The whole of the advantages possessed by the New-Jersey manufacturing company are under a law of that state. This act has exempted their workmen from military duty, except in cases of invasion or imminent danger: but this will not benefit them *in the smallest degree*, because it exempts only from calls under the laws of **the state, which are now annulled by the militia act of congress.** From militia duty they are not, nor can they be, exempted by the legislature of New-Jersey; and the law of the united states requires the same militia services, duties, fines, &c. from each of the New-Jersey society and its workmen, as from any other citizen or manufacturer.

The company have powers relative to canals and inland navigation, similar to those of the Susquehannah and Schuylkill canal companies, in Pennsylvania, and the Hudson and Mohawk river companies, in New-York. Corporations for inland navigation exist also in New-England, Maryland, Virginia, North and South Carolina; though it should seem that "the Farmer's" objections to this kind of association is such that it is doubtful whether he would consent to a *corporation* for making a turnpike road, or an inland navigation, or a religious society, or a public school. The yeomanry of the united states, however, it is presumed, have no such fanciful objections to these common and necessary means of obtaining such desirable and useful objects. We have city, town, and borough corporations in Pennsylvania, and in all the states, with civil powers to preserve the peace and order of the society, and the health of the people, and to facilitate the settlement of accounts and debts in places where there is usually some dealing and trade. Burlington, Amboy, Elizabeth-town, and Brunswic, in New-Jersey, being so incorporated, the legislature of that state appear to have thought it no great favour to provide for the establishment of the district which should become the principal seat of the manufactory in the same manner, but probably deemed it, as it really appears, highly expedient. It seems to be very wise to have a well regulated police, in a place where there will be many strangers. The assembly of New-Jersey have exhibited a cautious delicacy, worthy of the legislators of a free and intelligent people, in expressly providing, that the place or district shall not become incorporated, if a majority of the taxable inhabitants thereof shall signify to the governor their dissent from, or disapprobation of the incorporation.—So that if, on mature deliberation, they do not like the powers of the corporation, or if they are, on general principles, against any sort of incorporation, they have ample power to prevent it. This, fellow-citizens, is the law which the Farmer tells you, " grants to a few wealthy men the exclusive jurisdiction of six miles square, and a variety of unconstitutional privileges." It is not a few wealthy men, but all the taxable inhabitants, who are to be incorporated; and nearly all of the principal owners of the stock of the company actually reside in other townships, counties, and states, and a few in foreign countries; and not being " inhabitants" cannot be members of the territorial corporation, which is to possess the civil powers. Besides, it really is not true that they are to have ' exclusive' jurisdiction: for they are as much under the controul of the legislature, the governor, and the judiciary of the state and of the government of the united states, as any other city or corporate town, or any county or village in New-Jersey. It is not necessary to comment upon the indecorum and abuse (or art, design, criminality, folly, and breach of duty, which the Farmer bestows upon the government of New-Jersey, the secretary of the treasury, and the company, in the very page wherein he thus mistakes and misrepresents their proceedings. He speaks of danger from a capital stock of £.140,000 sterling in the hands of a great number of persons, when there are and have been these thirty years several individuals in the united states, who are each worth a greater sum. This stock is owned by manufacturers from abroad, who are employed by the company, by farmers, merchants, lawyers, physicians, women, minors, landed men, and monied men, members of the general government, officers of the state of New-Jersey, citizens of various states, and foreigners resident here and in other countries. Can it be expected, that combinations and devices, dangerous to liberty or honest industry, can take place in so mixed a society, or can it be supposed, that such perversions of the countenance of a state will be permitted to exist a single month?

The legislature of New-Jersey have exempted the workmen from their state poll taxes, taxes upon their persons and occupations; but it is not likely that such exemption will be of any value; because poll taxes are so generally disapproved, that they are always made very light in this country, and are seldom or never imposed on any but single men; and taxes on occupations are a very trivial part of the resources of any state: and the exemption will be of little avail, because congress will lay the principal taxes, and they have not given any exemption. The great taxes of all known countries are consumption duties, custom house duties, and the land tax. All these the workmen will have to pay, whether laid by the state or by the united states. But as the finances of New Jersey are in a very good situation and daily improving, and it is a state of great economy in its public expenses, its legiflature will have very little occafion to collect money from the people. The federal cuftoms, duties, and taxes, are thofe which the citizens of New-Jerfey will principally have to pay; and the workmen, raw materials, manufactures, lands, and tenements of the workmen and of the company, will be as liable to thofe cuftoms, duties, and taxes, as if the exemption had not been made by the ftate. For it is plain that a ftate act cannot exempt from taxes of any kind due to the general government.

The Farmer feems difpofed to alarm the hatters and tanners with fuggeftions, that their raw materials may be engroffed and made objects of fpeculation; but the third fection of the New-Jerfey law forbids the company to deal or trade in any raw materials, but fuch as are fit and neceffary for the articles it manufactures, and fuch as fhall be really and truly obtained therefor. It is known, that they have determined upon thofe branches which require water fpinning machinery, (a cafe peculiarly happy, as there are not two hundred water fpindles in the united ftates); and the imports of the goods they propofe to make, are ten times as great as their whole capital ftock, much of which they will inveft in buildings, lots, implements, machinery, working carriages, and cattle. It is plain, that bark, lime, and hides for the tanners, could not be fupplied to them from any diftance, and they could not engrofs either. In refpect to wool and our country furs, the fame remarks apply to them; and if they were to purchafe foreign furs, they would, no doubt, do it abroad. But they do not appear to have in view any thing but the fpinning and cotton dying and calico printing bufinefs, and are appropriating their funds to thofe branches.

The Farmer declares the grants of privileges, even fuch as they are, to be unconftitutional. Surely, then, there is no danger from them, as they muft be void and of no effect. If he will examine the civil lift of New-Jerfey, he will find no reafon to apprehend, from the gentlemen who compofe it, the enforcing of unconftitutional laws.

The origin and defign of the New-Jerfey manufacturing fociety has been frequently mifunderftood and mifreprefented. In the year 1791, feveral months after the government had been transferred from New-York, the fecretary of the treafury formed the plan. It was reprefented, that one of the great objections to manufactures in the united ftates, was the want of money; and although there was manifeftly a greater abundance of active capital in 1791, than for years before, yet there was no profpect of an early application of a fum equal to a moiety of the capital of any one of the firft fifty manufacturers and traders in Great Britain, France, Holland, Germany, or Flanders, by any individual. An union of many individuals was the only mode that could be adopted; and as there was fuppofed to be fome rifque, it was certainly a prudent method, as each would take care not to fubfcribe fo largely as to hurt himfelf, if a failure fhould take place. The fubfcribers, to avoid rifquing more than their fubfcriptions, were, of courfe, to apply for an incorporation; and it was not at all probable the fum would have been fubfcribed without one. The feveral banks in Philadelphia, New-York, Bofton, Baltimore, Providence, &c. had been made up in the same way; and the inland navigation companies and turnpike road companies in the Carolinas, Virginia, Maryland, Pennsylvania, New-York, &c. have been compofed in like manner. Moreover, as it was manifeft, that active capital was flowing into and arifing in the united ftates very rapidly, there was a fincere and ferious apprehenfion, that evils would arife from it, particularly a profufe confumption, unlefs objects to employ it were provided: and it appeared therefore a reafonable belief, that the want of capital, after one well-devifed and fuccefsful plan, would ceafe to be among the objections to manufactures. It is earneftly wifhed, that the body of *the fhip owners* in the united ftates may not furnifh an inftance of an over application of capital, in one of the old modes, which, as it is a cafh bufinefs, muft be accompanied with an actual excefs of money. The recent banks, canals, and turnpike roads, demonftrate, that without new objects, large fums of money muft have lain unemployed. With the impreffions above ftated, the plan of the manufacturing company was adopted: and it would be happy for Pennfylvania, if her Farmer would promote the eftablifhment of fuch an inftitution on that great interior canal, the river Sufquehanna, under the aufpices of the ftate legiflature. The yeomanry would find, that the capital and induftry of the manufacturing citizens would be wifely directed to a fpot, where their cattle, grain, wood, hemp, flax, wool, and iron, would be demanded by confumers, without encountering the coftly charge, and, in fome inftances, the infupportable expenfe of tranfportation to a fea-port. Several of thefe inftitutions would give a front to American manufactures which is neceffary in their competition with foreign commodities. But to return to the New-Jerfey manufactory: it appeared prudent to take a pofition in that ftate for the purpofe of interefting New-York and Philadelphia: and as New-Jerfey has very little foreign commerce, it was prefumed, that both her legiflature and her citizens would promote fo valuable a branch of internal trade. The latter have accordingly fubfcribed handfomely: and the ftate, knowing that thefe new enterprifes are attended with great expenfes at the commencement, with rifque, and fometimes with lofs, authorifed the company to raife by lottery 100,000 dollars, as an indemnification. Their real eftate was exempted from ftate taxes for ten years, and their ftock, or perfonal property, altogether. Thefe taxes, however, as before obferved, will be very fmall under the ftate laws: and they will be fo remote, that the manufacturers in other parts of the union cannot be fenfible of their effects.

The Farmer's fuggeftion that the company will be enabled, by the temporary advantage of a lottery, to underfell, is not even plaufible; for we know that merchants and manufacturers do not ufe their occafional advantages for the abfurd purpofe of underfelling their neighbours for a fhort time, but to increafe their own fubftance and ftock: and if they were to fell the cheaper for it, the purchafers and confumers, that is, "*the great body of the yeomanry,*" about whofe intereft the farmer wifhes to appear very anxious, would be benefited by it.

It will be perceived by every reader, that the letters which are under examination are not confined to the meafures which have been contemplated in regard to American (or national) manufactures. The fisheries, the navigation laws, the banks, the public credit, and the revenues of the united ftates, have each fuftained his efforts to wound them. The New-Jerfey manufactory has occupied but a part of his letters. A concluding number will therefore be employed in the examination of his four articles of impeachment againft the prefident of the united ftates, the majorities of the two houfes of congrefs, and the fecretary of the treafury, in behalf of each of whom the plea is *" not guilty,"* and the appeal is to the people.

A FREEMAN.

John Beale Bordley

Intimations on Manufactures (1794)

Most of the many late eighteenth-century tracts on the debate over manufactures supported either the farmer or the manufacturer. This pamphlet, attributed to the agriculturist John Beale Bordley (1727–1804), makes an unusual argument for a balanced economy. Let the farmers farm and the mechanics manufacture, and let there be free trade to provide luxuries, argued Bordley, and "the comforts of life will be certain."

Bordley modeled his economy on England's and supported it with arguments from an English writer, Arthur Young (1741–1820). Young believed agriculture to be the base on which manufactures and commerce rested and the basis of England's power. Manufactures balanced and strengthened agriculture; a powerful country needed both.

John Beale Bordley was a Maryland judge and gentleman farmer and an advocate of scientific farming techniques. He was a founding member of the Philadelphia Society for Promoting Agriculture.

[John Beale Bordley], *Sketches on Rotations of Crops and Other Rural Matters. To which are annexed Intimations on Manufactures; On the Fruits of Agriculture; and on New Sources of Trade interfering with products of the United States of America in Foreign Markets* (Philadelphia, 1797), pp. 69–72.

Somewhat has been said, in public, of manufactories in *America*; whether it be adviseable to promote them in this early stage of her political existence, or to depend on procuring them from other countries, with the produce merely of her own lands? Have we not " room for looms and the various arts?" Why then should not this nation, in its present youthful vigor, begin to apportion her employment between husbandry and manufactories? which in experience prove to be so coincident, so promotive of wealth and independence, as to have rendered *Britain* rich in all comforts, with a purse powerful in war; but which some on both sides of the Atlantic think has unwarily admitted of a degree of pride in her, that, according to what is common to that vice, bodes an approaching reverse in the current of her affairs. Besides, in the course of a great influx of emigrants to *America*, many, if not the greater number, are mechanics. When these land on the sea coast, and find little or no employment for them in the way of their profession, will they generally go to country labour? Past experience says they will recross the Atlantic, or travel farther westward, and sit down on lands easier obtained, and where they can live on less labor than they could among the old settlements in the hither country. But if manufactories were on foot among us, it would be natural that they should generally prefer the employment they had been ufed to; and by fitting down to their trades, they would gradually advance the arts in America, whilft the more rapid increafe of hufbandry would be the means of fupplying them with bread in payment for their goods, and leave an overplus to be exported to foreign markets. "It however is material to the vigor and worth of manufactories, that they be not difperfed." They are more or lefs advantageous, according as they are carried on in towns, or in detached habitations in the country. In general, the manufacturer in the country has his farm, or a lot of ground, which divides his attention with that of his fhop, whereby both crafts fuffer; and, certain it is, fays Mr *Young*, " their hufbandry is always execrable—the fhop and the field are conducted with little fpirit: both are mean in the quantity and the quality of the productions; and the living of the *farmer-tradefman* is according to it. But in towns the trade is alone depended on, and the productions are more and better: fo of the *thorough-farmer*, from whom he buys his bread, and to whom he fells his goods."

When our employment fhall be duly apportioned between hufbandry and manufactories, the comforts of life will be certain; as they will be procured within our country, independent of the caprice of foreign countries: with the overplus of thefe we are to obtain exotic delicacies, luxuries, and bullion.

" From a well chofen employment are derived the riches, the ftrength, the independency, and the happinefs of nations." If the employment be in things neceffary and convenient, it is infinitely better than when applied in producing luxuries. With neceffaries plentifully produced at home, we may be independent of other nations. An abſolute independency, which ſhuts out commercial and in effect ſocial intercourſe, is not meant. Nations do not all yield the ſame productions; and few, if any, properly divide their employment between huſbandry and manufactories. Britain is the neareſt to it. Even where the beſt proportion prevails, luxuries and trifles will have ſome ſhare of attention among the artiſts, although common ſenſe directs that, eſpecially for the intereſts of a young country, the firſt and principal application ſhould be to procure *neceſſaries* as well for *ſtaples of commerce* as for domeſtic uſes; ſuch as food, clothing, ammunition, &c. Yet legiſlators will not over buſily warp employment againſt its natural bent. They may invite and gently incline it; avoiding dogmatical inhibition or command, unleſs it may be on very extraordinary national occaſions. Nor will they erect monopolies, directly or indirectly, or give undue preferences. Temporary patent rights for inventions are not meant. * To ſet about making *fine* goods before we are full of *neceſſary* comforts, ſeems a beginning at the wrong end.

The manufactures wiſhed to be firſt promoted are eſpecially of plain *clothing and blankets, arms and ammunition.* Manufactures of *woolen* goods are full in our

---

* Perhaps it were better to grant *rewards* proportioned to the uſefulneſs of diſcoveries or inventions, than *excluſive* patent rights. There are conſiderable objections to the latter, in experience, however fair it ſtands in theory; and infinite advantages would ariſe from an immediate free uſe of the invention, at large.

view—In promoting thefe, we increafe the quantity of meat and fkins as well as wool. They are not exotic; but precious materials furnifhed by our hufbandmen. A bounty on the exportation of arms and ammunition *made within the nation*, would foon caufe thofe effentials to abound in the country for its neceffary defence. Yet it is in a fpirited and *flourifhing hufbandry that the foundeft* health and comfort of nations is found. It is a *plenty* of food and cloathing, that are plain and good, rather than fine things, which gives content and cheerfulnefs to a people; and it is the great mafs of the people that are induftrious, rather than the idle poor or the luxurious few, who are principally confidered by legiflatures.

Reproduced courtesy of Merrimack Valley Textile Museum.

George Logan

Constitution of the Lancaster County Society for Promoting of Agriculture, Manufactures, and the Useful Arts (1800)

The Constitution of the Lancaster County Society is attributed to George Logan, defender of the American farmer. In it he demands all the benefits of manufacturing, especially freedom from British merchants and suppliers, with none of the disadvantages. He wants, in short, manufacturing without manufacturers. Only when the farmer provides for his own needs, says Logan, will he be free, and only then will the United States be truly free of England. Reprinted here is the preamble to the Constitution. The body of the document comprises details of organization and rules for meetings.

The Constitution of the Lancaster County Society, for Promoting of Agriculture, Manufactures, and the Useful Arts, in George Logan, *A letter to the citizens of Pennsylvania, on the necessity of promoting agriculture, manufactures, and the useful arts* (Lancaster, Pennsylvania, 1800), pp. 15–20. By permission of the Houghton Library, Harvard University.

THE
CONSTITUTION
OF
The Lancaſter County Society,

For promoting of Agriculture, Manufactures, and the uſeful Arts.

PREAMBLE.

INDEPENDENT Communities do not owe their Characters to the Soil which they occupy; but to their Progreſs in the uſeful Arts. To thoſe Cauſes are to be attributed, not only the difference in the Characters and the Manners of Nations, but their Proſperity, Strength, and Happineſs. On this account, Political Juſtice requires, that every Individual, in becoming a Member of a particular Society, ſhould adopt a mode of conduct conſiſtent with his relative ſituation to ſuch Society. Men would never have aſſociated together, if they had not expected, that, in conſequence of ſuch Aſſociation, they would mutually conduce to the Advantage and Happinefs of each other. This is the *real Purpofe* the *genuine Foundation* of civil Society; and, as far as this Purpofe is anfwered, fo far does civil Society anfwer the end of its inftitution.

Upon the Emancipation of our Country from the Political Yoke of Great Britain, we deemed and called ourfelves a free and independent Commonwealth: But there are means of inferior and indirect Subjugation, from which our Country is not yet emancipated.

The Citizens of Pennfylvania are yet beholden to the Britifh, for the determinations of her Courts, for her Maxims of Commercial Policy, and for many Political Prejudices.

We are dependent on Great Britain for almoft every Article of Clothing we wear, for a great part of the Furniture of our Houfes, for the Inftruments of our amufements, and for the means of our Defence.

Nor is it only for Articles of immediate import from that Country, that we are dependent on Great Britain. The dangerous pre-eminence of her Navy: A Navy that Domineers, with impunity, on the Ocean; that inceffantly threatens the Peace of the Earth; that carries the devaftations of War, upon every pretended infult, from the Shores of that Ifland, to the remoteft parts of the Globe; that imperioufly forces every Maritime Country to take part in, or fuffer by her quarrels; and prohibits, at pleafure, the Commercial Intercourfe of the World. From the dangerous pre-eminence of this Navy we, alfo, are perpetually liable to be interdicted from thofe Articles of Confumption, which Expedience or Neceffity have induced us to feek for from other Countries: And yet our Clothing, imported from Great Britain, is made tributary to fupport this very Navy; which is daily committing the moft wanton Depredations on our Commerce. Nor is it an Evil of trifling magnitude, that the Credit, almoft forced upon our Merchants by the cupidity of the Britifh Traders, has overwhelmed our Country, with Britifh Merchandize, far beyond the real Wants of the Confumer. It has excited our Farmers to needlefs Expenfe, and involved them in Difficulties, for Articles of mere Luxury; It has rendered the plain, but comfortable, Manufactures, which employed the leifure hours of their Wives and Daughters, disreputable, because unfashionable. It has made the Farmer tributary to the Storekeeper; the Storekeeper, to the Merchant, of Philadelphia; the Merchant of Philadelphia, to the Merchant of Great Britain. The Credit thus given, can, at any time, be withdrawn; the Debts thus contracted, can, at any time, be demanded; and the Peace and Comfort of a numerous Body of American Citizens are now, and have long been, at the Mercy of British Merchants, and of the British Court. Hence are our Commercial Towns filled with British Subjects, who conduct our Trade; with British Agents, who drain our Wealth; with British Politics, British Interests, and British Influence. To lessen, in part, these enormous Evils; to render our Citizens, in their private as well as in their public Capacities, really as they ought to be, *independent of Foreign Countries,* for Articles which the Necessities, or the Comforts, of Life require; and to suppress the Temptations to improvident Expense; *We propose* a general encouragement to Agriculture, Manufactures, and the useful Arts. An encouragement that shall make *the use of* our own Productions and Manufactures the fashionable Articles of Consumption; at least in every Circle of *American Republicans.*

Not that it is our Defire to make this, in the commom acceptation of the Word, a Manufacturing Country: Nor do we contemplate any Manufactured Article for an export Trade, nor any Manufacture among ourfelves, which the natural Refources of our Country may not make profitable.

Still lefs are we defirous of introducing in this happy Country, that baneful fyftem of European Management which dooms the human Faculties to be fmothered, and Man to be converted into a Machine. We want not that unfeeling plan of Manufacturing Policy, which has debilitated the Bodies, and debafed the Minds, of fo large a Clafs of People as the Manufacturers of Europe.

Nor are we ambitious to fee a Manufacturing Capitalift, as in the great Manufacturing Towns of Europe, enjoy his Luxuries, or fill his Coffers, by paring down the hard-earned Wages of the laborious Artifts he employs.

But the Object of our Affociation is, to procure, from the fertile Soil of Pennfylvania, every Production it is capable of affording; and, from the Labour and ingenuity of independent Citizens, every Article of Manufacture and of the ufeful Arts, neceffary to render our Country happy, profperous, and truly independent.

The Philosophy of Manufactures

David Humphreys

A Poem on the Industry of the United States of America (1804)

David Humphreys (1752–1818) was a brilliant soldier of the Revolution, diplomat, poetaster, and early manufacturer. His consular assignment to Spain lent him the opportunity to import into the United States the famous Merino sheep and thus to build the woolen mills of Humphreysville, near Derby, Connecticut, now Grosvenordale. In spite of this earnest poem's title and its explicit endorsement of manufactures and machinery, the social vision of the poem is overwhelmingly agrarian, its landscape pastoral. Humphreys stated as plainly as any the ingenuous assurance of early industrial promoters that "agriculture will probably, for a succession of ages, be the chief employment of the citizens of the United States." For Humphreys, the "rural arts" and the "industrial arts" had hardly begun to differentiate themselves, and textile manufacture still meant "homespun." Yet Humphreys' vision of a pastoral empire rested on two points often found in arguments in favor of manufactures. He was strongly committed to federal sponsorship of "improvements," and he was vigorous in his belief that civilization required firm social restraints. In this latter regard, Humphreys' poem and its prefatory matter bear comparison with the argument of Walton Felch in his equally tasteless verse and the report of civilizing influence of factories along the Brandywine published in *Niles' Register,* both reprinted in this collection. Humphreys's landscape does already include the waterpowered spinning mill run by child labor: ". . . water labors in his forceful fall!/ Teach tiny hands with engin'ry to toil. . . ." In general the abstract quality of

David Humphreys, *Miscellaneous Works* (New York, 1804), pp. 89–114. By permission of the Houghton Library, Harvard University. Reprinted with an introduction by William K. Bottorff (Gainesville, Fla.: Scholars' Facsimiles and Reprints, 1968).

Humphreys's poetic diction obscures the extent to which the concrete realities of industrial life were already apparent to him. His prose "Address" and "Argument'" are rather more straightforward. Humphreys's complacency about the industrial future is unqualified by any contradictory evidence of the kind that troubled Zachariah Allen when he offered his version of a commonwealth of rural industries a quarter of a century later (see the Allen text in this volume).

The Philosophy of Manufactures

ADDRESS

TO THE

PEOPLE OF THE UNITED STATES OF AMERICA.

THE main scope of the author's principal productions in verse, has been to indicate to his fellow-citizens, in a connected manner, the measures best calculated for increasing and prolonging the public felicity. He deemed the success of our revolution the broad basis on which this superstructure was to be built. The first thing to be done was to establish our independence; the second to prepare the national mind to profit by our unusual advantages for happiness; and the next to exhibit in perspective those numberless blessings which Heaven has lavished around us, and which can scarcely be lost but by our own folly or fault. Having attempted to furnish his countrymen with some seasonable arguments and reflections on these subjects, in his "Address to the Armies," in his "Poem on the Happiness of America," and in the "Prospect of the Future Glory of the United States," he proposes now to show the prodigious influence of national industry in producing public and private riches and enjoyments.

One of the primary objects of a good government is to give energy and extent to industry, by protecting the acquisitions and avails of their labour to the governed. This industry is the cause of the wealth of nations. It hastens their advancement in the arts of peace, and multiplies their resources for war. Under such a safeguard, mankind, engaged in any lawful and productive profession, will advance, at the same moment, their own interest and that of the commonwealth. Universal prosperity must ensue. With us, the successful issue has been the best panegyric of such a system. Could industry become generally fashionable and prevalent, indigence, and the calamities that flow from it, would be confined within very narrow channels. With a few exceptions, such as are offered by the bee, the ant, and the beaver, social toil, which accomplishes works truly astonishing for their contrivance and magnitude, distinguishes the human race from every species of the animal creation. A reciprocation of wants and aids, as it were, rivets man to his fellows. What isolated person can perform for himself every act which his helpless and feeble state requires? By a combination of well-directed efforts, what miracles of improvement, what prodigies in refinement, may be effected! The expediency, and even the necessity of concerted and persovering operations, have a natural tendency to confirm and augment, through the medium of mutual services and benefits, fidelity, kindness, valour and virtue, among the members of civil society. Who, then, will envy the indolent and comfortless lot of the solitary savage, or the thinly scattered tribes of the desert?

The influence of industry is not less efficacious in procuring personal advantage and fruition for individuals. It commonly gives health of body and serenity of mind, together with strength of resolution and consistency of character. It thus furnishes a kind of moral force for overcoming the sluggishness of matter, which constantly inclines to repose. Influenced by a desire of being free from humiliating dependence and degrading penury, every man, who is not visited by sickness or prevented by disaster, will be enabled, in his youthful days, to provide a plentiful subsistence for his old age; so that, in the last stages of infirmity and decrepitude, distress and mendicity will seldom, if ever, be seen. Such is now the condition of the people of the United States of America. To flatter the idle and worthless, by perpetually declaiming on the duty of the industrious and wealthy to dispense largely their contributions and charities, is the insidious language often used in Europe by many vociferous demagogues and revolutionary scribblers. To prevent poverty as much as possible, by presenting employment to protected and provident industry, is the high office of a wise and just government. In our country that policy has been successful beyond all former example. The traveller may journey thousands of miles without meeting a single beggar. And herein a striking difference will be remarked between our country and most of the countries in the world.

That industry is capable of speedily changing a dreary wilderness into a cheerful habitation for men, the history of the progress of society in the United States of America has sufficiently proved. It is at present generally understood, that an unequalled share of happiness is enjoyed by the inhabitants of this newly discovered continent. This is, perhaps, chiefly attributable (under the benediction of Providence) to their singularly favourable situation for cultivating the soil. May we not fairly calculate that this effect will continue co-existent with the cause; namely, the abundance and cheapness of land? An almost unlimited space of excellent territory remains to be settled. Freehold estates may be purchased upon moderate terms. Agriculture will probably, for a succession of ages, be the chief employment of the citizens of the United States.

Notwithstanding the beauties and pleasures of rural life have so frequently been happily described in poetry, it was presumed the settlement and cultivation of a new hemisphere might supply some new topics and allusions. There many things wore a novel appearance, when examined in their process and result. The agricultural character was presented in action, with more than usual effect and felicity. The changes were, in some respects, like those in a garden of enchantment. Upon the introduction of civilization into those rugged and inhospitable regions, whose barbarity was coeval with the world, forests fell, houses rose, and beautiful scenery succeeded. It was not intended, by deviating from the beaten track of describing old establishments, to run unnecessarily into the bye-path of innovation and singularity. Many American prospects rose before the author's transported imagination, when he was far absent from his native land. How frequently did he wish for a magic pencil to make them equally present to the mental sight of his European friends! How often, and with how much ardour, did his fancy dwell on the humble and unvarnished blessings of peace, when contrasted with the proud and dazzling miseries of war! In thus ruminating on the walks of still life, he hoped he should at least be permitted, without incurring the displeasure of any ill-natured critic, to proceed in a course so amusing to himself, picking here and there a wild or cultivated flower, and attempting to delineate such landscapes as he might occasionally find, interspersed with scenes of romantic grandeur or domestic simplicity.

*

ARGUMENT.

The Genius of Culture invoked—prodigious effect of toil in changing the face of nature—state of our country when it was first settled by our ancestors—their manly efforts crowned with success—contrast between North and South-America—the latter remarkable for mines, as the former is for agriculture—in what manner labour embellishes the land—different branches of cultivation recommended—the fabrication of maple-sugar dwelt upon, as having a gradual tendency to the abolition of slavery—commerce to succeed—strong propensities of the people of the United States for extensive navigation—effeminate nations are always in danger of losing their independence—several specified which have experienced the debilitating consequences of sloth—its destructive influence on states—Congress called upon to encourage industry in the United States; and Washington, as President, to protect manufactures—machinery for diminishing the operations of manual labour—the loom—wool—sheep—flax and hemp—remonstrance against suffering our manufacturing establishments to be frustrated by an unreasonable predilection for foreign fabrics—the fair sex invited to give the example of encouraging home manufactures—their province in the United States—their influence on civilized society—deplorable condition of savage life—moral effect of industry on constitution and character—bold and adventurous spirit of our citizens—prepared by hardiness to distinguish themselves on the ocean and in war—allusion to our contest with Britain—happiness of our present peaceful situation—the Poem is concluded with the praises of Connecticut as an agricultural State.

A POEM

ON THE

INDUSTRY

OF THE

UNITED STATES OF AMERICA.

———

GENIUS of Culture! thou, whose chaster taste
Can clothe with beauty ev'n the dreary waste;
Teach me to sing, what bright'ning charms unfold,
The bearded ears, that bend with more than gold;
How empire rises, and how morals spring,   5
From lowly labour, teach my lips to sing;
Exalt the numbers with thy gifts supreme,
Ennobler of the song, my guide and theme!

    Thou, toil! that mak'st, where our young empire grows,
The wilderness bloom beauteous as the rose,   10
Parent of wealth and joy! my nation's friend!
Be present, nature's rudest works to mend;
With all the arts of polish'd life to bless,
And half thy ills, Humanity! redress.
On this revolving day, that saw the birth   15
Of a whole nation glad th' astonished earth;
Thee I invoke to bless the recent reign
Of independence—but for thee how vain
Each fair advantage liberty has giv'n,
And all the copious bounties show'r'd by heav'n?   20
Hail, mighty pow'r! whose vivifying breath
Wakes vegetation on the barren heath;
Thou changest nature's face; thy influence such,
Dark deserts brighten at thy glowing touch;
Creation springs where'er thy plough-share drives,   25
And the dead grain, an hundred fold, revives.
Thy voice, that dissipates the savage gloom,
Bade in the wild unwonted beauty bloom:

By thee and freedom guided, not in vain,
Our great fore-fathers dar'd the desert main : 30
O'er waves no keel had cut they found the shore,
Where desolation stain'd his steps with gore,
Th' immense of forest! where no tree was fell'd,
Where savage-men at midnight orgies yell'd;
Where howl'd round burning pyres each ravening beast, 35
As fiend-like forms devour'd their bloody feast,
And hoarse resounded o'er the horrid heath,
The doleful war-whoop, or the song of death.
Soon our progenitors subdu'd the wild,
And virgin nature, rob'd in verdure, smil'd. 40
They bade her fruits, through rifted rocks, from hills
Descend, misnam'd innavigable rills:
Bade houses, hamlets, towns, and cities rise,
And tow'rs and temples gild Columbian skies.
Success thence crown'd that bold, but patient band, 45
Whose undegen'rate sons possess the land;
Their great fore-fathers' principles avow,
And proudly dare to venerate the plough.

*

   Sages, conven'd from delegating states,
Who bear the charge of unborn millions' fates;
From early systems states their habits take,
And morals more than climes a difference make:
Then give to toil a bias, aid his cause 195
With all the force and majesty of laws;
So shall for you long generations raise,
The sweetest incense of unpurchas'd praise!

   Thou, Washington, by heav'n for triumphs nurs'd,
In war, in peace, of much lov'd mortals first! 200
In public as in private life benign,
Still be the people heav'n's own care and thine!
While thou presid'st, in useful arts direct,
Create new fabrics and the old protect.
Lo! at thy word, subdued for wond'ring man, 205
What mighty elements advance the plan;
While fire and wind obey the Master's call,
And water labours in his forceful fall!
Teach tiny hands with engin'ry to toil,
Cause failing age o'er easy tasks to smile; 210
Thyself that best of offices perform,
The hungry nourish and the naked warm;
With gladness picture rescued beauty's eye,
And cheek with health's inimitable dye;
So shall the young, the feeble find employ, 215
And hearts with grief o'erwhelm'd emerge to joy.

The Philosophy of Manufactures

First let the loom each lib'ral thought engage,
Its labours growing with the growing age;
Then true utility with taste allied,
Shall make our homespun garbs our nation's pride.   220
See *wool*, the boast of Britain's proudest hour,
Is still the basis of her wealth and pow'r!
From her the nations wait their wintry robe,
Round half this idle, poor, dependant globe.
Shall we, who foil'd her sons in fields of fame,   225
In peace add noblest triumphs to her name?

Shall we, who dar'd assert the rights of man,
Become the vassals of her wiser plan?
Then, rous'd from lethargies—up! men! increase,
In every vale, on every hill, the fleece!   230
And see the fold, with thousands teeming, fills
With flocks the bleating vales and echoing hills.
Ye harmless people! man your young will tend,
While ye for him your coats superfluous lend.
Him nature form'd with curious pride, while bare,   235
To fence with finery from the piercing air:
This fleece shall draw its azure from the sky,
This drink the purple, that the scarlet dye;
Another, where immingling hues are giv'n,
Shall mock the bow with colours dipt in heav'n: 240
Not guarded Colchis gave admiring Greece
So rich a treasure in its golden fleece.

Oh, might my guidance from the downs of Spain,
Lead a white flock across the western main;
Fam'd like the bark that bore the Argonaut,   245
Should be the vessel with the burden fraught!
Clad in the raiment my Merinos yield,
Like Cincinnatus fed from my own field;
Far from ambition, grandeur, care and strife,
In sweet fruition of domestic life;   250
There would I pass with friends, beneath my trees,
What rests from public life, in letter'd ease.

To toil encourag'd, free from tythe and tax,
Ye farmers sow your fields with hemp and flax:
Let these the distaff for the web supply,   255
Spin on the spool, or with the shuttle fly.
But what vile cause retards the public plan?
Why fail the fabrics patriot zeal began?
Must nought but tombs of industry be found,
Prostrated arts expiring on the ground?   260

The Philosophy of Manufactures

Shall we, of gewgaws gleaning half the globe,
Disgrace our country with a foreign robe?
Forbid it int'rest, independence, shame,
And blush that kindles bright at honour's flame!

    Should peace, like sorcery, with her spells controul   265
Our innate springs and energies of soul;
To you, Columbian dames! my accents call,
Oh, save your country from the threaten'd fall!
Will ye, blest fair! adopt from every zone
Fantastic fashions, noxious in your own?   270
At wintry balls in gauzy garments drest,
Admit the dire destroyer in your breast?
Oft when nocturnal sports your visage flush,
As gay and heedless to the halls ye rush,
Then death your doom prepares: cough, fever, rheum,   275
And pale consumption nip your rosy bloom.
Hence many a flow'r in beauty's damask pride,
Wither'd, at morn, has droop'd its head and died.*
While youthful crimson hurries through your veins,
No cynic bard from licit joys restrains;   280
Or bids with nature hold unequal strife,
And still go sorrowing through the road of life.
Nor deem him hostile who of danger warns,
Who leaves the rose, but plucks away its thorns.

*

    Of savage life so spring the bitter fruits,
For savage indolence the man imbrutes.
From industry the sinews strength acquire,   345
The limbs expand, the bosom feels new fire.
Unwearied industry pervades the whole,
Nor lends more force to body than to soul.
Hence character is form'd, and hence proceeds
Th' enlivening heat that fires to daring deeds: 350
Then animation bids the spirit warm,
Soar in the whirlwind and enjoy the storm.
For our brave tars what clime too warm, too cold,
What toil too hardy, or what task too bold?
)'er storm-vex'd waves our vent'rous vessels roll,   355
Round artic isles or near th' antartic pole;
Nor fear their crews the fell tornado's ire,
Wrapp'd in a deluge of Caribbean fire.

---

* This, it is wished, may be received as a useful warning by young persons against exposing themselves, when too thinly clad, to the winter air. Many deaths have been occasioned by imprudencies of this nature.

The Philosophy of Manufactures

The wonders of the deep they see, while tost
From earth's warm girdle to the climes of frost: 360
Full soon to bid the battle's thunder roar,
And guard with wooden walls their native shore.

*

Hail favour'd state! CONNECTICUT! thy name
Uncouth in song, too long conceal'd from fame;
If yet thy filial bards the gloom can pierce,
Shall rise and flourish in immortal verse.
Inventive genius, imitative pow'rs, 405
And, still more precious, common-sense, is ours;
While knowledge useful, more than science grand,
In rivulets still o'erspreads the smiling land.

Hail, model of free states! too little known,
Too lightly priz'd for *rural arts* alone: 410
Yet hence from savage, social life began,
Compacts were fram'd and man grew mild to man.
Thee, Agriculture! source of every joy,
Domestic sweets and bliss without alloy;
Thee, friend of freedom, independence, worth, 415
What raptur'd song can set conspicuous forth?
Thine every grateful gift, my native soil!
That ceaseless comes from *agricultural toil:*
This bids thee, dress'd, with added charms appear,
And crowns with glories, not its own, the year. 420
Though, capp'd with cliffs of flint, thy surface rude,
And stubborn glebe the slothful race exclude;
Though sultry summer parch thy gaping plains,
Or chilling winter bind in icy chains;

Thy patient sons, prepar'd for tasks sublime, 425
Redress the rigours of th' inclement clime,
Clothe arid earth in green, for glooms supply
The brightest beauties to th' astonish'd eye.

What though for us no fields Arcadian bloom,
Nor tropic shrubs diffuse a glad perfume; 430
No fairy regions picturesque with flow'rs,
Elysian groves, or amaranthine bow'rs,
Breathe sweet enchantment—but still fairer smile,
Once savage wilds now tam'd by tut'ring toil.
The rolling seasons saw with rapture strange, 435
The desert blossom and the climate change.
Roll on, thou sun! and bring the prospect bright,
Before our ravish'd view in liveliest light.
Arise in vernal pride, ye virgin plains!

The Philosophy of Manufactures

With winning features which no fiction feigns. 440
Arise, ye laughing lawns! ye gladd'ning glades!
Poetic banks! and philosophic shades!
Awake, ye meads! your bosoms ope, ye flow'rs!
Exult, oh earth! and heav'n descend in show'rs!

Where the dun forest's thickest foliage frown'd, 445
And night and horror brooded o'er the ground;
While matted boughs impenetrably wove
The sable curtains of th' impervious grove;
Where the swart savage fix'd his short abode,
Or wound through tangled wilds his thorny road; 450
Where the gaunt wolves from crag-roof'd caverns prowl'd,
And mountains echoed as the monsters howl'd;
Where putrid marshes felt no solar beams,
And mantling mire exhal'd mephitic steams;
See, mid the rocks, a Paradise arise, 455
That feels the fostering warmth of genial skies!
While gurgling currents lull th' enchanted soil,
The hill-tops brighten and the dingles smile.

Then hail for us, ye transatlantic scenes,
Soul-soothing dwellings! sight-refreshing greens! 460
And chiefly hail, thou state! where virtue reigns,
And peace and plenty crown the cultur'd plains.

Nor lacks there aught to soothe the pensive mind,
Its taste on nature form'd, by truth refin'd:
For pure simplicity can touch the heart, 465
Beyond the glitter and the gloss of art.
Not wanting there the fountain's bubbling tide,
Whence flows the narrow stream and river wide,
With gladsome wave to drench the thirsty dale,
Or waft through wond'ring woods the flitting sail. 470
Not wanting there the cottage white-wash'd clean,
Nor town with spires that glimmer o'er the green:
Nor rich variety's uncloying charm,
The steeds that prance, the herds that graze the farm;
The flocks that gambol o'er the dark-green hills, 475
The tumbling brooks that turn the busy mills;
The clover pastures deck'd with dappled flow'rs,
Spontaneous; gardens gay with roseate bow'rs;
The tedded grass in meadows newly shorn,
The pensile wheat-heads and stiff Indian corn; 480
The grafts with tempting fruit, and thick-leav'd groves,
Where timid birds conceal their airy loves:
Along th' umbrageous walk, enamour'd meet

The artless pairs, in courtship chaste as sweet,
In wedlock soon to join—hail, sacred rite! 485
Delicious spring! exhaustless of delight!
No poor, for wealth withheld, accuses heav'n,
Nor rich, insulting, spurns the bounties giv'n.
No wretched outcast—happy, till beguil'd—
Pollution's sister, and affliction's child! 490
Shivering and darkling strays through wintry streets,
And lures (for bread) to brothels all she meets;
Or tir'd and sick, with faint and fearful cry,
At her betrayer's door lies down to die.
No scenes of woe the pleasing prospect blight, 495
And no disgusting object pains the sight;
For calm *content*, the sunshine of the soul,
With bright'ning *ease*, embellishes the whole.

'Tis rural innocence, with rural toil,
Can change the frown of fortune to a smile. 500
Ah, let the sons of insolence deride
The simple joys by humble toil supplied:
Not him whose breast with false refinement pants,
Factitious pleasures, artificial wants,
Such scenes delight—nor boasts that state a claim, 505
For man's or nature's grandest works, to fame.

Of life sequester'd, fond and frequent theme!
Th' instructed few with higher reverence deem:
For o'er its moral part a lustre shines,
That all around enlivens and refines. 510
'Twas there the joys of wedded love began,
And health and happiness there dwelt with man:
The city's palaces though man has made,
The country's charming views a God display'd—
Still the best site from art derives new charms, 515
In villas fair and ornamented farms.

*

The Philosophy of Manufactures

Reproduced courtesy of Merrimack Valley Textile Museum.

Robert Southey

Espriella's Letters (1807)

The English romantic poet Robert Southey (1774–1843) couched this extensive, critical prose commentary on the state of British society in the form of a pseudonymous travelogue made up of a series of letters purportedly written by a Spanish aristocrat. As a humanitarian and moderately reformist attack on the domestic ills of an old oppressor, the book was popular in the United States. It was reprinted here the same year it appeared in England (1807) and went through at least seven editions between then and 1820. Southey's commentary on the squalor of industrial Manchester was a common point of reference for Americans concerned by the effects of manufactures. Note, however, Southey's conclusion: the introduction of the "utmost" in machinery in a new nation could be of great social benefit. This argument probably did as much to encourage American manufacturing interests as Southey's condemnation of child labor in Manchester did to arouse opposition.

[Robert Southey], *Letters from England: by Don Manuel Alvarez Espriella*, first American edition (Boston, 1807), pp. 163–167. By permission of the Houghton Library, Harvard University.

LETTER XXXVIII.

Manchester.—Cotton Manufactory.—Remarks upon the pernicious Effects of the manufacturing System.

J. had provided us with letters to a gentleman in Manchester ; we delivered them after breakfast, and were received with that courtesy which a foreigner when he takes with him the expected recommendations is sure to experience in England. He took us to one of the great cotton manufactories, showed us the number of children who were at work there, and dwelt with delight on the infinite good which resulted from employing them at so early an age. I listened without contradicting him, for who would lift up his voice against Diana in Ephesus!—proposed my questions in such a way as not to imply, or at least not to advance, any difference of opinion, and returned with a feeling at heart which makes me thank God I am not an Englishman.

There is a shrub in some of the East Indian islands which the French call *veloutier ;* it exhales an odour that is agreeable at a distance, becomes less so as you draw nearer, and, when you are quite close to it, is insupportably loathsome. Alciatus himself could not have imagined an emblem more appropriate to the commercial prosperity of England.

Mr. —— remarked that nothing could be so beneficial to a country as manufactures. " You see these children, sir," said he. " In most parts of England poor children are a burthen to their parents and to the parish ; here the parish, which would else have to support them, is rid of all expence ; they get their bread almost as soon as they can run about, and by the time they are seven or eight years old bring in money. There is no idleness among us :—they come at five in the morning ; we allow them half an hour for breakfast, and an hour for dinner ; they leave work at six, and another set relieves them for the night ; the wheels never stand still." I was looking, while he spoke, at the unnatural dexterity with which the fingers of these little creatures were playing in the machinery, half giddy myself with the noise and the endless motion : and when he told me there was no rest in these walls, day nor night, I thought that if Dante had peopled one of his hells with children, here was a scene worthy to have supplied him with new images of torment.

" These children, then," said I, " have no time to receive instruction." " That, sir," he replied, " is the evil which

The Philosophy of Manufactures

we have found. Girls are employed here from the age you see them till they marry, and then they know nothing about domestic work, not even how to mend a stocking or boil a potatoe. But we are remedying this now, and send the children to school for an hour after they have done work." I asked if so much confinement did not injure their health. "No," he replied, "they are as healthy as any children in the world could be. To be sure, many of them as they grew up went off in consumptions, but consumption was the disease of the English." I ventured to inquire afterwards concerning the morals of the people who were trained up in this monstrous manner, and found, what was to be expected, that in consequence of herding together such numbers of both sexes, who are utterly uninstructed in the commonest principles of religion and morality, they were as debauched and profligate as human beings under the influence of such circumstances must inevitably be ; the men drunken, the women dissolute ; that however high the wages they earned, they were too improvident ever to lay-by for a time of need ; and that, though the parish was not at the expense of maintaining them when children, it had to provide for them in diseases induced by their mode of life, and in premature debility and old age ; the poor-rates were oppressively high, and the hospitals and workhouses always full and overflowing. I inquired how many persons were employed in the manufactory, and was told, children and all about two hundred. What was the firm of the house ?—There were two partners. So ! thought I,— a hundred to one !

"We are well off for hands in Manchester," said Mr. — ; "manufactures are favourable to population, the poor are not afraid of having a family here, the parishes therefore have always plenty to apprentice, and we take them as fast as they can supply us. In new manufacturing towns they find it difficult to get a supply. Their only method is to send people round the country to get children from their parents. Women usually undertake this business ; they promise the parents to provide for the children ; one party is glad to be eased of a burthen, and it answers well to the other to find the young ones in food, lodging and clothes, and receive their wages." "But if these children should be ill-used ?" said I. "Sir," he replied, "it can never be the interest of the women to use them ill, nor of the manufacturers to permit it."

It would have been in vain to argue had I been disposed to it. Mr. —— was a man of humane and kindly nature, who would not himself use any thing cruelly, and judged of others by his own feelings. I thought of the cities in Arabian romance, where all the inhabitants were inchanted : here Commerce is the queen witch, and I had no talisman strong enough to disenchant those who were daily drinking of the golden cup of her charms.

We purchase English cloth, English muslins, English buttons, &c. and admire the excellent skill with which they are fabricated, and wonder that from such a distance they can be afforded to us at so low a price, and think what a happy country is England! A happy country indeed it is for the higher orders; no where have the rich so many enjoyments, no where have the ambitious so fair a field, no where have the ingenious such encouragement, no where have the intellectual such advantages; but to talk of English happiness is like talking of Spartan freedom, the Helots are overlooked. In no other country can such riches be acquired by commerce, but it is the one who grows rich by the labour of the hundred. The hundred, human beings like himself, as wonderfully fashioned by Nature, gifted with the like capacities, and equally made for immortality, are sacrificed body and soul. Horrible as it must needs appear, the assertion is true to the very letter. They are deprived in childhood of all instruction and all enjoyment; of the sports in which childhood instinctively indulges, of fresh air by day and of natural sleep by night. Their health physical and moral is alike destroyed; they die of diseases induced by unremitting task work, by confinement in the impure atmosphere of crowded rooms, by the particles of metallic or vegetable dust which they are continually inhaling; or they live to grow up without decency, without comfort, and without hope, without morals, without religion, and without shame, and bring forth slaves like themselves to tread in the same path of misery.

The dwellings of the labouring manufacturers are in narrow streets and lanes, blocked up from light and air, not as in our country to exclude an insupportable sun, but crowded together because every inch of land is of such value, that room for light and air cannot be afforded them. Here in Manchester a great proportion of the poor lodge in cellars, damp and dark, where every kind of filth is suffered to accumulate, because no exertions of domestic care can ever make such homes decent. These places are so many hot-beds of infection; and the poor in large towns are rarely or never without an infectious fever among them, a plague of their own, which leaves the habitations of the rich, like a Goshen of cleanliness and comfort, unvisited.

Wealth flows into the country, but how does it circulate there? Not equally and healthfully through the whole system; it sprouts into wens and tumours, and collects in aneurisms which starve and palsy the extremities. The government indeed raises millions now as easily as it raised thousands in the days of Elizabeth: the metropolis is six times the size which it was a century ago; it has nearly doubled during the present reign ; a thousand carriages drive about the streets of London, where, three generations ago, there were not an hundred ; a thousand hackney coaches are licensed in the same city, where at the same distance of time there was not one ; they whose grandfathers dined at noon from wooden trenchers, and upon the produce of their own farms, sit down by the light of waxen tapers to be served upon silver, and to partake of delicacies from the four quarters of the globe. But the number of the poor, and the sufferings of the poor, have continued to increase ; the price of every thing which they consume has always been advancing, and the price of labour, the only commodity which they have to dispose of, remains the same. Work-houses are erected in one place, and infirmaries in another ; the poor-rates increase in proportion to the taxes ; and in times of dearth the rich even purchase food, and retail it to them at a reduced price, or supply them with it gratuitously : still every year adds to their number. Necessity is the mother of crimes ; new prisons are built, new punishments enacted ; but the poor become year after year more numerous, more miserable, and more depraved ; and this is the inevitable tendency of the manufacturing system.

This system is the boast of England,—long may she continue to boast it before Spain shall rival her ! Yet this is the system which we envy, and which we are so desirous to imitate. Happily our religion presents one obstacle ; that incessant labour which is required in these task-houses can never be exacted in a Catholic country, where the Church has wisely provided so many days of leisure for the purposes of religion and enjoyment. Against the frequency of these holydays much has been said ; but Heaven forbid that the clamour of philosophizing commercialists should prevail, and that the Spaniard should ever be brutalized by unremitting task-work, like the negroes in America and the labouring manufacturers in England ! Let us leave to England the boast of supplying all Europe with her wares ; let us leave to these lords of the sea the distinction of which they are so tenacious, that of being the white slaves of the rest of the world, and doing for it all its dirty work. The poor must be kept miserably poor, or such a state of things could not continue ; there must be laws to regulate their wages, not by the value of their work, but by the pleasure of their masters ; laws to prevent their removal from one place to another within the kingdom, and to prohibit their emigration out of it. They would not be crowded in hot task-houses by day, and herded together in damp cellars at night ; they would not toil in unwholesome employments from sun-rise till sun-set, whole days, and whole days and quarters, for with twelve hours' labour the avidity of trade is not satisfied ; they would not sweat night and day, keeping up this *laus perennis** of the Devil, before furnaces which are never suffered to cool, and breathing-in vapours which inevitably produce disease and death ;—the poor would never do these things unless they were miserably poor, unless they were in that state of abject poverty which precludes instruction, and, by destroying all hope for the future, reduces man, like the brutes, to seek for nothing beyond the gratification of present wants.

How England can remedy this evil, for there are not wanting in England those who perceive and confess it to be an evil, it is not easy to discover, nor is it my business to inquire. To us it is of more consequence to know how other countries may avoid it, and, as it is the prevailing system to encourage manufactures every where, to inquire how we may reap as much good and as little evil as possible. The best methods appear to be by extending to the utmost the use of machinery, and leaving the price of labour to find its own level : the higher it is, the better. The introduction of machinery in an old manufacturing country always produces distress by throwing workmen out of employ, and is seldom effected without riots and executions. Where new fabrics are to be erected it is obvious that this difficulty does not exist, and equally obvious that, when hard labour can be performed by iron and wood, it is desirable to spare flesh and blood. High wages are a general benefit, because money thus distributed is employed to the greatest general advantage. The labourer, lifted up one step in society, acquires the pride and the wants, the habits and the feelings, of the class now next above him.† Forethought, which the miserably poor necessarily and instinctively shun, is, to him who earns a comfortable competence, new pleasure ; he educates his children, in the hope that they may rise higher than himself, and that he is fitting them for better fortunes. Prosperity is said to be more dangerous than adversity to human virtue ; both are wholesome when sparingly distributed, both in the excess perilous always, and often deadly : but if prosperity be thus dangerous, it is a danger which falls to the lot of few ; and it is sufficiently proved by the vices of those unhappy wretches who exist in slavery, under whatever form or in whatever disguise, that hope is as essential to prudence, and to virtue, as to happiness.

* I am informed by a catholic, that those convents in which the choir service is never discontinued are said to have *laus perennis* there.—Tr.

† This argument has been placed in a more forcible light in the first volume of the Annual Review, in an article upon the Reports of the Society for bettering the Condition of the Poor, attributed to a gentleman of Norwich. It is one of the ablest chapters upon this branch of political economy that has ever been written.—Tr.

George W. P. Custis

An Address on the Importance of Encouraging Agriculture and Domestic Manufactures (1808)

This address was given by Custis before the Arlington [Virginia] Sheep-Shearing, an annual event founded by him in 1803. At these meetings, held to celebrate and promote agriculture and the wool industry (as well as to shear sheep), awards were presented for the best sheep, best wool, and best wool cloth. In these excerpts from his address, Custis touches on reasons why true patriots should support domestic manufactures. Note that Custis supports manufactures while opposing corporations and machinery.

George Washington Parke Custis (1781–1857) was the foster grandson and adopted son of George Washington. In addition to his work in promoting the wool industry, he was a playwright and author of a biography of Washington.

George W. P. Custis, *An Address to the People of the United States on the Importance of Encouraging Agriculture and Domestic Manufactures: Tending to shew that by a due encouragement of these essential interests, the nation will be rendered more respectable abroad and more prosperous at home. Together with an Account of the Improvements in Sheep at Arlington* . . . (Alexandria, Virginia, 1808).

ADDRESS.

FELLOW-CITIZENS,

IN the present important crisis of our national affairs, when the public mind is bent with concern upon our relations abroad, and turns with anxiety to our welfare at home, it is becoming to urge for your consideration, those subjects, which may be promotive of peace to the one, or prosperity to the other.

In the war of politics which has long unhappily raged among us, I have ever beheld the scene without a desire to participate any further than necessity, or a sense of duty, might warrant. But in the humble walks of rural life, in the promotion of Agriculture and Domestic œconomy, I have endeavoured to take as active a part as my situation, and means of knowledge would admit. Convinced that Agriculture is the great basis of all national prosperity, the key-stone that centers the arch of government, and the great impetus that moves the machine of national wealth and individual industry, I have been a laborer in its cause. The importance of manufactures, in all countries, but particularly in one which calls itself independent, must be present to every view, and the situation of European affairs renders it doubly necessary to provide for those wants at home, the supply of which may be denied us abroad.—— The rising population of the United States will in a few years demand employment for its numerous offspring, and the necessities of the country will of course encrease with its enlarged people. The old, the infirm, and those as yet too young to bear the labors of the field, are all candidates for manufacturing employment: and a manufacturing establishment founded upon a liberal scale, would at once employ and give bread to hundreds, who are now objects of want and disgust. Among the praiseworthy attributes of European countries are the vast works of public benevolence, where thousands are rescued from vice, and bred up in industry. The produce of so much labor, gives wealth to the nation, and does honor to the national character. Again, we now occupy the very humble rank in the commercial world, of being the carriers for others, and thereby deriving emolument from their exertions—but how much more characteristic would it be, of a great people like ourselves, rather to bear to distant quarters of the globe, the produce and ingenuity of our own workshops than to wait at the doors of our neighbours, to receive their burdens for hire. 'Tis true, the abundance of our fields load many vessels with the necessaries of life, and returns the Agriculturist a reward for his labors, but their produce is only borne to those countries, where no harvest bless the land, or where the season has denied the usual supply of bread. Were all countries able to fill their own garners America would want a trade.— Those who deal alone in the necessaries of life, appeal to the distresses of their neighbours for a market. Thus we find that the greatest commercial nation on the earth exports the industry and ingenuity of her workshops, and receives the bounteous gifts of the soil in return.

In some branches of foreign manufacture fabrics are formed so costly by the force of improvement and skill, that the labour of a day will purchase the labour of one of our citizens for half a year. So very helpless are the people of some parts of the United States, that they actually import articles, the material of which they possess in a better degree at home. It may be urged that the expence of labour, is as yet too great in America for works of art to flourish; but the reason is very plain—we are mostly engaged in one occupation, tilling the soil; and not having an alternative, those who possess no land will naturally require a high price, to work for those who have; but very many of these people would take a much lower price, where they could be employed at all seasons, and receive a settled stipend for the support of their families. The labours of the field are precarious from the seasons, and the condition of the farmer to give employment, whereas the manufactory which supplies the wants of the people, can only cease with those wants being supplied. The vast impetus which manufactures would give to the national machine, can only be imagined from beholding the fiscal resources of other countries. As collieries and fisheries form the basis of naval strength, so agriculture and manufactures are the foundation of commercial importance: so essential are they to each other, that agriculture may be termed the heart, which diffuses life to the manufacturing system, and regulates the pulse of the national constitution.

The situation of the United States possessing a vast maritime coast, and rivers visiting the internal parts of every state, shews its disposition by Providence, for a great agricultural, manufacturing and commercial commonwealth. Again, the almost boundless latitude of our vast possessions, embracing a portion of every clime, and every soil, shews that nature has done every thing, and we as yet but little, for our prosperity. Our mountains contain ores of all the useful kinds, (happily but little gold) which would tempt our avarice and weaken our industry, for the hand that would seek gold, will find it in honest industry. Our vallies smile with glowing pastures, and the hills produce at once the timber for machinery and naval defence. In very many parts of our northern states, great improvements have taken place in agriculture, and the dawn of our manufacture is fast progressing to a happy day. From the encreased white population, and the more confined space of these portions of our country, a more rapid improvement was to be expected. But if in these latitudes, where the seasons are by no means so propitious to the labours of the field, the inhabitants shall improve in art to supply what nature has denied, certainly the southern country, blest with a more genial sun, and possessing a much greater space for action, may blush for its appearance. Were societies formed in every state, of virtuous, patriotic, and well informed men, the knowledge necessary for improvement in agriculture, might be diffused to the people at large, for the benefit of the community, and the honor of the school. Were every useful patent for improvement in husbandry, or domestic manufacture, purchased by government, and published for the use of the country, much benefit would accrue to both. Were laudable incitements offered for discoveries and improvements, or for the practice of useful methods in agriculture, or manufacturing systems of affairs, a praiseworthy ambition would result in the happiest effects. The soldier receives a sword and vote of thanks, for gallantly performing the duties of his station. Does not he who performs a like duty in guiding the plough, or causing two " spears of grass to grow now, where but one grew before," deserve equally well of his country.

In a word, let government protect and cherish the infant agricultural establishments for the benefit of our country, and hold up the meed of honor to him who shall worthily serve her cause : the citizen will do the rest. The introduction of canals, of roads, and bridges, and all means which shall facilitate the communication of the various parts of our country, would result in vast benefit to the state, and nobly aid the cause of domestic œconomy. If we are to be denied the ocean, the great thoroughfare assigned by Providence for the use of mankind, and on which the little bark of the poor Indian has the same right to navigate, as the magnificent vessel of the prince.——
I say if this right, derived from a source, whose authority no laws ought to change, nor any civilized being question. If our intercourse cannot be defended by a like violence, and protected by means alike those of our aggressors, we must abandon the ocean, and within ourselves, form a great mart, for all the world to visit. China affords an instance which is precisely similar to this, but China affords no annals like ours—no epoch in her history can sound like '76—a great people struggling for their liberties, and nobly daring to proclaim their freedom. May the ocean never be abandoned, may the sovereign of the seas dread the prediction to Macbeth, and fear that " Birnam wood shall come to Dunsinane," and may our

forests descend to guard the soil which gave them birth, and protect the people who cherished their growth, and, as an humble individual, I pray God that American Oak, and Iron, may bear the flag of our country in virtuous pride, to all quarters of the globe.

But, my fellow-citizens, let not the aspect of foreign relations relax your endeavours to provide for your domestic wants, or to promote the cause of your infant manufactures. Remember, that thirty years since at the formation of the present government, its illustrious chief, when inaugurated to the highest gift of his country, *was wholly cloathed in American Manufactures.* Was not this an example worthy of imitation, and ought not its remembrance, at this late period, to cause a patriotic emotion in every American breast.— It ought! I trust it will! The great, the virtuous Washington lived but for his country; each action of his long and glorious life, was a lesson to his countrymen, and an example to the world.—— That example is yet fresh in your remembrance, and by following its wholesome dictates, you discharge a patriotic duty, and do honor to the memory of the FIRST OF MEN.

In times of maratime warfare, a large and very useful part of our community, become either liable to be seized and pressed into a foreign service, or obliged to remain and starve at home. Were a spirit of manufacture existing to a liberal degree, a great proportion of this useful and meritorious class of citizens, would either be engaged in the works, or transporting their produce on our, various inland waters. The mechanic also, a no less useful member of society, will want the materials of his art, if there is no intercourse with abroad, and no manufactures at home. And last, not least, the agriculturist comprising by far the greatest proportion, of the population of all countries, must want the reward of his toil; if no manufactures demand his produce.

The boundless tracts of southern country, devoted to the production of cotton, imperiously demand the labor of machinery, and skill, to convert this valuable article into use. And ere long, it may be hoped the improvements in wool, will alike call for manufacturing aid.

No country was ever more liberally gifted by nature with scites for manufactories, the streams possessing sufficient planes of water to force the wheels of any works. The elevated situations where the winds could be called in to aid the labor for the public good, and the vast forests which would contribute the steam, so necessary and useful in modern improvements.

I trust the genius of our countrymen will not be found wanting, in aid of this great undertaking, as the various models which the office of state is daily patronizing and will justly show.

In the event of an establishment of manufactures, new discoveries will daily be made, and thus a vast school of useful improvement established—new treasures, now hidden in the bowels of the earth, will daily see the light, and new systems be the results of these encreased means of knowledge. Europe may soon learn to respect our genius, our industry, and our patriotism, and no longer believe that we are to derive from her alone, those gifts which Providence has placed within our reach, and intended for our use. No fears need be entertained of want of market, for the fabrics which we may form. In the revolutions of empire which almost daily occur, new marts will always be afforded, and those countries which are suffering the scourge of war, must look to the peaceful regions of industry, for a supply of their wants.

Necessity, which is termed the mother of invention, will certainly aid our offset into manufacturing life, and supply in many cases, the knowledge derived from years of study and improvement.—— Our necessities during the revolutionary war, supplied many wants which at first appeared far beyond our reach. In that holy cause we put our trust in him, who alike visits with his divine protection, the most worthy, and nobly overcame every obstacle to freedom.— Let us, my fellow-citizens, but exert a small share of our fortitude, and perseverance, and we shall alike remove all impediments to manufacturing importance.

Is it not a reflection that even the *flag of our country*, is made of Foreign manufacture, and our legislators and patriots, while delivering the most dignified and national sentiments, are clothed in the produce of a foreign land? It is—We shall ever bear a secondary grade, in the rank of nations, if we are not independent of all. We shall ever feel our insignificance, if we are dependent on others for what we most want, and what we can best supply. Among the many systems whose effects result in happiness to mankind, that in which a due employment is given to every class of society, is the best: for idleness being the source of national as well as individual evil, injures the state as much as it does the man. Every citizen, owes to his country certain duties, such as he can best perform, within the compass of his knowledge or ability, and the due performance of these duties will always give wealth and happiness to the people, and greatness to the state.

*

The first and most material article now wanting, for our use, is the blanket, which is more appropriate to every condition of citizens than any other fabric whatever—Next the flannels, which the vicissitudes of our climates render particularly necessary—Lastly, strong cloths, which will at once repel the weather and give warmth to the labourer in winter. Added to these are hose and all the varieties of flaxen manufacture. Linens, which seem indispensable at present, for our use, may be greatly obviated by the introduction of cotton shirting, which is better adapted to warm climates, than the linen, and which may be made at small expence, there being already a considerable quantity manufactured within our country.— Any thing which will relieve the weight of foreign obligations, will be meritorious and patriotic ; and very many substitutes may be called into use, which will answer good purposes and remove the necessity of importation. Carpeting being an article of luxury, it does not come within the observation of this address, which only comprehends articles that are useful and necessary. It may be urged as an objection to manufactures, that they require many workmen who would otherwise be employed in agriculture. To this I answer —That the superfluous population of the country, not employed in agriculture, are obliged to resort to casual means of livelihood, whereas manufactures would employ them at home, and thereby contribute to retain the strength of the nation within our own doors. In respect to machinery which may be necessary for the works, the genius of the American people has never yet been found wanting, when called into action, and I believe will be fully adequate to the task. It will certainly be adviseable in the commencement of manufactures to lessen the labor of machinery and encrease the demand for workmen, since this will give the citizens a confidence in the utility of these establishments, and an opportunity of comparing the respective merits of labor and machinery. Expensive machines can only be erected by companies, which soon form monopolies.— This prevents competition, which is the life of all infant establishments. The employment of many citizens in the manufactories, will encrease the population of our country, by giving a zeal to emigration. Indeed, nothing can be weighed against manufactures on a limited scale ; for the abuses of the system are only felt in very populous countries, where monopolizers engross the profits, and the community are deprived of their due share. But in America, for a century yet to come, nothing can be derived from manufactories, but their immediate and important benefits. When the nation becomes entirely independent, it will be proper to restrain these systems, within due bounds, and not deprive agriculture of any part of its usefulness by a want of labour or patronage. Agriculture being the broad basis of the nation's wealth and prosperity, must be the sun from whence the other systems derive their light, and as they mutually depend upon this primary source for their origin or importance, they must give every essential aid to support its influence and promote its utility. It is said in Europe, that the most manufacturing parts of the country, are the most indifferently cultivated. Granted.—On the principle of monopoly only. The farmer and manufacturer should live as neighbours and not the one entirely monopolize and exclude the other.— The great manufacturing county of Lancaster in England, is said to be worse cultivated than the agricultural county of Kent ; but the one is the vast emporium for commerce, while the other is engaged in tillage and improvement.

Charles Brockden Brown

Address on the Utility and Justice of Restrictions upon Foreign Commerce (1809)

Charles Brockden Brown (1771–1810) is best remembered as the first professional literary man in the United States, whose sensational novels, inspired imaginatively and politically by William Godwin, gained notoriety. *Weiland* (1798) is his most successful fiction. He was also a prolific Jeffersonian propagandist. His unsolicited address to the United States Congress, excerpted here, is especially interesting in its refusal to make absolute judgment about a preference for agricultural or industrial occupations. After some survey of circumstances around the western world, Brown concluded, "Thus embarrassed and qualified are all reasonings and comparisons about the eligibility of different callings and conditions in human life."

Charles Brockden Brown, *An Address to the Congress of the United States, on the Utility and Justice of Restrictions upon Foreign Commerce, with Reflections on Foreign Trade in General, and the Future Prospects of America* (Philadelphia, 1809), pp. 76–84.

THE domestic evil effects arising from the total stop of exportation in the United States, are quite beyond the reach of human calculation. No penetration can foresee the resources, balances and compensations which human society affords in all such exigencies. Imports would necessarily cease with exports, and here would be opened a new and enormous source of change and privation. This would doubtless be somewhat balanced by the application of a great capital to domestic manufactures. The distribution of society into classes, would be somewhat altered. The manufacturers and townsmen would increase, but the farmers would not diminish. The internal sources of property would still impart to our population a rapid progress, but would this progress be as rapid as in the ordinary state of things? How much would the farmers be augmented by the addition of merchants out of business? How much would cities lose by the emigration of merchants?

In a country which has hitherto depended so extensively for cloathing, ornaments, food, tools, furniture, and all the apparatus of civilized life on importation ; which has hitherto exported in value, at least, one-third of all the produce of its land and labor, a total and sudden cessation of this import and export, must needs produce a vast and wide spread revolution. By what means, in what time, and to what degree the chasm would be filled up, is a problem no human wit can solve. What would be the immediate effects? What the ultimate? Volumes might be filled with attempts to explain these points, but the arguments, however shrewd, the theory however plausible, would gain credit with no reasonable man. He would obstinately withhold his faith from any testimony but experience, and *that* will never be granted him, because this revolution is impossible. Laws are possible; considerable but temporary checks to commerce are possible; but an absolute and permanent extinction of foreign trade, is plainly impossible.

It is of no importance to the nation, whether its governors approve of a commercial system of society or not. These topics being necessarily remote from practice, they may enjoy their sentiments in quiet, and without meriting suspicion or reproach. On this head, they are surely at liberty to think and talk as freely as on the best proportion of oxygen in the atmosphere, or the advantage of excluding all eccentric bodies from the solar system. They can never make their own country, an example of a purely commercial, or a mere agricultural state. They can do nothing, in short, but what their creators, the people, approve either beforehand or after, but in this respect, the whole society is under the influence of laws, which the abstract opinion of the largest majority cannot shake.

Ought we to be all foreign traders? All farmers? Or partly traders and farmers and foreign traders? Or partly farmers and partly manufacturers, without foreign traders. If the whole society be made heartily to agree in one of these opinions, no one finds it easier on that account to change the path in which his destiny has placed him. The merchant, the artificer and the farmer, has been such before he begins to reason on these subjects. Is there a magic in a mere opinion of this kind, which will supercede the influence of habit; supply the want of education; and create a farmer where there is no room for him; an artizan where all the custom is already engrossed; a merchant, where all the current business is already done by others. He must continue in the track, he must acquire a living, and raise a fortune in the way in which he is fixed by the complex and motly system of a great and civilized society, and may amuse himself with theories and visions, as to a better order of things, with his book and his closet, if he has the time and the taste.

When we reflect on the perpetual jealousies and bickerings, the hardships and oppressions, the frauds and cruelties to which an extensive foreign trade necessarily exposes a state; on the gigantic evils of war, to which the clashing interests of our own, and of other nations make us liable; on the intestine broils and factions which derive their existence and their venom from this source; on the seeming necessity there is of protecting trade, if we have it, by frigates and by squadrons; on the enormous and insupportable expense which any efforts of this kind require; on the utter hopelessness of affording adequate protection to it in the present state of the world; on the mortification and shame of submitting to injustice and oppression, when we cannot revenge or repel it; on the absolute misery which this injustice and oppression diffuses through a part of the community; and which arises from the fluctuations of war and peace among foreign powers: When we think on our helpless dependence, for the comforts and decencies of life, upon nations three thousand miles off, we may, without a crime, be disposed to wish that all intercourse of this kind, were at an end; that we should sit, quiet spectators of the storms tnat shake the rest of the world, secure in our solitude and in the waste that rolls between them and us; employing all our vigor in building up an empire here in the West; and in cementing the members of our vast and growing nation, into one body.

There is something charming too in the picture of a world within ourselves; of bringing within our limits, all the sources of comfort and subsistence; of supplying all our wants with our own hands; of gaining all the functions, occupations and relations of a polished nation; of being a potent political body, complete in all its members and organs, and in which no chasm or defect can be found. We catch likewise an imperfect notion that we should be richer and more populous by this means. We should go on multiplying persons and towns and cottages faster; and thus become much greater and more wealthy, if all our surplus products were consumed by mouths at home, and not abroad. If the millions who now weave and sow and hammer and file for us, were members of our own body, swelling by their gains and their expences, the tide of circulation in our own community.

We cannot be blamed if we ponder with pleasure on such splendid images, if we are reluctant to pursue any path of reflection, which appears to lead us away from them. They charm us, because they are visions of national felicity. They doubly charm us, because the nation around which they hover, is our own. And are we, on more deliberate reflection, obliged to relinquish them?

Certainly our confidence is somewhat shaken, when we come to recollect, that most of this formidable train of evils is hitherto in prospect merely; that though we are a nation, whish wrested its independence from Great Britain by force of arms; though pursuing the commercial path with a zeal and to an extent unexampled; though our commercial interests have been perpetually interfering with hers; though her maritime power be ALL and ours NOTHING; though for twenty years, superior to all naval rivalship in Europe, and engaged in a war, in consequence of which we were able to gather immense profits and mount to enormous opulence, yet we have hitherto escaped the grosser evils of that war, which seemed but now the necessary consequence of foreign trade. That navy which seemed so necessary for the guardianship of such a trade as ours, we have never had, and yet the molestations and vexations it has suffered, has not disarmed us of our caution, or made us rashly incur greater evils, in resentment of these injuries. What is the amount of these vexations. If they are great, they have still borne so small a proportion to the benefits of the trade, growing out of the war itself, that our advances in population and riches, have been truly wonderful; the mass of our gains has been enormous; and what we have gained by eluding the allowed rights of belligerents has been infinitely more than what we have lost by the lawless and unauthorized exertion of their power. *Remember that.*

Now, indeed, the prospect begins to lour. The insults and vexations we meet with begin to urge us to hostilities. We are ready to bring invasion on our own shores; we are willing to give up the remnants of foreign trade, allowed us by the lords of the ocean, because they have denied some of that which we have been accustomed to enjoy: We are willing sullenly to abandon all, because we cannot retain all, especially as this proceeding is adapted to annoy and perplex the injurer. But whether we have now arrived at the point at which a foreign war is to grow out of foreign trade, is not yet quite certain. Hope still fondly believes that the government will still recall its fiery menaces; and annul its avenging embargoes; that we may once more stand on the brink of war, and yet escape it; that since we are to fight for gain,—embargoes and interdicts, being only modes of fighting, to which we have recourse in the want of armies and squadrons—Since we aim at regaining the commerce of foreign states, for no end but the profits it affords, may we not find that we destroy our own ends, by our proceedings, and make this discovery while the evil is reparable?

With regard to intestine feuds, they have certainly appeared to draw most of their ferocity from our intercourse with foreigners. But this intercourse, like all other political conditions and transactions, produces many effects, and some of them must necessarily be evil. We are bound only to consider the evil which it excludes, or the proportion of its good and evil. But what is the extent of this evil. Faction has not hitherto shaken or overturned, or seriously endangered the public peace. The war has been a verbal one, and its victories are only seen in numerical changes at elections; nor in truth do we owe the existence or ferocity of faction to foreign *trade*. If there be any foreign ingredient at all in the cup, it must chiefly be ascribed to our intellectual intercourse with foreigners. Their political conduct and opinions only can have an influence on us affecting our internal policy, the frame and constitution of our government, and no one, I suppose, seriously imagines it either possible or eligible to stop this kind of communication.

Faction, however, belongs to human nature. Forms of government create it not; extinguish it not. They only vary or regulate the mode of its action, the field of its contests; ours are regulated by our manners and political forms, the most memorable effect of which is to make it talkative, loud, noisy, vociferous; to make it harangue without fear, and scribble without end. It is a spirit necessarily active, and it matters little how it is employed, since it must be busy; and happy shall we be, if it continue to prate and to bluster with as little injury to peace and order, as it has hitherto done.

With regard to the beauty of the spectacle afforded by a nation in which all the departments of human society are filled up by its own citizens; whose farmers feed and furnish its citizens and traders only, and whose artisans and traders supply each other and those farmers only, little need be said. It is in reality commended by nothing but a certain shape of order and completeness, adverse and not friendly to the dignity and happiness of mankind. Human society is a complex body, the members of which are not equal in use or value; in their faculties of self enjoyment; in the power they exert over the commonweal; in the immediate benefits which, as individuals, they derive from their employments. There must be some to till the ground and raise bread, but they must produce more than they consume, otherwise they can merely eat. They demand clothes, shelter, and domestic accommodation. Tools and machines to till the ground, even for their own use, they must necessarily have; and to procure these, they must raise a surplus produce, and give it to the artisan, who makes, and the merchant who fetches and carries. Artizans must be numerous in proportion as the farmer's taste and habits require much accommodation, and as his surplus produce is great. They must be divided into numerous classes, in proportion to the refinement which the arts have attained. As their numbers are greater, they are obliged to work for lower wages, and with more industry. The fruit of their work is cheaper, and those whom they supply get better things, and more of them, for the same quantity of produce, or money which represents it; but the subsistence of the artisan is more precarious, his enjoyments fewer, his drudgery more painful, all the causes of human misery and distress operate more cruelly upon him, as his wages are scantier. And thus the condition of the artisan is worse, as the produce of his labor is cheaper, and of consequence the ease and luxury of those who buy that produce greater.

Now what is our present condition? We are a nation of farmers, traders, and artisans, but our wants are not entirely supplied by our artisans. The produce of manufacturing labor, which is annually consumed is immense, but a vast proportion is produced by the labor of distant nations. We are clothed, and adorned, and supplied with tools, in a great degree, by artisans beyond the ocean. Our principal employment is to catch fish, to cut timber, to reap corn, to feed cattle, and to carry what we do not consume of these away, in exchange for the viands of foreign climates, or those articles of foreign manufacture which they will buy, and which we want. From circumstances peculiar to the countries that supply these, their workmen are many; they are obliged, therefore, to work cheap: powerful machines are in use which do more than human hands at a less cost. Hence, the misery and want of the artisans of these countries; hence the cheapness and abundance of their work to us; and hence more than half of the population of these countries are cooped up in towns; while only one-fourth or one-fifth of our population is thus cooped up.

Considering mutual dependance as a principle of unity, there is a body of farmers, traders and artificers, who live by each other, amounting perhaps to seven millions. Six millions live in America; one million in Europe. The six millions here are chiefly cultivators, woodmen, and carriers. The million at a distance are chiefly day laborers in towns and factories. Now to a benevolent eye, it may be of little consequence how they are dispersed. The mutual relations being the same, and the effects upon their happiness the same, it matters not, in such a view, how they are distributed; whether all the artificers are here, and all the farmers there; but, with those who have national feelings, it may be of some importance. How far is it eligible that six millions of Americans should be classed as at present, or a change take place, in consequence of which one million of the six* are employed in manufactures?

The intricacy of this subject is well known. The delusiveness of all comparisons between different classes and professions, as to their enjoyments, is apparent. It is evident how much the real condition of a human being is disconnected with the nature of his employment, and is modified by his climate, the products he cultivates, his social relations, his education, his property, and even his personal character flowing from religion or morals.

Take an Highland cottager and a Kentish one. They are equally poor and their kind of labor is the same, but the Highland man has the worst soil, the worst fare, and the worst house: Yet education and manners make him a better and happier man. Take the Irish peasant into this comparison. He is poorer, worse clothed, and worse housed, more unlettered and improvident, and more degraded in his political

---

* These numbers are taken for the sake of a clear statement. They are not, and need not be accurate. The conclusions are equally valid whether they be precisely true or not.

condition than either, but he is better fed, because he has learned the value of potatoes; and is healthier in course. He is likewise happier by temperament, and because he feels an inward impulse to laugh, when the others are inclined to weep.

Take the linen manufacturer in Lancashire, in Silesia, and in Glarus. Their employments are the same, and their gains not very different; but the first passes the day in the work room, crowded, hot, noisome, pestiferous, and his nights in the ale-house. The others are country cottagers, and work alone amidst rural scenes and mountain airs.

Take the plowman of Kentucky or Connecticut, and of Yorkshire. The former is a citizen, a proprietor, and calls no man master. The other is an half starved hireling, and meditates no asylum in age or sickness but the alms-house. But add to these the plowman of Roanoke and the seedsman of Santee. What are *they* compared to the Yorkshire peasant, or their *countryman* of Connecticut? They are negroes, slaves; worse, far worse than the serfs of Hungary or Poland; yet the Polish serf, the Yorkshire cottager, the Connecticut farmer, and the slave that hoes corn and plants rice beyond the Chesapeak, are all of one vocation; they till the earth and enjoy the country. In like manner the manufacturers of Lancashire, of Antrim Glarus, or Geneva, follows all the same calling, but it leads to very different degrees of health, ease and happiness. He that tills the earth in Georgia is far inferior to the pin-maker of Birmingham. The Newark shoemaker is much more happy and respectable than the Yorkshire plowman or the Teveot dale shepherd, though the former is a townsman, an artisan of the lowest class, and works for exportation. And all of them are, by many degrees, higher than the race of men called sailors, which it is the peculiar tendency of foreign trade to propagate and multiply.

Thus embarrassed and qualified are all reasonings and comparisons about the eligibility of different callings and conditions in human life. There may be a solid ground at bottom, and we might possibly find it, were it our present business to search for it; but this would lead us into volumes, and is not to our present purpose. We are at present a community of merchants, artificers and farmers. Some of our artificers even work for exportation. If things are left to their own course, the period will inevitably arrive when most of our own wants will be supplied by our own hands, except such as peculiarities of soil and climate deny us. But we shall not stop here. We shall become the manufacturers of other nations. Such we are, even now, to a certain degree, but our manufactures will multiply in a larger proportion than our husbandmen. Less and less of the produce of mere handicraft will be imported from abroad, and more and more will be exported. The latter will depend, in some measure, on the state of foreign nations; but the former will be chiefly influenced by internal and domestic causes. It will even be our lot to import the raw materials of our manufactures, even for our own consumption, nor can any thing but the real state of foreign countries prevent us, in fine, from importing bread and meat itself. All this will come. Eternal and immutable causes will bring us finally to this point, without the aid of government to push us forward, and in spite of all its efforts to hold us back; but the impossibility of any internal regulation to suspend our commercial intercourse with foreigners, will be as evident when that intercourse consists in carrying cloths and fetching flour and cotton, as now when it consists in carrying cotton and fetching cloths. This is a point not susceptible of serious argument.

Henry Clay

Speech on Domestic Manufactures (1810)

Many measures to promote domestic manufactures were debated in the Senate and the House of Representatives in the first half of the nineteenth century. Most of the speeches made in favor of or opposition to these measures were confined to the specific measures under discussion. In this speech, given March 26, 1810, Henry Clay, speaking before the Senate in favor of a proposal that the Navy prefer American supplies to foreign goods, discusses some of the general issues of concern to those regulating manufactures.

Henry Clay (1777–1852), senator, representative, and two times a presidential candidate, was one of the most fervent supporters of laws to protect American manufactures. In 1808, for example, as a member of the Kentucky legislature, he introduced a resolution urging that its members wear homemade clothes. He coined the phrase "the American System" to describe the system of tariffs and internal improvements that would favor American industry and was largely responsible for the strong tariffs passed in the 1820s that did much to benefit American manufactures.

Henry Clay, "Speech on Domestic Manufactures [March 26, 1810]," *The Papers of Henry Clay*, ed. James F. Hopkins (Lexington, Ky.: University of Kentucky Press, 1959), vol. 1, pp. 459–463.

The local interest, Mr. President, of the quarter of the country which I have the honor to represent will apologize for the trouble I may give you on this occasion. My colleague has proposed an amendment to the bill before you, instructing the secretary of the navy, in providing supplies of cordage, sail cloth, hemp, &c. to give a preference to those of American growth and manufacture. This part of the amendment is moved by the gentleman from Massachusetts (Mr. Lloyd) to be stricken out. And in the course of the discussion which has arisen, remarks have been made on the general policy of promoting manufactures. The propriety of this policy is perhaps not very intimately connected with the subject before us, but is nevertheless within the legitimate and admissible scope of debate. Under this impression I offer my sentiments.

In inculcating the advantages of domestic manufactures, it never entered the head, I presume, of any one to change the habits of the nation from an agricultural to a manufacturing society. No one I am persuaded ever thought of converting the plough share and the sickle into the spindle and shuttle. And yet this is the delusive view too often taken of the subject. The opponents of the manufacturing system transport themselves to the establishments of Manchester and Birmingham, and perceiving the indigence, vice and wretchedness prevailing there, by pushing it to an *extreme*, argue that its introduction into this country will be attended by the same mischievous consequences. But what is the fact? That England is the manufacturer of a great part of the world, and even there the numbers thus employed bear an inconsiderable proportion to the whole mass of population. If we were to become the manufacturers of other nations, effects of the same kind might result. But if we *limit* our efforts by our own wants, the evils apprehended would be found to be chimerical.—The invention and improvement of machinery, for which the present age is so remarkable, dispensing in a great degree with manual labour; & the employment of those persons who, if we are engaged in the pursuits of agriculture alone, would be either unproductive, or exposed to indolence and immorality, will enable us to supply our wants without withdrawing our attention from agriculture; that first and greatest source of our wealth and happiness. A judicious American farmer, in the household way, manufactures whatever is requisite for his family. He squanders but little in the gewgaws of Europe. He presents in epitome what the nation ought to do. Their manufactories ought to bear the same proportion, and effect the same object in relation to the whole community that the part of his household employed in domestic manufacturing does to the whole family. It is certainly desirable that the exports of the country should continue to be the surplus production of tillage, and not become those of manufacturing establishments. But it is important to diminish our imports—to furnish ourselves with clothing made by our own industry—and to cease to be dependent for the very coat we wear upon a foreign and perhaps inimical country. The nation that imports its cloathing from abroad is but little less dependent than if it imported its bread.

The fallacious course of reasoning urged against domestic manufactures, the distress and servitude produced by those of England, would equally indicate the propriety of abandoning agriculture itself. Cast your eyes upon the miserable peasantry of Poland. Revert back to the days of feudal vassalage, and you may thence draw copious arguments of the kind now under consideration against the pursuits of the husbandman! What would become of commerce, the favorite theme of some gentlemen, if assailed with this sort of weapon? The fraud, perjury, cupidity and corruption with which it is unhappily too often attended would at once produce its overthrow.—In short, sir, take the black side of the picture and every human occupation will be found pregnant with fatal objections.

The opposition to manufacturing institutions recalls to my recollection the case of a gentleman of whom I have heard. He had been in the habit of supplying his table from a neighbouring cook and confectioner's shop, and proposed to his wife a reform in this particular. She revolted at the idea. The sight of a scullion was dreadful, and her delicate nerves could not bear the clattering of kitchen furniture. But the gentleman persisted in his design; his table was thenceforth better and cheaper supplied, and his neighbour the confectioner lost one of his best customers. In like manner Dame commerce will oppose domestic manufactures. She is a flirting, flippant, noisy jade, and if we are governed by her fantasies we shall never put off the muslins of India and the cloths of

Europe. But I trust that the yeomanry of the country, the true and genuine landlord of this tenement, called the U. States, disregarding her freaks, will persevere in reform until the whole national family is furnished by itself with the cloathing necessary for its own use.

It is a subject no less of curiosity than of interest to trace the prejudices in favor of foreign fabricks. In our colonial condition we were in a complete state of manufactural and commercial, as well as political dependence on the parent country. For many years after the war, such was the partiality for her productions, that a gentleman's head could not withstand the influence of solar heat unless covered with a London hat—his feet could not bear the pebbles or frost of this country unless protected by London shoes—and the comfort or adornment of his person was only consulted when his coat was cut out by the shears of a tailor "just from London." At length, however, the wonderful *discovery* has been made that it is not absolutely beyond the reach of American skill and ingenuity to provide these articles, combining with equal elegance greater durability. And I entertain no doubt that in a short time the no less important fact will be developed, that the domestic manufactories of the United States fostered by government, and aided by household exertions, are fully competent to supply us with at least every necessary article of cloathing. I, therefore, sir, *for one* (to use the fashionable cant of the day) am in favour of encouraging them, not to the extent to which they are carried in England, but to such extent as will redeem us entirely from all dependence on foreign countries. There is a pleasure—a pride (if I may be allowed the expression, and I pity those who cannot feel the sentiment) in being clad in the productions of our own families.—Others may prefer the cloths [of] Leeds and of London, but give me those of Humphreysville.

Aid may be given to native institutions in the form of bounties and of protecting duties. But against bounties it is urged that you tax the *whole,* for the benefit of a *part* only, of the community; and in opposition to duties it is alledged that you make the interest of one part, the consumer, bend to that of another part, the manufacturer. The sufficiency of the answer is not always admitted, that the sacrifice is merely temporary, being ultimately compensated by the greater abundance and superiority of the article produced by the stimulus. But, of all practicable forms of encouragement, it might have been expected that the one under consideration would escape opposition, if every thing proposed in congress were not doomed to experience it. What is it? The bill contains two provisions—One prospective, anticipating the appropriation for cloathing for the army, and the amendment proposes extending it to naval supplies also, for the year 1811.—And the other, directing a preference to be given to home manufactures & productions whenever it can be done *without material detriment to the public service.* The object of the first is to authorize contracts to be made before hand with manufacturers, and by making advances to them, under proper security to enable them to supply the articles wanted in sufficient quantity.—When it is recollected that they are frequently men of limited capitals, it will be acknowledged that this kind of assistance, bestowed with prudence, will be productive of the best results. It is in fact, only pursuing a principle long acted upon, of advancing to contractors with government, on account of the magnitude of their engagements. The appropriation contemplated to be made for the year 1811 may be restricted to such a sum as, whether we have peace or war, we must necessarily expend. The discretion is proposed to be vested in officers of high confidence, who will be responsible for its abuse, and who are enjoined to see that the public service receives no *material detriment.* It is stated that hemp is now very high, and that contracts made under existing circumstances will be injurious to government. But the amendment creates no obligation upon the secretary of the navy to go into market at this precise moment. In fact, by enlarging his sphere of action, it admits of his taking advantage of a favorable fluctuation, and getting a supply below the accustomed price, if such a fall should occur prior to the usual annual appropriation.

I consider the amendment under consideration, of the first importance in point of principle. It is evident that, whatever doubt may be entertained as to the general policy of the manufacturing system, none can exist as to the propriety of our being able to furnish ourselves with articles of the first necessity, in time of war. Our maritime operations ought not in such a state to depend upon the casualties of foreign supply. It is not necessary that they should.

With very little encouragement from government, I believe we shall soon not want a pound of Russia hemp. The increase of the article in Kentucky has been rapidly great. Ten years ago there were but two rope manufactories in the state. Now there are about 20 & between 10 and 15 of cotton bagging; and the erection of new ones keeps pace with the annual augmentation of the quantity of hemp. Indeed the Western country alone is adequate to the supply not only of whatever of this article is requisite for our own consumption, but is capable of affording a surplus for foreign markets. The amendment proposed possesses the double recommendation of encouraging at the same time the manufacture and growth of hemp. For increasing the demand for the wrought article, you increase the demand also for the raw material, and consequently present new incentives to the cultivator.

The three great subjects that claim the attention of the national legislature are the interests of agriculture, commerce and manufactures. We have had before us a proposition to afford a manly protection to the rights of commerce, and how has it been treated? Rejected!—You have been solicited to promote agriculture, by increasing the facilities of internal communication through the means of canals and roads, and what has been done? Postponed! We are now called upon to give a trifling support to our domestic manufactories, and shall we close the circle of congressional inefficiency by adding this also to the catalogue?

1. [John Pope, senator from Kentucky 1807–1813]

James Mease

Address to the Cultivators, the Capitalists, and Manufacturers (1811)

This unsigned address was published in the *Archives of Useful Knowledge,* a short-lived Philadelphia periodical edited by Dr. James Mease. Mease (1771–1846) was a much-published student of Benjamin Rush who wrote frequently on medical and agricultural subjects. In this address, attributed to Mease, the patriotic urgency of domestic manufactures in anticipation of hostilities with Britain echoes the appeal of Rush's speech in Carpenter's Hall thirty-five years earlier. The address continues at length beyond this excerpt with detailed speculations on the possibilities for American manufacture of wool, skins, metals, wood products, and the like.

[James Mease], "Address to the Cultivators, the Capitalists, and Manufacturers of the United States," *Archives of Useful Knowledge,* volume 1 (January 1811), pp. 229–233, 237–243.

ARCHIVES

OF

USEFUL KNOWLEDGE.

Vol. I.　　　　JANUARY, 1811.　　　　No. 3.

ADDRESS

TO THE CULTIVATORS, THE CAPITALISTS, AND MANUFACTURERS OF THE UNITED STATES.

THE times require your enlightened co-operation in the improvement and extension of *the national industry.* An unprecedented course of foreign misconduct, demands equally of your interests and of your virtues, a deportment corresponding with the new situation into which we are drawn. The subject of WAR, as a possible remedy for the disorder of the times, is studiously left to other pens, and is respectfully submitted to the proper authorities of our neutral country. But it is a consideration of the utmost importance, in favour of the public views, which it is proposed to suggest in these papers, that they may tend to manifest our vast resources for a just and necessary contest with any foe; and to indicate the means of their sure and economical improvement and increase.

Under the strongest impressions from these painful, injurious, and awful times, it is proposed to take the necessary view of certain divisions of our internal industry, in which the interests of the cultivators, the manufacturers, and the capitalists, well understood, appear to unite. We shall commence with the great and novel article, the importance of which has been so miraculously increased in the United States, by measures like these, and which is rapidly advancing to the character of the greatest material for clothing, furniture, and diversified equipment, on the face of the earth.

COTTON.

The woollen manufactures of Europe constitute one of the greatest rivals of the cotton cultivation of the United States, and of the increasing manufacture, at home and abroad, of that illimitable raw material. It is an instructive fact, that the woollen manufactures of Great-Britain have been steadily computed at a little more than sixteen millions sterling per annum, for the whole period between the peace of 1783, and the beginning of the present war in the year 1803.—The weight of the wool annually consumed there, varies little from the weight of our whole *surplus* cotton, as exported in the greatest year. If our cotton shall be impeded by the belligerents in its way to foreign markets, we must and shall manufacture many cotton goods, so as to rival foreign woollens. The American will not be uncomfortable in his own cotton velvets, velverets, corduroys, swanskins, and cotton blankets. The scarcity of hands is no objection to ginning, carding, spinning and plate printing by horse, by water, and by steam. An effect of their cotton manufacture has been to keep the woollen manufacture stationary in England for twenty years, during which the cotton branch has been raised from less than one million sterling to more than ten millions.

The presence of the raw material will provoke to, excite and produce the manufacture.—This is a plain and a sound maxim.— We find it proved by the course of events, even in the instances of the most unnecessary, luxurious and difficult branches. Thus the presence of the *Grecian* marbles occasioned their formation into the most costly, exquisite, difficult and unnecessary manufactures in the world—*their marble statues.* Italy followed in this superfluous branch, because, *there also was the native raw material, and the Greek marble was adjacent.* France and Great-Britain have not offered to the world *the statuary's wares,* because they had not the marble; although, it is observable, that they have respectively made a distinguished figure in the more exquisite and complicated relative manufactures of the painter, because importations constantly occasioned *the presence of the raw materials*. So of the people of the whole seventeen provinces of the Netherlands. They could make paintings, but not marble statues.

The presence of the most considerable mass of the best hemp, in Russia, has produced *an unrivalled excellence* in the finer sail cloth, and in the common sheetings of that country.

The various fossils of Great-Britain have produced an immense number of useful manufactures from mineral substances.

The moist climate and rich soil of Ireland produce the best flax in abundance, and the presence of that excellent raw material has occasioned the most celebrated manufacture of linens, which have been offered in the markets of the world.

The American cotton, in like manner, will *surely* produce the home manufacture with a celerity, *proportioned to foreign interferences with its rightful sale in external marts*. A strong collateral ground for this presumption is, that we have insensibly attained the actual manufacture of *all* the other raw materials, which are either the spontaneous productions of the earth, or the fruits of cultivation. The American *metals, wool, hemp, flax,* and *skins*, have, by their mere presence, produced the gradual rise of a body of regular and qualified manufacturers, actually competent to the manufacture, not only of all of them, which we can obtain from the landed interest, but all we can procure by means of importation. We have nothing to do in regard to those last five raw materials, but to increase their quantity, and to ameliorate their manufacture. It is the presence of cotton only, as a *redundant* raw material, which is to produce in the United States a new proof the truth of the maxim, that *the presence of the raw material will excite and produce the corresponding manufactures*.

Many important articles of public supply, for which the country paid excessive prices in the revolutionary war, can be made of the cotton, which Europe interrupts in its way to foreign markets. We were not aware, that we possessed the sources of this profuse raw material; neither had we the gin to free it from the seed; nor the carding and spinning machinery, with which to manufacture it. We were not aware, that it would make blankets, of which (of foreign wool) there are twenty-five millions of pounds weight in daily use within our country. Its utility in girth webbing for military saddles, for belts of various kinds, for pantaloon, vest, jacket, trowser, frock, shirt, sheet, tent, knapsack, wagon, and sail cloths, twilled and untwilled, would ensure and cheapen that great mass of army and navy supplies, which have been made of ravens duck, drilling, sheeting, ticklenbergs, dowlas, girth webbing, and even sometimes of leather, as in the case of belts. If the military blanket, by land and sea, ought to be of wool, except in the hospitals and garrisons, still the manufacture of cotton blankets facilitates and insures the command of woollen blankets in time of war. For there could be no difficulty, in a war of virtuous and sound principle, to obtain from the private owners of three millions and one half of woollen blankets, now and always in use, the exchange of a sufficient quantity for the soldiers and mariners, to be paid for in new, clean and good cotton blankets, on fair terms.*

The charges on the exportation of cotton wool to Europe, and upon importing it in manufactures from thence, amount to 50, 55 and 60 per cent. on the sterling cost of blankets, common velvets and corduroys, cotton drilling, girth webbing, and other goods of constant and general consumption. It is evident and certain then, that our good house-wives, and cotton manufacturers, are protected by a difference of more than 50 per cent. in their favour. It is a similar difference, which has created and

* That this opinion is well grounded, will be inferred, when it is known that during the revolutionary war, the patriotic citizens of Philadelphia parted freely, and without a call of government, with their blankets for the use of our army destined to invade Canada. If then they voluntarily gave these articles, and without any equivalent,—at a time too when, from the existence of a war, the means of replacing them was cut off, and when the method of manufacturing them in the United States was scarcely known, it cannot be doubted that the present generation will freely exchange their used woollen for new cotton blankets, should the increasing rapacity of the European powers force us to appeal to arms to recover and perpetuate our rights.— EDITOR.

protected the coach-maker, the cabinet-maker, the gold and silver-smith, the paper-maker and stainer, the hatter, the plumber and the other manufacturers of *all* our hides, skins, flax, hemp, wool, and metals.

The various foreign invasions of our personal rights, and of our merchant vessels and ships of war, against all decorum, reason and justice, are powerful considerations in favour of every proper attention to the internal walks of our national industry. Our honest exertions, capital and skill, employed in the production of raw materials and subsistence for manufacturing families and their cattle, and in manufactures themselves, will give us *a great mass of certain, cheap and excellent supplies*, necessary in war, comfortable in peace, and profitable in both seasons.

*

HEMP.

EVERY movement of the belligerents is calculated to force the United States into the manufacture of their own supplies. The blockade of Elsineur and Sweden by the British government, the acts of the Danes, and of the Swedes in Pomerania, oblige us to find substitutes for the cordage, linens, steel, and iron, which we have heretofore drawn from the numerous ports of the Baltic. We have imported in a single year 6,500 tons of hemp, and 210 tons of cordage and cables, besides vast quantities of hempen drilling, sheeting, duck, and other linens. The mere unmanufactured hemp for these goods, at the present price of 400 dollars per ton, is worth in our markets two millions eight hundred thousand dollars, and weighs nearly fifteen millions of pounds. Our planters must supply the place, by American hemp, flax and cotton, of the sheetings, drillings, ravens duck, sail cloth, table cloths, and towels, which we procured from the Baltic. All those goods can be made also of flax; and they, as well as twine, cables, and cordage, can be well made of hemp. It will be our interest to cultivate more flax and hemp, and much less tobacco. A small quantity of this latter plant is consumed by our army and navy; but flax and hemp are necessary for the public supply, military and naval. The modification of our agriculture, so as to suit our own manufacturing purchasers, is necessary to the profit of our cultivators, and to the certainty and sufficiency of the public supplies on sea and land. The rich swamps in the eastern counties of North-Carolina and Virginia, and great quantities of the better class of lands in the western states, are well adapted to the production of hemp: and it is plain, that the linen manufacturers of middle and western Pennsylvania, and other places, could readily execute, by close imitations, linens like the best we receive from Russia. Twine, hempen yarns, white rope, and tarred cordage and cables, are extensively made on the Ohio and the Mississippi. All these sustain the expenses of transportation better than the less valuable raw materials of which they are made. The rival foreign articles sustain the transportation of the raw material, by a long inland navigation to St. Petersburgh, in Russia: and hemp and flax, as well as linens, twine, yarn and cordage, are burdened with a heavy freight and insurance from that port to those of our country.

Steam engines can now be employed where mill-seats are wanting. Machinery to obviate the deficiency of hands is obtained. Capital has become abundant. Persons acquainted with every kind of manufacture are to be found in our country. Employment in foreign trades is so much reduced, that men of the best business, habits and talents, are at leisure to engage in manufactures. The more bulky is the native raw material, and the nearer to the place of its growth is the American manufacturer, the more sure and considerable is the profit.

As hemp and flax are universal productions of the United States, it cannot but happen, that the manufacture of them must be general, increasing, and, before long, very great. Ireland has lost, from various causes, much of her linen manufacture, and has not given us, for a long time, very large quantities of those kinds of linens, of which we are deprived by the blockades affecting the Baltic sound. There is therefore the greater necessity for us to make our own linen supplies, and especially those demanded by the army and navy.

The linens of Germany, (especially for coarse shirts, frocks, and trowsers,) with some from the Dutch and Flemish manufactories, have constituted our principal supplies for several years, particularly since the great agitations of Ireland, during the modern revolutions of France. All these German, Dutch and Flemish linens, are liable to interruption in their exportation, by the French exclusion of our vessels from the continent of Europe, and the British captures of our vessels in going and coming. These things must give a spring to the production of flax and hemp; and they must occasion American substitutes from those two raw materials and from cotton, to be made in our families and manufactories. The universal instruction of females in the art of weaving, as a domestic employment, would contribute greatly to such manufactures, and to the comfort and happiness of the sex. It should be a condition of all female's indentures, that the girls should be taught to weave during their time of service. It can be easily effected in a few days, or in the evenings of a very few weeks. In many of the counties of England the women weave, as large numbers do in our eastern, northern and southern states. By this excellent practice the men are left for the duties of the farm, and other employments, requiring exposure and strength. If by water, horse and steam, we card and spin, and females weave, our manufactures will the better support the agriculture of the men. It may be safely assumed, that, if the women of our country, now unemployed in weaving, were to work in that way during one half of six days in the week, they would make enough of woollen, linen, and cotton cloths, and hosiery, to furnish our houses and clothe the inhabitants, including a war establishment of the army and navy. In this way may our American matrons and their daughters, contribute to correct the authors of the indignities and injuries, which we suffer from foreign violence and improbity. They would furnish the hammocks of the honest tars, who are to bear our thunders over the main, and the beds of the soldiers, who are to defend our shores, our altars, and our homes. It is impossible for us to foresee how soon our supplies may be cut off: how soon our means of paying for them, in our own produce, may be terminated: nor how soon a reluctant commitment in the outrageous contest of the old world, may require of us, that *all our men, and women too, should do their duty.* Let us then turn our agriculture, our business, our capitals, and our minds, in prudent timely anticipation, to the beneficial and necessary increase of manufactures, requisite to comfort and defence.

In considering the subject of hemp and flax and their manufactures, cut off from our markets by foreign violence and improbity, we are not to forget, that, if they amount to seven or eight thousand tons, our *surplus* cotton exported in one year, has been sixty-four millions of pounds, which exceeds 28,500 tons. Since the invention of the American cotton gin, by which the powers of agriculture in the southern states have been set loose, there is no limit to our capacity to bring cotton to market, but the demand for manufacture or exportation. Every year does and will add to our list of goods made from cotton. In no branch will these goods be more frequent, more easy, or more useful, than in that which furnishes substitutes for goods of foreign hemp and flax. It is conceived, for example, that one fiftieth part of our ordinary cotton crop would make as much American twilled cotton drilling (in lieu of the hempen drilling, ravens duck and sheetings of Russia) as would furnish two summer suits for each man of an army of one hundred and fifty thousand men. Considering the redundance and steady increase of our cotton, and the numerous substitutes for foreign goods, which may be made of it, too frequent occasion cannot be taken to notice its applicability to old and new uses. When we remember, that cotton is much more easy and less costly to raise or procure than flax, hemp, wool, silk, or leather, we cannot doubt, that it will fall in price, till the cheapest cotton shall be as low as the cheapest of those materials. To increase our cotton manufactures, and either to duty or to banish those of the countries beyond the Cape of Good Hope, *which are never made of our cotton*, would obviously be to encourage and support the American agriculture. Upon the whole subject of this paper, as it relates to public supplies and those for private use, it behoves us ingeniously and deliberately to consider the most easy, prompt, effectual and profitable modes of procuring, through the means of our various raw materials, substitutes for the hemp and flax, and the hempen and flaxen goods, of which foreign cupidity and usurpation are likely to deprive us by orders and decrees, or captures and seizures. The illegitimate conduct of the belligerent powers has happily placed our manufactures, as it were, in a hot-bed.

The obvious tendency of the present systems of the belligerent powers is:—

1st. To decrease or rather to destroy the profits of the agriculture and land of the United States, by preventing the sales of our produce in some foreign countries, and by loading those sales with various heavy and debasing charges in other foreign countries.

2dly. To enhance the cost of American supplies by preventing their transportation hither from some countries, and by increasing the charges, equally heavy and debasing, of the transportation from others.

The natural efforts of the American capitalists (both in land and in money) and of our manufacturers, and even of our circumscribed merchants, are to buy, raise and manufacture American produce and consumable articles.

No production of the United States has sustained a happier impulse than

WOOL.

THE fine cloths of England are advanced more than one hundred per cent. upon the prices of 1790. Those of France and Holland are prohibited by English usurpation, and those of Britain by French usurpation. The manufacture of cloth in Spain is destroyed or suspended. The Merino sheep are straying through the world, to us and all nations. In England they were sold in the year 1790, at the most excessive prices : even 800l. and 1000l. sterling, have been paid there for the use of a single breeder for *a single year.** In America, they have been procurable on various terms, at the highest for less than a quarter of this sterling money for the full price of the finest ram, as the momentary, occasional and highest cost of the animal. The carding of wool has become an operation of the utmost facility, by water, cattle, or steam. The spinning is done by water, horse and steam. Steam engines have been introduced, improved and rendered familiar. The manufactory of cloths in private families has been much improved. Chemistry has thrown prodigious light on the art of dyeing in Europe and America. The stocking loom has been greatly improved, as also the dressing and fulling of stocking webs for hosiery, and for a new, curious, and excellent species of woollen cloth. The convulsions of Europe have driven hither every description of artist employed in the woollen manufactory. Sheepskins, formerly an offal, often left to rot, or only used to make glue, cannot now be procured in sufficient quantities for the skin-

---

* This refers to Bakewell, who let a ram of his breed for 1200 guineas, for one year.—EDITOR.

dressers, book-binders, wool card-makers and glovers—a great saving or rather profit to the sheep raiser, on whom the foundations of the woollen manufacture depend. The United States are intelligently bent upon the increase of sheep, as well the *heavy* fleeced sheep, as the *fine* fleeced sheep. Every kind of sheep is propagated, nursed, and spared to procure *fine* wool and *much* wool. The Merino and other breeds are secured, and cannot ever be lost. The mixture of our wool and our redundant cotton aids the clothing manufacture. The two branches work well together —a great and peculiar advantage in the clothing business of the United States. The women, as before observed, weave in many parts of America, and may be happily taught to weave in all the states. The increase of manufactures gives women full, safe and profitable employments, which formerly they had not. It is estimated by the Secretary of the Treasury, that our family manufactures exceed in value *forty* millions of dollars per annum, in the woollen, flaxen, and cotton branches alone. It is plain and certain then, that our army, navy and militia, in any possible war, could be supplied with clothing, far below the high costs of the late revolutionary war. The importance of our clothing manufactures is manifest and great. Their increase is rapid and incessant. The household branches alone, at forty millions, are worth double the value of all our exports at the adoption of our present blessed union and constitution. All our ordinary stock of wool is consumed in manufactures, and all we can procure by our import trade, and by all the foreign breeds of sheep. The great efforts of the farmers, *to procure wool*, do not keep pace with the increasing demand, though none is exported, and hence the prices of wool are kept up and for the best payments. Such are the new advantages of the American woollen manufacture. Shall we then continue to raise tobacco, to be dutied, plundered, rejected, restricted or excluded, according to their criminal will, by foreign governments, or shall we decrease it and raise more sheep, procure more and better wool, and provide a market at home, free from injury, insult and vexation, by the industry of our women and children, and the power of machinery, with a little aid from regular and skilful male manufacturers? Is it better to get from

45 to 150 cents per lb. for wool at a *sure* American market, or to raise tobacco to be plundered, dutied or driven from port to port, in distant seas ? Let us consider *well* the past course, the present condition, and the apparent prospect of our cultivation and sales of tobacco, and all other productions. Let us consider, on principle, whether it is not better often to avoid articles of mere foreign consumption, the sales of which must be effected under all the disgrace and evils of these times on the ocean; and to raise or extract from the earth, wool, hemp, flax, cotton, leather, iron, copper, and lead, to arm, supply and clothe our soldiers, sailors, and militia, and to provide for the raiment of our families and the furniture of our houses. Let us observe well the state of Rhode-Island. With wool, flax, hemp, and leather of her own, she demands the iron and hemp of the Baltic and of her Southern sisters, for her manufacturers, and the cotton of Virginia, the Carolinas, Georgia, and the western states for her *pre-eminent* cotton mills. In that state there are already erected a number of cotton mills to card and spin, equal to one for every township, and her enterprising merchants and cotton millers have erected others in the adjacent townships of the more wealthy and populous states of Connecticut and Massachusetts. The native country of the illustrious Greene, has already made effectual provision for the sure and cheap supply of all the cotton goods, which can be necessary for any force of army, navy and militia, which we can want in any war: and this too, *in certain support of the Southern agriculture of the United States*. The population of the whole state of Rhode-Island is less than that of the English town of Manchester. Shall we cast the eye of fellowship and attachment upon our neighbour state, or upon our foreign rivals in British Lancashire, or Gallic Rouen? Shall we consume the provisions of the agricultural states by the cotton manufacturers in Rhode-Island and its vicinity, subject to the light expense of a coasting voyage, or shall we send them to Liverpool or Rouen, for foreign cotton spinners, under heavy charges and more intolerable indignities? Let us use the *home* market, since the roads to the foreign are blockaded, or subjected to a scandalous tribute.

Thomas Cooper

Prospectus of the Emporium of Arts and Sciences *(1813)*

Thomas Cooper (1759–1839) was an English emigrant to the United States in 1795, an associate of Joseph Priestly, a Jeffersonian pamphleteer, a scientist, and one of the most influential educators in the young republic. He taught chemistry, law, mineralogy, and political economy and professed a strong brand of anticlericalism. For a brief period in the middle of his career he edited the *Emporium of Arts and Sciences* in Philadelphia, which published the latest information in applied sciences. His prospectus for that publication is notable on two counts. It stresses "6thly" the intellectual benefits of encouraging manufactures, which "would extend knowledge of all kinds, particularly scientifical." It also shows the embarrassment of an advocate of free trade, who, nevertheless, has to acknowledge that government intervention (the protective tariff) is the only route to industrial development in the United States. Cooper came to oppose manufactures stoutly. His condemnation of machinery as a "dreadful curse" (1823) is also reprinted in this volume. Cooper's devotion to laissez-faire led him to support South Carolina in the Nullification Act of 1832, in protest against federal tariff legislation. The excerpt reprinted here includes all but the introductory and concluding remarks.

Thomas Cooper, "Prospectus," *Emporium of Arts and Sciences*, New Series, volume 1 (June 1813), pp. 4–10.

Whether it be worth while to *encourage* manufactures in this country, or to turn aside a part of the capital from the immediate employment of agriculture is a question of great moment. All bounties and protecting duties, are taxes upon the rest of the community, in support of that employment of capital, which, without them, would be injudicious and unproductive. While so much land remains uncultivated, there can be no want of opportunities of employing capital in America. Generally speaking also, the interference of government is sadly misplaced, when it attempts to direct the capitalist what he shall do with his money. *Laissez nous faire*, is the proper reply. Still, there are considerations of great weight with me, in opposition to this general reasoning, that I have never seen urged.

1st. Our population is becoming scattered over such an extent of territory, that the nation is really weakened by it; defence is more difficult and expensive; active hostility almost impossible; the communication of society, and of course of knowledge, is greatly retarded; many of our citizens are tempted to live in a half savage state; and even the administration of law, and the maintenance of order and necessary subordination, is rendered imperfect, tardy, and expensive.

2dly. Our agriculturists want a *home* market: manufactures would supply it. Agriculture at great distances from seaports, languishes for want of this. Great Britain exhibits an instance of unexampled power and wealth, by means of an agriculture greatly dependent on a system of manufacture: and her agriculture, thus situated, is the best in the world, though still capable of great improvement.

3dly. We are too much dependent upon Great Britain for articles that habit has converted into necessaries. A state of war demands privations that a large portion of our citizens reluctantly submit to. Home manufactures would greatly lessen the evil.

4thly. By means of debts incurred for foreign manufactures, we are almost again become colonists: we are too much under the influence, indirectly, of British merchants and British agents: we are not an independent people. Manufactures among us would tend to correct this, and give a stronger tone of nationality at home. I greatly value the intercourse with that country, of pre-eminent knowledge and energy, but our dependence upon it is often so great, as to be oppressive to ourselves.

5thly. The state of agriculture would improve with the improvement of manufactures, by means of the general spirit of energy and exertion which no where exists in so high a degree as in a manufacturing country; and by the general improvement of machinery, and the demand of raw materials.

6thly. The introduction of manufactures would extend knowledge of all kinds, particularly scientifical. The elements of natural philosophy and of chemistry, now form an indispensable branch of education among the manufacturers of England. They cannot get on without it. They cannot understand or keep pace with the daily improvements of manufacture without scientific knowledge : and scientific knowledge is not insulated; it must rest upon previous learning. The tradesmen of Great Britain, at this day, can furnish more profound thinkers on philosophical subjects, more acute and accurate experimenters, more real philosophers thrice told, than all Europe could furnish a century ago. I wish that were the case here; but it is not so. I fear it is not true, that we are the most enlightened people upon the face of the earth; unless the facility of political declamation be the sole criterion of decision, and the universal test of talent. We should greatly improve, in my opinion, by a little more attention to mathematical and physical science ; I would therefore encourage whatever would introduce a general taste for such pursuits.

7thly. Because the home trade, consisting in the exchange of agricultural surplus for articles of manufacture produced in our own country, will for a long time to come, furnish the safest, and the least dangerous, the least expensive, and the least immoral—the most productive and the most patriotic employment of surplus capital, however raised and accumulated. The *safest*, because it requires no navies exclusively for its protection : the least *dangerous*, because it furnishes no excitement to the prevailing madness of commercial wars : the least *expensive*, for the same reason that it is the safest and the least dangerous : the least *immoral*, because it furnishes no temptation to the breach or evasion of the laws; to the multiplication of oaths and perjuries ; and to the consequent prostration of all religious feeling, and all social duty: the most *productive*, because the capital admits of quicker return; because the whole of the capital is permanently invested and employed at home; because it contributes directly, immediately, and wholly, to the internal wealth and resources of the nation : because the credits given, are more easily watched, and more effectually protected by our own laws, well known, easily resorted to, and speedily executed, than if exposed in distant and in foreign countries, controuled by foreign laws and foreign customs, and at the mercy of foreign agents : the most *patriotic*, because it binds the persons employed in it, by all the ties of habit and of interest, to their own country ; while foreign trade tends to denationalize the affections of those whose property is dispersed in foreign countries, whose interests are connected with foreign interests, whose capital is but partially invested at the place of their domicil, and who can remove with comparative facility from one country to another. The wise man observed of old, that " where the treasure is, there will the heart be also," and time has not detracted from the truth of the remark.

Nor can there be any fear that for a century to come, there will not be full demand produced by a system of home manufacture, for every particle of surplus produce that agriculture can supply. Consider for a moment what are the articles that may fairly be regarded as of the first necessity, that an agricultural capitalist will require either to conduct his business, or for his reasonable comforts. 1st. The *iron manufacture* in all its branches from the ore to the boiling pans, the grate, the stove, the tire, the ploughshare, the spade, the scythe, the knife and fork, the sword and the gun : the *copper manufacture*, for his distilling vessels ; for the bolts and sheathing of ships : the *lead manufacture*, for his paints and his shot : the *tin manufacture*, for his kitchen utensils ; the manufacturing of powder for blasting and for fire arms : he cannot dispense with the wheel-wright, the mill-wright, the carpenter, the joiner, the tanner, the currier, the sadler, the potter, the glass maker, the spinner, the weaver, the fuller, the dyer, the shoemaker, the hatter, the maker of machines and tools, and very many trades and handicrafts not enumerated. Of all these occupations, every one of which may be employed in furnishing articles either of immediate necessity of reasonable want, or of direct connection with agriculture, we have in abundance the raw materials of manufacture, and the raw material, uninstructed man to manufacture them. Is it to be pretended that these occupations when fully under way at home, will not furnish a market for the superfluous produce of agriculture, provided that produce be, as it necessarily will be, suited to the demand ? Or ought this variety of occupation, and above all, the mass of real knowledge it implies, to be renounced and neglected for the sake of foreign commerce—that we may not interfere with the profits and connexions of the merchants who reside among us, and that we may be taxed and tolerated and licensed to fetch from abroad, what we can with moderate exertion supply at home ? And yet this is the doctrine not merely advocated and recommended among us, but likely to become the fashionable creed of political economy, wherever mercantile interests and connections prevail. It appears to me of national importance to counteract these notions. As a source of national wealth, I would no more encourage manufacture than I would encourage commerce—I would encourage or discourage neither : for I am persuaded that the aggregate of individual, constitutes national wealth; and that a government is conceited and presumptuous, when it attempts to instruct an individual how he can employ his industry and his capital most beneficially for his own interest.

Every treatise on political economy ought to have its

first page occupied with the answer to Colbert, LET US ALONE.

But as a mean of national defence, and national independence—as a mean of propagating among our citizens the most useful and practical kinds of knowledge—as a mean of giving that energetic, frugal, calculating and foreseeing character to every branch of our national industry, that does not exist but among a manufacturing people—as a mean of multiplying our social enjoyments by condensing our population—and as a mean of fixing the consumers and the producers in the immediate neighbourhood of each other—I would encourage the commencement at least of home manufacture. Not the manufacture of gold and silver—not the velvets of Lyons or the silks of Spitalfields---the laces of Brussels and the lawns of Cambray---not the clinquaillerie and bijouterie of Paris and Birmingham, but such as we feel the want of in time of war; such as may fairly be regarded as of prime necessity, or immediately connected with agricultural wants and pursuits.

8thly. I would remark, that nature seems to have furnished the materials of manufacture more abundantly in Pennsylvania in particular, than in any country I know of. The very basis of all profitable manufacture, is plenty of fuel, easily, cheaply, and permanently procurable: the next desirable object, is plenty of iron ore; iron being the article upon which every other manufacture depends. It is to the plentiful distribution of these two commodities, that Great Britain is chiefly indebted for the pre-eminence of her manufactures and her commerce. I have not a doubt on my mind, but both pitcoal, and iron ore, are more plentifully distributed in Pennsylvania than in Great Britain; and that both the one and the other can be gotten at more easily and cheaply in this country, than in that. Moreover we have a decided superiority in the raw materials of Cotton, Hemp and Flax; in our alkalies for glass works; in the hides and the tanning materials of the leather manufactory; and we can easily procure that advantage, so far at least as our own consumption requires it, in the woollen manufactory. Other branches might be enumerated wherein our advantages of internal resource are undeniable; but I cannot see why we should neglect or despise these. Nothing but a stimulus is wanted to induce and enable us to make a proper use of our domestic riches. But men of skill and men of capital, fear to begin; lest on the return of peace, they should be exposed in the weakness and infancy of their undertaking, to contend with the overwhelming capital, and skill of the European powers, particularly of Great Britain.

For these reasons, I think it would be expedient so far to aid the introduction of manufactures in this country, by protecting duties, as to afford a reasonable prospect of safety to the prudent investment of capital, and the industrious pursuit of business; but no bounty to wild speculation, to negligent workmanship, or to smuggling.

Daniel Webster

Debate on Repeal of the Embargo (1814)

Daniel Webster (1782–1852) was a young congressman from rural New Hampshire when he spoke in support of a bill repealing the embargo and nonimportation acts. At this point in his career Webster opposed the double tariff imposed by those acts. Webster was not opposed to manufactures per se but to government support of industry at the expense of other parts of the economy. Later in his career, after moving to Boston, he became spokesman for the Boston manufacturing community, as other documents of his reprinted in this collection indicate, but even early on he saw the direction American politics would take him. Not long after this 1814 speech, he observed portentously, "The manufacturing interest has become a strong *distinct political party*. This you may rely upon. In short, I believe we are all to be thrown into new forms, & new associations" (letter to Samuel Bradley, April 21, 1816).

In this excerpt from his 1814 speech, Webster warns against the problems that extensive manufacturing, supported by the continued tariff, would cause. He contrasts a Jeffersonian image of agrarian society to the evils of industry as it had developed in England. Webster mixes patriotic, economic, and moral arguments in his plea for a lower tariff. "Their" and "them" at the opening of this passage refers to manufacturers.

Daniel Webster, *The Debates and Proceedings of the Congress of the United States*, 13th Congress, first and second sessions (Washington, 1854), pp. 1972–1973.

I am not, generally speaking, their enemy. I am their friend, but I am not for rearing them, or any other interest, in hot-beds. I would not legislate precipitately, even in favor of them; above all, I would not profess intentions in relation to them which I did not purpose to execute. I feel no desire to push the capital into extensive manufactures faster than the general progress of our wealth and population propels it.

I am not in haste to see Sheffields and Birminghams in America. Until the population of the country shall be greater in proportion to its extent, such establishments would be impracticable, if attempted, and if practicable, they would be unwise.

Whatever manufactures can be conducted in the household, where children may be occupied under parental guardianship and protection, are useful in the highest degree. Many others, cotton and woollens for instance, of which the material constituting the article of chief value, is or may be the produce of our own soil, are likewise worthy of protection and care. But of those products of which the chief ingredient is the labor bestowed, which can be made profitable only by the employment of vast capital, by the minutest division and subdivision of labor, and by the toil of children of both sexes, drawn together in great numbers, and put out of sight of those who have a natural interest in the preservation of their health and morals, one can hardly speak in terms of so decisive approbation. Habits favorable to good morals and free Governments, are not usually most successfully cultivated in populous manufacturing cities. It is one of the consequences of such employments to render the laborer altogether dependent on his employer. This arises from the extent to which the division of labor is carried in great manufacturing establishments. He whose occupation it has been for his whole life to perform only one of the many operations necessary to the production of a single article, is necessarily among the most dependent of human beings. The trite example of the number of persons—sixteen or eighteen—who have all their several labors and operations in the production of a common brass pin, sufficiently illustrates my idea.

One of these laborers, utterly incapable of making and carrying to the market on his own account the smallest entire article, is necessarily at the mercy of the capitalist for the support of himself and family. Any cause which deprives him of that particular occupation for which only he is fit, by habit and education, throws him a burden on society. As such causes must occur often, it is in the neighborhood of such establishments that hands unemployed or ill employed will be found in greatest numbers.

It is in her manufacturing districts that England recruits her armies. It is there, principally, that those are found who have the least hold on society, and where necessities, or where habits, force them to the camp. I am not anxious to accelerate the approach of the period when the great mass of American labor shall not find its employment in the field; when the young men of the country shall be obliged to shut their eyes upon external nature, upon the heavens and the earth, and immerse themselves in close and unwholesome work-shops; when they shall be obliged to shut their ears to the bleatings of their own flocks, upon their own hills, and to the voice of the lark that cheers them at the plough, that they may open them in dust, and smoke, and steam, to the perpetual whirl of spools and spindles, and the grating of rasps and saws. I have made these remarks, sir, not because I perceive any immediate danger of carrying our manufactures to an extensive height, but for the purpose of guarding and limiting my opinions, and of checking, perhaps a little, the high-wrought hopes of some who seem to look to our present infant establishments for "more than their nature or their state can bear."

It is the true policy of Government to suffer the different pursuits of society to take their own course, and not to give excessive bounties or encouragements to one over another. This, also, is the true spirit of the Constitution. It has not, in my opinion, conferred on the Government the power of changing the occupations of the people of different States and sections, and of forcing them into other employments. It cannot prohibit commerce any more than agriculture; nor manufactures any more than commerce. It owes protection to all. I rejoice that commerce is once more permitted to exist; that its remnant, as far as this unblessed war will allow, may yet again visit the seas, before it is quite forgotten that we have been a commercial people. I shall rejoice still further, when I see the Government pursue an independent, permanent, and steady system of national politics; when it shall rely for the maintenance of rights and the redress of wrongs on the strength and resources of our own country, and break off all measures which tend, in any degree, to connect us with the fortunes of a foreign Power.

American Society for the Encouragement of Domestic Manufactures

Address to the People of the United States (1817)

The American Society for the Encouragement of Domestic Manufactures was one of the more important organizations established in the early nineteenth century to promote American manufactures. It was based in New York, and its president, Daniel Tompkins, was vice-president of the United States. The society helped to establish many smaller societies that worked in favor of the same ideas, and it sponsored a national convention to advance the cause of industry.

The arguments put forward in the address are the same ones that advocates of industry since Alexander Hamilton had presented, but they are tempered by the War of 1812 and its economic dislocations. The most interesting part of this document is its call, at the end, for a network of societies bound together to "bring important truths to light, dispel prejudice, refute sophistry, excite patriotism, cherish industry, and, above all, give to public opinion that expansive swell that will harmonize with the rising tide of our country's prosperity." To the writers of this phamphlet, the encouragement of American manufactures was a matter of great moral and patriotic importance, not just a simple economic concern.

Address of the American Society for the Encouragement of Manufactures, to the People of the United States (New York, 1817), 32 pp.

ADDRESS, &c.

The Committee charged to report an Address to the public, from the Society for the encouragement of Domestic Manufactures, have cheerfully complied, for if there be any interest dear to the patriot's heart, and precious in the eyes of humanity, it is that of a nation's industry, advancing hand in hand with her civilization, glory, and independence. National industry is the true source of imperishable riches, the means of pure enjoyment, the support of good morals, the natural ally of social prosperity and individual happiness. In its effects, and in its causes, it is identified with the advancement of the sciences and the progress of the human mind.

*

Some minds, deserving of a better direction, have, from long habit of a particular mode of dealing, associated the idea of commerce with that of a ship from abroad, loaded with stuffs of foreign manufacture. And they cannot see how another branch of industry can bear any competition. Yet a little attention to the progress of man's civilization will show, that without reference to national advantage, to be a manufacturer is a law of man's nature; witness his attitude, his structure, those limbs which are not destined to support his body, but supple, flexible with motion and articulation, suited to every operation that the will of the most improved intelligence can exact. And if he cannot assure his own preservation, nor procure food, raiment, or habitation, without manufacturing implements for defence.

or for the chase; nor fell a tree in the forest, or turn a furrow in the field, till he has manufactured the plough and the axe, then we may say with Franklin, whose wisdom spoke in smiles—in any one of whose sallies there is concentrated more profound thought than in volumes of common place, " That man is a tool-making animal," or, in words less lively or emphatic, that he is by nature a manufacturer.

But we cannot help regretting, that not only the objects of our commerce, but our moral and political opinions, have been too long of foreign manufacture. And we think they treat us unfairly; for the opinions they force upon our credulity are such as they never use themselves. They are manufactured for exportation, not for home consumption. If we adopt them they will profit willingly, but, in return, smile at our credulity.

In a word, all the arguments used by the partisans of foreign manufactures, are resolved into one point, Shall we manufacture for ourselves, or shall Britain manufacture for us? This is the question; and now, having stated it fairly, we shall meet it boldly, and argue it candidly.

On the part of the adversary, the following objections are relied upon as insurmountable:

1. That this ought to be a commercial and agricultural, and not a manufacturing country.

2. That manufactures are unfriendly to commerce and agriculture.

3. That they cannot be carried on to advantage, because labour is higher than in Europe.

4. That they demoralize and deprave those employed in them.

5. That they should be left to themselves, and not forced into premature existence by government patronage.

6. That such patronage would diminish the revenue and resources of government.

True to her interest, when Great Britain cannot force a market by the bayonet, she does it by circumvention. It was this policy, exercised towards these states whilst colonies, that, with other aggressions, led to resistance. It was the continuance of this policy, and the influence of her manufactures, that lately went near to prostrate our government, sever our union, and overturn our independence. And this policy, as long as it is fed with any hope of our ruin, will leave no means untried to injure us. Such is the policy that carries despotism round the globe; that whispers in our ears, and would instil into our hearts, pernicious counsels.

And now to our argument:

1st. That this ought to be a commercial and agricultural country.

If this position were not the entering wedge for other sophistries, we should have nothing to do but to agree: but when they go the length of saying, " Give up manufacturing that you may be commercial and agricultural," we say, no! but we will manufacture, that we may be agricultural and commercial. And we tell them, read your history, and see how England's commerce has depended on, and grown out of, her manufactures.

If England's commerce has depended upon her manufactures, and without any agricultural resources she has risen to wealth, we may well say, having a resource the more in the abundance of our soil, " Do you give up all competition, let us manufacture for you." Great Britain would surely think this an arrogant pretension, and she would think rightly. Why, then, presume that we should be her dupe?

Does any one seek to be convinced, by a single fact, that the settlement of the lands, and the prosperity of the country, depend, essentially, upon manufacturing establishments, let him go to the western part of this state, the rapid growth of which is without a parallel in the history of nations, and he will find that mills and manufactures formed the first rudiments of those almost countless villages and towns which spangle that fertile and beautiful country, emphatically styled, the Eden of the state.

2d. That our manufactures are noxious to our commerce and agriculture.

This is little else than so many empty words. How can that which widens the field of commerce be said to injure it? Will these logicians assert that British manufactures have injured British commerce? No; but they speak with two tongues; one for themselves, and one for us. We have three resources; they have but two: abandon one, they say, that we may be equal. When did they set us the example of such complaisance? And as to any pretended injury to agriculture, by the absorption of labour, we find that out of 200,000 persons formerly employed in our factories, in two branches alone, more than 120,000 were women and children. Was agriculture benefited when, on the stopping of the cotton and woollen manufactures, these women returned to idleness, the children to the poor house, and the men, not to the farms, but to the cities from whence they came?

3d. That manufactures cannot be carried on here to advantage whilst labour is so much higher than in England.

This may be plausible to those who are as ignorant of that country as its partisans are, or affect to be, of this. Our labour is, indeed, numerically higher; but taxes and impositions are so much lower, that we can afford to pay more, because our goods are charged with little else. It is true that in England the labourer receives less, because what he earns by his industry is paid away, before it reaches his hands, in tithes, pensions, taxes, poor-rates, and a thousand exactions to pamper the pride and luxury of those who live but to consume the fruits of the earth—who neither work, nor add to the stock of national wealth.

But it proves nothing for the lowness of wages, that this

poor man's substance is eaten up by so many that had no share in earning it. And there is another answer worth attention : If our fabrics are upheld for a time, a power will develop itself which will sink this formidable objection into nothing; that of labour-saving machinery; a power of which no man can at present foresee the limit or extent; a power indigenous in this country, where men, by the free exèrcise of their will and faculties, have acquired a characteristic aptitude for mechanical inventions. Many instances prove this position, so honourable to our country.

And what field of competition is so desirable as that which calls into activity the finest powers and greatest energies of useful intellect; the powers that will make us strong in war, secure in peace, respected abroad, happy at home. But there is another motive, still nearer at hand: these manufactures give bread to many whom years, infirmities, or sex, disqualify from labours of a ruder cast, and make them rather a source of wealth to the community than an incumbrance. And so little does the depression of our manufactures depend upon scarcity of hands, that many are carried on by apprentices without wages. And since the peace, many persons have been obliged to return from them to the poor houses, and be again consigned to pauperism.

What we have said of machinery will be of more weight, when it is considered what abundance of mill-sites are to be had in this country, of which the fee-simple, and all other charges, would not cost the annual expense of a steam engine; and though in England wages are higher than on the continent of Europe, yet that has not prevented her from underselling all her rivals, except such as have lately adopted the counteracting policy we would recommend.

It is worthy also of notice, that all these labour-saving machines, and mechanical improvements, which would be hailed by us as new planets in the firmament, are, in that country, the signals of mobs, assassinations, and revolt; and are, in fact, at last established by the sole protection of the strong arm of government.

We refer on this head to Mr. Tench Coxe's "Statement of the Arts and Manufactures of the United States," who asserts that the diminution of manual labour in 1808 was estimated in England, in regard to the cotton business, at 200 to 1. And who observes further, that Mr. John Duncan, of Glasgow, an able writer and artist, considers it to be much more. In the same work Mr. Coxe instances the saw-gin, invented by Mr. Ely Whitney of Connecticut, as saving manual labour in the proportion of 1000 to 1. If it were consistent with our limits, or our present object, we could quote abundance of valuable matter from this authentic and useful work. We can only here recommend it to the perusal of all who take interest in their country's welfare.

4th. That manufactures degrade and demoralize.

We are inclined to believe that in the British factories are found disgusting exhibitions of human depravity and wretchedness. But we cannot believe that the exercise of industry could ever be the cause of demoralizing any race of men; although unequal laws and bad examples may have that tendency. In this country there are extensive manufactories, and yet no such consequences are observed.

The best account we have of the pollution of British manufactures is in a work entitled "Espriella's Letters." To judge from that work, British manufactories are objects of abhorrence. But, for the honour of humanity, we must suppose that picture something over-coloured.

Surely, we have not witnessed in our fabrics any of those fearful apparitions, flitting through the smoke of their dismal repairs, like the spirits of the damned, squalid and pallid, with green hair, red eyes, distorted members, and ghastly aspect. But whoever has travelled through the towns and cities of the Brtish Isles, during the last twenty-five years of war, must know that it is not alone in manufacturing districts, or manufacturing countries, that beggary and wretchedness are to be found. Whoever would describe depravity and immorality, may visit barracks, camps, and men-of-war; and, moreover, those nations which are not manufacturing will be found most to abound in profligacy and disorder. In those countries that enjoy the benefit of manufactures, their wholesome effect upon the morals of the people is too often defeated by the immoderate use of spirituous liquors, which, and not manufactories, are the most prolific source of poverty and immorality. Experience has shown that the persons employed in manufactories are as sober as any of the working class. A reason for which may be, that the employers have better means of watching over their conduct, and controlling their disorders ; or, where that cannot be effected, discharging those whose bad example might corrupt the rest.

And it appears, from the authentic treatise of Mr. Colquhoun, that before the present unparalleled state of distress in England, there were only seven paupers to every hundred inhabitants in the manufacturing districts, and in others, not manufacturing, there were twenty-one.

Was it manufactures that humbled Spain, whose power and pride stood once as high as England's ? What manufactures strew the streets of Naples with idle Lazaroni? What manufactures debase Portugal? Is it the manufacturing of tooth-picks at the university of Coimbra? or is it the stripping off the bark from the cork tree in the forest, to be carried to England, cut, and sent back to bottle their wine? Is it the encouragement of domestic manufactures that has degraded the children of Erin ? or is it that every demoniac effort has been used, to depress its industry, stifle its genius, and trample down its virtues ?

And why is Canada so different from the United States, although untaxed? Because, even the timber of their woods is sent to be made into ships, and returned, ready framed, to be launched on the lakes for their defence.

But at length, though late, the continental nations have taken the alarm, and combinations are formed, by both sexes, against the importation of these manufactures! Shall we be less quicksighted? If, in war, they could not overcome us, shall they in peace destroy us? If they feel now the effects of their ambition, they cannot complain: "They are the general challengers. We come but as others do, to try with them the strength of our youth."

We have, besides, none of those great manufacturing cities; nor do we wish for such. Our fabrics will not require to be situated near mines of coal, to be worked by fire or steam, but rather on chosen sites, by the fall of waters and the running stream, the seats of health and cheerfulness, where good instruction will secure the morals of the young, and good regulations will promote, in all, order, cleanliness, and the exercise of the civil duties. This, with the beneficial clauses usual in our indentures of apprenticeship, and the vigilant eye of the magistrate to enforce them, will obviate every apprehension. And we hazard nothing by the assertion, that some of the best educated of the poorer class, in this country, are those brought up in factories, and such as would otherwise have been destitute of education altogether; and those whose tenderness inclines them to make this objection are requested to reflect, that the paternal regard of the legislature is awake to this subject; and that, to every institution of this kind a school will be appendant. Then, if it please heaven to redeem the thousands, and tens of thousands, that groan in the land of bondage, and open them a passage through the waves, as to the Israelites of old, this shall be their land of promise. Here shall their industry find its reward; and if they fear sickness or decrepitude in our factories, there is no authority, power, or necessity, that can confine them for a day. They may shape their course to any part of a territory as expansive as the ocean they have traversed, find a thousand ways to bestow their industry to their advantage, with land, free and unoccupied, on which to settle; and under no circumstances need they fear the dreadful calamity of famine, from which they fled.

5th. That manufactures should be left to their natural growth.

To the friends of America, it will be argument enough that domestic manufactures are for the permanent interest of their country, and the only sure means of our independence. What would not wisdom and patriotism do to secure such objects?

We ask not one-third of the protection which Britain has bestowed upon her manufactures. We ask not more protection than our commerce has received by discriminating duties and navigation laws; and what we do ask, is but until our tender grizzle shall be hardened, and our joints knit. But under what protection British manufactures grew, and still maintain themselves, we shall now show; and then, in our turn, ask these advisers, why ours should be left to themselves rather than their own.

Coeval with the first dawn of English prosperity, we find in the British code laws for the protection of British manufactures. One of their ancient kings, the third Edward, is magnified in their history for his wise foresight in enacting these statutes, to which their increasing greatness is ascribed. To those acts is referred the consequence to which that little island has since attained; the bursting of the feudal chains; the growth of art and science; and that power, of which the abuse has at length recoiled upon the head of pride and usurpation.

We do not ask for such laws as the British code exhibits. We would not sacrifice to a golden idol the rights or feelings of humanity. We would not chain to the ground the harmless artificer; nor under accumulated penalties restrain his natural rights. Yet such are Britith statutes. The oppressor may trample on him; famine stare him in the face; his children cry for bread, when he has none to give them; be his disgust or his enterprise what it may, he " must abide the pelting of the storm ;" his native land is his dungeon, and his industry his crime. If a master of an American vessel offer to transport him to a country where his heart's hopes are centred, he, too, is condemned, as "*a seducer of artisans*," to like ruinous inflictions, and punished for his charitable ministry. The exporter of a tool or implement used in any art, or the master who receives it in his ship, is subject to similar pains and forfeitures.

Nor is this, like the feudal laws, or monastic institutions, an obsolete system; many of these statutes are of modern date, and some of the time of the reigning monarch.* We wish for nothing that can affect the personal right of any individual, citizen, alien, native, or foreigner; we claim only for our country the honourable protection of its very dearest interests. But, we think this argument may show how far Great Britain is from doing that herself which her emissaries never fail to preach to us—that is, letting her manufactures take care of themselves. Nor is it the king, nor his cabinet, nor his parliament, to whom this policy is to be ascribed. It is the public voice. So dearly do Englishmen prize that interest they would have us forego.

We would here notice two branches of domestic manufactures, the shoe and hat manufactures, which have, by the means of the protection of government, prospered to that degree that they, at this day, render us independent of foreign supply. But facts are so abundant that the details would lead to interminable length.

* Geo. I. c. 27. Geo. III. c. 13. Geo. III. c. 71. Geo. III. c. 37. Geo. III. c. 60.

We find a member of parliament, the celebrated Mr. Brougham, who brought about the repeal of the orders in council, by showing the effects of our non-importation law upon their manufactures, this energetic denouncer of the abuses of power, versed in the subject, and speaking for popularity, in arraigning as madness the excessive exportations to the continent of Europe, admits, nevertheless, " that it is well worth while to incur a loss on the first exportation, in order, by the glut, to stifle in the cradle those rising manufactures in the United States, which the war had forced into premature existence, contrary," as he is pleased to assert, " to the natural course of things." And a celebrated writer on the colonial policy of Great Britain, whose words are considered next to official, in a chapter on the relative situation of Great Britain and America, as manufacturing rivals, speaks thus : " This is the era (he says) of a systematic contest which must, eventually, endanger the safety of the manufactures of the one or the other." Now, though this is not a war of arms, yet it is a war more subtle and more deadly, a war that can deprive us of every means of future resistance, and insure success to some future invasion. It is that warfare, which, two years after victory, has left us worse than a conquered nation ; without a single piece of coined money in the purse of any individual. If we hesitate now, we deserve our adversary's scorn; if we will be deceived, why should he not deceive us ; if we are content to be undone, why should he feel remorse ; if we have no remedy, we are to be pitied and not blamed ; if we have, and want courage to apply it, we are to be blamed, but not pitied. If we do not make a stand upon this ground, we need defend no other post ; their interest, supported by the government, by their laws, by public patronage, and wealthy combinations, by export duties, and bounties on exportation, will prevail against our's, unsupported and neglected, and our interest will be more than *endangered*, in this *systematic contest*, if one gives all the blows, and the other passively receives them.

*

There is living testimony within the reach of this society, that, in certain British manufactories, the French marks were put upon their goods without any affectation of concealment, and the purpose openly avowed, as well as the connexion that subsisted between the real manufacture in Britain, and the fictitious one in France.

And, at the commencement of our woollen manufactures, for the purpose of degrading our fabrics, goods of the worst quality, but highly finished to the eye, were sent to this city from England, marked " Humphrey's Ville," that they might, by passing for the productions of that manufactory, injure its well-merited reputation.

It is well known to many, that, during the late war, British goods were smuggled into this country, and exposed to sale as American, Spanish, and Portuguese; it is quite of course, too, for their agents who have come out here since the war, in speaking of the glutting of the European markets, to say, that the speculation was not so unwise as unfortunate, for, if the government and people had not taken the alarm, they should have destroyed their manufactures, and afterwards had their own price.

In the beginning of the year 1792, when the report of General Hamilton, then secretary of the Treasury, made by orders of the House of Representatives, was published in England, it created such alarm, that meetings were called in the manufacturing towns, and Manchester alone, at a single meeting, subscribed 50,000 pounds sterling, towards a fund to be vested in English goods, and shipped to this country for the purpose of glutting our market, and blasting the hopes of our manufactures in the bud.

The lucrative speculations which the wars of Europe gave rise to, the examples of rapid fortunes made by foreign commerce, and the temporary advantages of our neutral state amongst so many powers, eager for each other's destruction, prevailed over the prophetic wisdom of that illustrious statesman; but things being now restored to their natural order, that important document which has been almost smothered in oblivion, and is of all his works that which has been least noticed or appreciated, must now be brought into full view. And we call upon the friends of American independence, upon those who raised to his memory a humble monument suited to be the record of private affection, and to number his days, to join with us in raising this fallen column of his true renown.

And before we despatch this important head of " leaving manufactures to themselves," we must advert to that phenomenon of art, the steam-boat, that proudest specimen of *American manufacture.* Had it been left to itself there would have been lost to the human race an inestimable benefit, and to this republic the proudest monument of its glory. It came forth with throes and pangs of travail like a giant's birth; and had not an enlightened legislature fostered its inventor with encouragement and hope, and renewed from time to time the period limited for its production, it would not now be seen stemming the current of our magnificent rivers, glittering like the enchanted galley on the tide of fate, topping the ocean's wave, or gliding like the pride of swans upon the lake.

6th. We come now to the last head of our argument, "the public revenue." And here we would remove that error which supposes that foreign importations pay the revenue to government. It is not so! they are barely the medium through which the government collects the revenue from the private purses of the private citizens. It is the citizen, and not the ship, that pays. It is the citizen, and not the foreign goods, that pay. It is the consumer, and not the importer. During the recent war, so far from supporting the revenue, these importations (too often carried on in partnership with treason) developed their characters, drained the country of its specie, and its bullion, and left the government in a situation too humiliating to be recollected without pain by any patriot.

But, happily for this country, fortune has brought this evil to a period. And few will be so headstrong as not to acquiesce in the change of times and circumstances.

It surely makes no difference to our citizens which way they pay the money that goes to support their government, and they can have no objection to pay it in the way most beneficial to their country, by raising it on the domestic manufactures. The necessity of a direct tax will be lessened, which will come in ease of the landed interest and of the merchant.

Mr. Isaac Briggs, in his Statement to the Chairman of Commerce and Manufactures, has proved, by exact calculations, founded on a *present* and *prospective* view of our population, wants, produce, and the foreign markets, that if our agriculturists depend, in future, upon any other market than that which domestic manufactures will afford, that their produce will lie upon their hands, or they must accept of whatever price the foreign merchant may be pleased to offer, for such portion as he will condescend to accept. For produce will no longer serve as payment where it is no longer wanted, and payment in specie will clearly be impossible.

For the tables and calculations we refer to the 9th volume of Niles' Weekly Register, where this valuable document will be found.

As the public may not be aware of the great interest, even *now* in jeopardy, we will barely mention, upon good authority, that there were, at the peace, 600,000 spindles employed in the cotton factories alone, the value of each of which, with the appendages, averaged 80 dollars, embracing, in capital, above forty millions, *besides* the capital employed in working the raw material, which amounted to twenty millions more; and the woollen factories, though of much more recent origin, a capital of about the same amount. all which appeared, from a report to the Representatives of the People of the United States, by the Committee of Commerce and Manufactures of the last session, founded upon authentic data, furnished by the agents of the manufacturing interest, who were examined before separate committees of senate and representatives. It has, moreover, been since ascertained, that preparations were made for the extension of both branches, which would have augmented the capital employed in them respectively to a much greater amount.

Let us now look back and see what this idol, foreign importation, was, and whether it is wiser to keep life in our own manufactures, or to struggle unnaturally to revive that unprofitable traffick.

It is a fact, which we assert on the authority of intelligent merchants, that the importing commerce has, in the two last years, (since peace has brought things to their natural course,) diminished the mercantile capital one-third, and, if continued, will result in the total impoverishment of every class. But what in its best days did it do for us? It corrupted our patriotism; domineered over our opinions; excited party spirit; embarrassed the government, and aimed a mortal blow at our union and independence. It carried the views of fortune of many good citizens from their own, to a foreign land, and brought among us a host of mischievous agents, whose business was, by night and by day, to irritate the public mind, fester every sore, and warp the measures of the government to a foreign interest. Instead of furnishing money, the sinew of war, it cut that sinew in the critical moment when its action was most wanted. Before a blow was struck on our part, it had stained our own waters with the blood of our countrymen; taught the nations of the earth to disrespect us, placed six thousand of our kidnapped citizens in British prisons, and forced others to shed the blood of their fellows and kindred in battle; and now, at the end of two years from the cessation of the war which it induced, although victory crowned our arms, bankruptcy stares us in the face. Is it, then, upon this rope of sand that government can rely in the event of any future war?

Happily the frauds of the foreign merchants have brought conviction home to the knowledge and sensibilities of our importers. Our merchants have found out that their order is no sooner executed by the English merchant, than other cargoes, of the like kind and quality, invoiced at reduced prices, are immediately shipped on their own account. And the duties being as much less as the invoice is lower, the revenue is defrauded of so much, and these goods are then thrown upon the market at this reduced price; added to which, the facilities afforded them by sales at auctions, (where the foreign merchant is exempt from license duty,) enable them to "*glut our markets*," as their term is, to the ruin of the merchant and manufacturer, and to the prejudice of the revenue. By all these means they reap the profits of smuggling without incurring any of its risks.

Mr. Brougham, indeed, has flattered them, that though these enterprises are desperate as regards the continent of Europe, where the merchants will not pay, that the American merchants will pay; and these practices of glutting and destroying may be safely adventured against them. Mr. Brougham could not have known that our merchants were already reeling under their balance-sheets of foreign commerce, uncertain whether the next assault of the unsteady element, on which they ride, may not send them to the abyss of ruin.

It is no time for jealousies between farmer, merchant, and manufacturer; one common bond of interest and patriotism unites them now. Let the government take advantage of this propitious crisis, stand firmly to its post and do its duty, as we trust it will; confidence will soon revive, capital be vested, machines improved, competition will bring our own goods to market at a reasonable price, and prevent those exactions which some affect to anticipate on the exclusion of foreign manufactures. On the other hand, if the foreign importations are ever again relied on as the means of revenue, what can ensue but a repetition of those vexatious embarrassments which our government experienced during the war, and which it cost the best blood of our country to surmount.

If it clearly now appears, that Europe will not take from us the produce of our soil upon terms consistent with our interest, the natural remedy is to contract as far as possible our want of her productions. And if there be no other way to independence than that of manufacturing for ourselves, at least for our own consumption, it is hoped that the prejudice against *home* is not so strong in the mind of any American, but that it may be overcome.

The encouragement, besides, of domestic manufactures will increase the capital of the country as the manufactured article exceeds the value of the first material; sometimes one hundred fold, without speaking of the saving of all extra charges of shipping and reshipping, increasing in proportion the value of the land, and easing the landholder of his burden in supporting the expenses of the Government. It has been exultingly asserted by a great statistical writer in England, that one man in a factory maintains four soldiers, and one steam engine subsidizes three hundred German mercenaries.

Having discussed the various topics of argument, as far as the time allotted to our labour would permit, we shall set forth the titles upon which we presume to solicit universal co-operation.

In the first place, we can safely affirm, that our society is not the diminutive offspring of selfish or party combination, nor the foundling of accidental caprice. It is the legitimate birth of circumstance and occasion, and has burst forth into existence spontaneously and full grown, like the Goddess of Wisdom from the brain of the great progenitor; for it is the child of mighty and irresistible necessity.

Its object is to give to national industry the impulse it is susceptible of, by all the means within our power, and to endeavour to discover what helps it most needs. We must solicit the patronage of an enlightened public, and the protection of a wise government. We must rescue opinion from the dominion of prejudice, and enlist in our ranks genius, knowledge, and experience. Our activity must depend less on the feelings of private interest than the more exalted sentiment of love of country. But when individual interest is blended with the general good, why should it not prosper?—how can it but succeed?

We must aim at acquiring extensive knowledge of all

useful facts that have relation to our subject; the power of generalizing will follow as of course. The artificer and philosopher must combine their efforts, and theory walk by the side of practice. Useful knowledge will thus be acquired and disseminated, like rays converged to one focus, and reflected wherever their application may be wanted. The head that conceives, will soon find the hand that can execute, and nothing of the stock of intellect will go to loss. Inventions already known will be improved, and their use rendered easy and familiar. All the powers of inquiry, experiment, and combination, will be in full activity. The embryo conception will not be chilled by neglect, but, cheered by timely attention, will exceed the hopes of the projector himself. If we have not a treasury to dispense pecuniary recompenses, yet, there are rewards more grateful to genius, because more worthy of its acceptance; and the most animating of all rewards to a free and noble heart will be the civic crown.

Our proceedings must be so squared with the public good as to be no more than echoes of the public wants and wishes. Servile fashion, and all the baleful prejudices that dedicate to foreign productions the tribute of their devotion, must fly before the majesty of the public voice, and the pride of national character rise on the ruins of prejudice.

Let nothing, then, check our onward march, nor the vigour of our efforts. Let genius and patriotism, from whatever quarter of the earth, be naturalized amongst us, and nothing be exotic in this generous republic that blooms and bears good fruit.

And we now respectfully invite our fellow-citizens throughout the union, to unite with us in this great national concern, to establish societies with as much promptitude as possible, and to correspond with us, and with each other. Such diversified and rapid communication will bring important truths to light, dispel prejudice, refute sophistry, excite patriotism, cherish industry, and, above all, give to public opinion that expansive swell that will harmonize with the rising tide of our country's prosperity.

It is not to one class, nor to one interest, that we address ourselves, but to the whole and each respectively.

We call on our manufacturing brethren, and artists of every description, to communicate directly, or through the medium of some affiliated society, all such facts or information as may be subservient to the prosperity of domestic manufactures in general, or of any in particular.

And you, agriculturists, owners and possessors of the soil, the standing pillars of your nation's independence, we conjure you, for yourselves, and for your country, to second us by all your energies. Explore, with new activity, and determine, by new inquiries, the nature and productions of your estates, and the adjoining territories. Every view, statistical, economical, geological, or topographical, is connected with this great national concern. You may find that you have been unconsciously walking upon hidden treasures, richer than the mines of Golconda. The three kingdoms of nature may have been long tendering to your acceptance the willing tribute which you have heedlessly disregarded. Who can have so much interest as you in the opening of canals and roads, the increase of national industry and capital, with all its ramifications, which must reach you like irrigating streams of living waters, and enhance the value of your possessions? The great improvements that must follow in the train of national industry, are too far beyond ordinary calculations to be readily conceived. You will have, not one, but a choice of markets for your produce, of which wars, blockades, or the casualties of foreign nations, cannot deprive you. You will have speedy returns of whatever you may want, and your approximation to the mart of exchange will put it in your power to be the comptrollers of your own fortunes, and the arbiters of your own concerns. Our southern agricultural brethren, in particular, would do well to reflect that Great Britain is now, and has been for some time, creating new sources for a supply of cotton, by encouraging its culture in India, on the Coromandel and Malabar coasts, Africa, Brazil, and other places; and will shortly render herself independent of any supply from this country, and probably prohibit the importation of American cotton into her market. When this event, which is not far distant, shall take place, you will be destitute of a vent for your cotton, unless a market can be found in our own country, by the establishment of domestic manufactures.

To you, merchants, now sinking by these foreign importations to ruin and bankruptcy, we appeal; by your dearest interests, and those of your country, we conjure you to contribute all the power of your intelligence and enterprise, and to aid in counteracting those frauds upon yourselves and the revenue, of which you, your fellow-citizens, and the government, are common victims. A new and unforeseen crisis has put an end to those delusions which heretofore arrayed agriculture and commerce against domestic manufactures. It is now demonstrated, that whatever adds prosperity to either of these modes of industry is beneficial to them all.

And of you, sons of science, who possess the rich treasures of cultivated intellect, and can teach their application to the useful arts of life, we claim the lights you can shed on this great subject. Too many of your former important communications have been lost to the public, from the inauspicious times in which they appeared, and have perished like seed sown by the way side. We entreat you to come forth anew in the pride of intellectual vigour, to break the spell of ignorance, and emancipate the genius of your country.

You who redeemed your fellow citizens from the barbarian's yoke and foreign captivity, who, mingling the battle's thunder with the cataract's roar, made Niagara's falls the eternal record of the well-fought field; and you, citizen soldiers, who re-echoed victory where Mississippi rolls her latest waves along—we invite you to participate in our civic triumphs. If your country's cause should call you forth hereafter, you will go girded with swords of native steel; and the arms you wield will be committed to you by the hands of your affectionate countrymen.

And you, fair daughters of Columbia, whose sway is most ascendant when the hearts of freemen do you homage, assert your dignity; disdain the fashions of foreign climes; let not the daughters of Belgium, Austria, or Russia, exceed, in patriotism, the free-born fair; let your dress be national; let your ornaments be of your country's fabric, and exercise your independent taste in suiting the array of your toilet to your own climate and your own seasons. You do not vote in the counsels of your nation, but your empire is everywhere where man is civilized. Let the power of beauty add impulse to the springing fortunes of the land which you adorn; and let the charms of your persons be ever associated with your country's love.

With this view of the past and present we might conclude; but, may we not look forward with anticipated delight to the prospect that bursts upon our sense: not through the vista of a long perspective, but which our children may enjoy in all its splendour; when a territory, vast as the European continent, shall pour its riches forth; when the protecting shade of equal laws, and the misery of another hemisphere shall have increased our population to the measure of our wide domain; when the genius of the republic, towering like the eagle on the Appalachian heights, shall, looking from the proud summit to either ocean's wave, survey the wealth of every soil, the fruit of every clime. Where the bear roams, and the wildcat prowls, flocks and herds shall pasture, and the savage's dreary repair out-bloom the gardens of Hesperia. There cities, towns, and villages, centres of intersecting orbits through which domestic commerce will revolve, shall rise and flourish. And whilst the plough shall trace the silent furrow, the mill shall turn, the anvil ring, and the merry shuttle dance. The exhaustless stores of mind and matter shall be this nation's treasury. Adventurous man, triumphing over the obstacles of nature, shall search the recesses of the stubborn mountain. The sounding tools, and the voice of human speech shall wake the echo in the vaulted space, where, from the beginning, silence and darkness reigned; and the rich ore shall quit its hidden bed, and sparkle in the upper day. Innumerable communications, by land and by water, shall bear, in all directions, the native produce of the soil and of its industry. Majestic rivers, enriched by their tributary streams, shall waft on their smooth tide the treasures of teeming abundance. And those proud cars to which magic genius has yoked the discordant elements of fire and flood, shortening the distance of time and space, shall stem the mighty current. The immeasurable coasts, with all their bays and inlets, shall invite the mariner to commerce, or beckon him to shelter from the storm. Those inland seas, memorable by the victories of freemen, the classic scenes of future Muses, shall be studded with barks which national industry has set in motion; the white canvass swelling to the breeze, the ensign of freedom waving to the sky. One people, one tongue, one spirit, grappled by ten thousand relations of interest or affinity—what factious demagogue, what ambitious usurper, will then find a spot to insert the wedge to sever such a union? A thousand heartstrings must be rent before the smallest member can be separated.

Let the world, then, in arms, assail this great Republic.

Like a proud promontory, whose base is in the deep, whose summit strikes the clouds; the storms of fate may smite upon its breast, the fretful ocean surge upon its base; it will remain unshaken, unimpaired—type of duration—emblem of eternity!

And who is he that is not proud of such a country—jealous of its prosperity? Who would be thought the subject of a king that could boast the title of citizen of this Republic?—countryman of Franklin and Fulton—child of Washington!

Signed,

THOMAS MORRIS,
SAMUEL L. MITCHILL,
ARTHUR W. MAGILL,
WILLIAM SAMPSON,
JONATHAN LITTLE,
THOMAS HERTTELL,
JAMES ROBERTSON,
THADDEUS B. WAKEMAN,
ISAAC PIERSON,
J. R. B. RODGERS,
EDWARD P. LIVINGSTON,
} Committee of Correspondence.

On Motion, Resolved, That the foregoing address be approved, and that the Corresponding Committee cause 5,000 copies to be printed; and that they transmit a copy to the President of the United States, to each of the members of Congress and Heads of Departments of the General Government, and to the Governor and Members of the Legislature of the States respectively.

DANIEL D. TOMPKINS, President.
STEPHEN VAN RENSSELAER, First Vice-President.
WILLIAM FEW, Second Vice-President.
JOHN FERGUSON, Third Vice-President.

DOMINICK LYNCH, Jun.
and
PETER H. SCHENCK,
} Secretaries.

N. B. Communications to the Society will be addressed to any of the members of the Corresponding Committee.

1. [similes]

2. Robert Southey, *Letters from England by Don Manuel Alvarez Espriella* (London, 1807). See excerpt reprinted in this collection.

3. Patrick Colquhoun, *A Treatise on Indigence, exhibiting a general view of the national resources for production labor* . . . (London, 1806).

4. Isaac Briggs, "Statement and Remarks . . . on the Subject of Agriculture, manufactures and Commerce," *Niles' Weekly Register*, vol. 9, no. 23 (February 3, 1816), p. 389.

5. Henry Peter Brougham, *An Inquiry into the Colonial Policy of the European Powers* (Edinburgh: D. Willison for E. Balfour, Manners and Miller, 1803).

James Swan

Address on the Question for an Inquiry into the State of Agriculture, Manufactures and Commerce (1817)

This address is one of the most vigorous anti-industrial writings of its time. Swan argues that manufactures are undemocratic, morally repulsive, and less profitable than agriculture; in short, "manufactures are a monstrous evil." Swan would permit only industries that were absolutely essential to the farmer, and these only in small units.

It is difficult to determine why Swan (1754–1831) was so opposed to manufactures. He was a Massachusetts Revolutionary War hero who spent most of his life in France. He made and lost several fortunes in commerce and land and spent many years in debtor's prison, where he wrote this address. Swan was also the author of several other pamphlets, mostly on political economy.

Colonel [James] Swan, An Address to the Senate and House of Representatives of the United States, on the Question for the Inquiry into the State of Agriculture, Manufactures and Commerce (Boston, 1817), 24 pp.

To the Senate and House of Representatives of the United States of America.

———

An inquiry into the state of the agriculture, manufactures and commerce of the United States, will no doubt occupy your attention in the course of the present session. It merits, and no doubt will receive the most steady, dispassionate and grave investigation; it demands that you should fully comprehend the peculiar nature and situation of our country, her wants and her advantages, her domestic and foreign relations. My weak efforts to that end, are respectfully submitted.

That the dignity and safety of the states depend upon the social comforts and happiness of the people, is more applicable to America, than to any other country, because there is not and there never was another nation which had not a long period of infancy; no one was ever in her position,—young in existence, yet ripe in intelligence, of an immense extent, and not one twentieth populated.

The light and information to be found in histories of the policy of ancient states, can very little contribute to the profit of America; because her condition is totally different from any, of which we have written or traditional accounts.

The writings of the ancients, it is true, contain many precepts and principles of liberty: but their habit of life, from the mixture of liberty and slavery; from the intrigue and turbulence of faction, and from the thirst of conquest and dominion, should serve only to teach us to repress those passions which are too much disposed to intoxicate the mind by enjoyments. The writings and habits of the ancients thus considered, may lead us to that tranquil and instructive philosophy, and to that practice of humanity so forcibly inculcated by our religion.

Our country I contend, is unique in her position, in her nature, and in all her relations, and totally different from all states that ever existed, or do exist at present: her like is not to be found either amongst the ancient or the modern nations. I speak thus generally, because speaking to men of information it is not necessary that I should describe the rise, progress or fall of the ancient nations, nor the rise, progress or present state of the nations which now exist, in order to compare them with America; for you gentlemen, can retrace those particulars in your own minds.

America is not a small country, she is not a small power, she is not a country to be protected by artifice, she does not require foreign possessions to support her: England cannot say this; she is a small country, her political power and dominion is the mere effect of artifice; therefore, she is, if I may use the term, an artificial nation, for she cannot exist as a great nation without foreign possessions, and she has wisely pursued colonization in places advantageous to her nature and her wants: but America has no such wants, possessing at home, all which England searches for abroad; therefore she is totally different from England.

America is a very fine country, of an immense, almost of an incalculable extent, embracing every climate in which the sugar cane, and lofty pine, receive their growth and acquire their maturity. The far greater portion of her domains remain still in a state of primitive nature, uncultivated and unprofitable, where the foot of the husbandman has never yet trod, although a land that invites the hand of the labourer, whose industry would make her flourish with every abundance desirable by man; a land that would give support and happiness to millions and millions of families to the latest posterity.

This gift, this bounty, this felicity to the human race, rest at the disposal of the United States of America: it is a deposit from Providence confided to their hands for the benefit of mankind; and it is the duty of America to discharge that trust with care, with activity, and with integrity. The trust is immense, the performance of the obligation is imperative on the government; and I hope presently to show, that the performance of that obligation claims a decided preference to either the acquisition of political power, or individual riches, and that the very prosperity of the country depends on the performance of it.—Let it not be supposed at the same time, that I mean to depreciate either political power, or individual riches: on the contrary, I hope to be able to shew, that, by the due performance of the trust confided to the American States, her political power will most rapidly, but silently grow to a most collosal strength ; and I contend, that whatever goes to impede the performance of the duty the United States owe to mankind, goes also to impede the real and permanent aggrandizement of her political power. And as to individual riches, I shall not be an enemy to it, if it shall be obtained in the progress of the great work; for it then will be gradual, and be surrounded in a great degree by the wealth of other persons: but I am decidedly the enemy of it when it is obtained in any pursuit injurious to this great national object; for it then not only becomes equally injurious to the national prosperity by the pursuit, but it may be too hastily obtained whilst surrounded by comparative poverty, which will give the rich a dangerous desire of power; and his ambition may do mischief even to the constitution of the country.

If it should be asked what it is that I consider as the primary duty of government?—I answer without reserve —to give every possible encouragement to the clearing and cultivating of American lands, be it by grants at long payments without interest, to real and active cultivators ; by premiums in lands, to such persons as shall have cleared and cultivated the greatest quantity, in a given time, and with an equal number of hands; by bounties to all manufactures which tend to consume the wood on our lands ; such as forges of iron, glass-works, &c.; and by discouraging all pursuits which shall tend to impede the cultivation of lands. Such encouragement, would not only stimulate the energies of the present people of America, but would draw in to it, millions of the inhabitants of Europe, who would become active and profitable subjects of the states.

It will not be denied by any one—for the proof and evidence is in our countrymen, that rural life produces a healthy and robust race of men, whose progeny rapidly increase, and will be equally healthy and robust: a being very different from the puny and decrepit city mechanic; and as fifty or sixty years, is but a very short space of time in the age of an empire; and as that time would commence the third generation of even the present children, giving twenty-one years before each marriage, the increase of the American population would be immense even in that time, and America having become numerous in a robust population, her aspect alone would be such a political power, that it would of itself bear an influence and a sway in the cabinets of Europe, more commanding than either fleets or armies, and carry an influence that would make every nation in the East court her friendship and alliance, and fear to give her offence. What nation, but the American has been composed of freemen, of whom nine in ten are proprietors, or sons of proprietors, of the soil? and what nation dare quarrel with such a people?—It is therefore with the view of the lands being inhabited by every possible facility and encouragement that I endeavor to draw your attention; and although the wit may say, that I am talking of the work of centuries, I shall not be discouraged, and, I trust you will not be by such a remark, as the progress, of even half a century, is more than he can calculate, when the American mind and her resources are directed to the point.

It has been advanced by persons whom I respect and esteem, that it will be extremely beneficial to America to give every possible encouragement to manufactures, so as to prevent the necessity of importing worked goods from abroad. I am well aware of the flattery of this proposition, in support of which we hear so much of the immense advantages that manufactures have been to Great Britain, the great source of her wealth, and of her high political character.

I have no difficulty in admitting what is said of England in regard to her manufactures; and I would even go farther and say, that they are her sole dependence: but I deny, and I deny it most firmly, that manufactures are desirable for America, because they are so essentially useful for England. The two countries are totally different, and the law of nature and of society demand that their pursuits should be equally different; and I hope I shall be able fully to convince you of that most essential difference, and that what in one country is laudable from necessity, would in the other be a gross error.

Every nation governed by a wise policy will encourage that sort of industry which is permanent, which is immovable, and which tends to procure subsistence to the people. Agricultural industry is permanent, because every season assures returns equal to the labour given. To realize the produce of that labour depends not on fancy and taste, as in manufactures, (which one year are the mode and sold with rapidity, whereas the following year no one will even look at them): the existence of man calls for its consumption; and whether the crop be great or small, the revenue is nearly equal: for if great, the price lowers, if small, the value rises.—Agricultural industry is immovable: it holds to the soil and cannot be transported abroad: hence the durable riches of that nation, where agriculture is the principal occupation of the people; whereas riches placed in, or arising from manufactures, are personal to the proprietors and form not the real riches of the nation; for if a manufacturer in England, for instance, (and we see it daily) be assured of greater profits in France or in Germany, than those he has in England, he sells off his stock and transports himself and his industry thence. Manufactures, then, form not national riches, but agriculture does. It is then for the interest of America, to encourage the latter, and to do nothing to promote the former.

England being a small country, an island, and seeking to maintain a rank amongst nations, incompatible with her extent of territory, requires other means than territory to support that rank;—in consequence her policy has been directed to maritime affairs, commerce and manufactures. To further these objects, she has obtained possession of different points on the globe, as places of rendezvous, and to give vent to her manufactures: she is therefore maritime and commercial of necessity: America is not so. England is surcharged with inhabitants, whilst America is in want of population. It is an invariable truth, that no country can wisely have manufactures, until her lands be peopled, and there be a surplus of population, who seek for employment in manufactures, for want of it in agriculture.

England has long since become a monarchy, and has long supported all the pomp and pageantry of ancient monarchies. America is a young republic of aggregate states, where pomp and ostentation is unsuitable, as it is unnecessary to the character of America and to the real dignity of man. She should be guided and governed by the plain and simple principles of rectitude and justice.

I have as much respect for England as any of you, gentlemen, and I give her every credit for the great merit she has possessed for centuries, especially for that extraordinary stand during twenty years, against the most atrocious usurpations and crimes, that ever blackened the page of history. When liberty was banished from Greece and from Rome, and had no resting place on the continent of Europe, Providence seems to have directed her steps to Britain, where has been consecrated to her, a temple, that has preserved her sacred fire: that fire, that illumined America, and which seems to begin to illuminate some parts of Europe. I think it a glory to say, generally speaking, that we Americans have sprung from that free people, and that doubtless the principles of freedom preserved in Britain, engendered the principles of freedom which exist in America: a freedom, which I hope, the sacred bond of our constitution, the wisdom of our legislators, and the virtue of our people, will preserve till time shall be no more.

With the constitution of England, I have nothing to do. It may be the best for her under her circumstances, and perhaps the people are as happy as the situation of the country may admit of. I am not going to look into their situation, in that respect: but I think it my duty to expose to you, what effect manufactures have upon the people, because the same effect is inseparable from manufactures wherever they exist.

Manufactures have existed so long in England, that she is enured to the habit, and has long lost her moral sensibility of the human sacrifice she offers daily to her manufactures. Habit has so blinded her, that she does not see it: but let a stranger, let an American, view the frightful picture within a Manchester manufactory. Men, women and children grouped together, breathing foul and unwholesome air, in unhealthy and cramped postures, with a species of exercise, that instead of animating the human frame, tends to give it a fixed and determined distortion; and thus they remain for 14, 16, or 18 hours a day, and every day all the year round, Sundays excepted, and sometimes not excepted, and not only all the year round, but all their lives, which renders them miserable objects, and entails that misery upon their posterity: their children, their grand-children, are a race of pitiful objects, barely a race of men, deformed and withered before they arrive at the years of puberty. Pregnant women in those manufactories have the foetuses even distorted in the womb, by means of their miserable occupation. It is not possible for the mind to take fully into its consideration the whole of the sufferings of the people employed in different manufactories in England: their wretched appearance shows more than language can describe.

It will not be mal-apropos to quote here, what Lady Morgan says in her *France*, lately published, when speaking of the happiness of rural life compared to that of manufacturer—" One turns their eyes with disgust and
" pity from that pale and lean population, who trains their
" existence amidst the unhealthy vapours of a mine, or
" who sickly languishes within the walls of a manufacture,
" and who degraded in moral and physic, pass their lives
" between the extremes of want and intemperance"—and by contrast, she adds—" But can one see, without a sen-
" timent of envy, a country of which the inhabitants are
" invited by a rich soil, to give to its culture all their en-
" ergy? There is in the pastoral life a perpetual moving
" scene, something picturesque and very particularly in-
" teresting. Grazing countries present to the heart of a
" philanthrope, images, much more attractive for him,
" than all the details of commerce and manufactures."—
Speaking of the peasants in France, become proprietors in the revolution, she says—" Each peasant has some cat-
" tle ; there are few who have not a cow, and many add
" to that a hog, a mule, and an ass, according to their

"means. One may well think that amongst these little
" proprietors, there are many who have not lands enough
" to cultivate so as to employ a plough and oxen: but of-
" ten a certain number of neighbours unite, and have one
" in common. After all, that man is a proprietor, he is
" independent; the little field he cultivates is his; it is
" for himself that he sows it, and the produce belongs to
" him. His children eat the fruits of the tree which he
" planted, and that little portion of land maintains him and
" his family in independence." Who can hear of such a
scene of happiness and independence from such small
means, and not congratulate our countrymen on that
which they possess, I mean those who follow agriculture?
and what regrets must we feel to think that thousands employed in manufactures in the United States, are not only deprived of such blessings, but that they are from habitude, doomed to the miserable end that such find in England?

You will naturally ask, if manufacturing has such a dreadful pernicious tendency, how comes it that people apply themselves to such occupations? Gentlemen, I will tell you: the origin of the manufacturer, is generally from the work-house; if he himself did not come from it, his father did, and if his father did not, then almost to a certainty, his grand-father did. And the women in the same manner: for it is the constant practice to send children from the parish work-house in scores, to be apprenticed at these grand manufactories; and when once a child gets into the trade, he is so tied to it, and is so ignorant of every other occupation, that he is fit for that employment only: hence, he must not only stay in it himself, but he must also bring up all his children to follow the same work; and the women in like manner. Thus the poor creatures and their progeny, have the curse of manufacturing regularly entailed upon them.

That the habits of manufacturing bring on disease, misery and immorality, will hardly be denied: but if it should, look at the hospitals in every manufacturing town; for an hospital must always be built close to a manufactory as a necessary appendage, to relieve as much as possible, the certain and unavoidable evil; look at a late publication in London, wherein it is calculated, from the number imprisoned for crimes, that at Manchester there were 1 in 140; in London, 1 in 800; in Ireland, 1 in 1600, and in Scotland, 1 in 20,000. This is caused by a difference in education, and occupation, in the people. Manchester, a manufacturing town, gives the greatest number. These calculations were on the average of nine years.

England being a mere commercial and manufacturing country, is subject to variations in her domestic economy, according to the degree of prosperity or depression of foreign markets. The continental nations, after a desolating war of 20 years, found their pecuniary sources dried, and their specie in the pockets of their invaders. America, after her late war, was hardly more at her ease. England, pressed to find a sale for her merchandize, overwhelmed all countries with her goods: no returns could be made; the manufacturers from necessity stopped working, and drew with it the most afflictive distress ever known in any country. I shall give some particulars, taken from proceedings in the British parliament.

* [Swan here quotes several pages of statistics from proceedings of the British parliament enumerating the afflictions of industrial workers in England.—Eds.]

Can any honest American wish to have these scenes exhibited in his country?—If he does support manufactures, they certainly will happen, for all these horrors are only found amongst the unfortunate people who depended on manufactures for subsistence. If manufactures continue to be encouraged amongst us, such evils as are just recited, are as necessarily attached to them, as the shadow is to the body; and if that part of our population now employed in manufactures continues in them, they will as certainly meet the horrors the English feel, as that the sun is vertical at noon. The interest of the British manufacturers, and that of the government, urge the ruin of every fabrick in the world, which interferes with theirs, and the glow of wealth, which warms some of our manufacturers will ere long be changed into the cold sweat of certain ruin, or expiring prosperity.

It is well known that in England individuals have made large fortunes by manufactures; that they have become legislators, and statesmen, by means of such fortunes, and that government have had the advantage of their talents. America has need of no such recourse to obtain wise and able legislators or statesmen. Can any man tell me how many victims have been sacrificed to raise such statesmen in England? How many men, women and children have had their lives abridged, themselves and posterity deformed, to make one plump secretary of state for Ireland, from the fruits of their unhealthy labours?—In plain terms, from the fruits of their slavery? I am aware that I may be legally answered, that work performed for hire, and which a man has the liberty either to follow or quit, is not slavery: But I am not speaking of slavery in a strict technical sense,—I am speaking to the heart of man, to his feelings, in a moral and humane sense; for what is the situation of the helpless parish boy when he is put to be a manufacturer? (or a girl either?) Can he quit his occupation when his apprenticeship is served? Certainly not. He is wedded to it for life, for he knows no other pursuit, his spirit is bowed to it; he thinks his lot is hard, but thinks he is bound to bear it as he sees others do: hence he follows this debasing drudgery and entails this slavish servitude on his issue, if he has any. But as the laboring manufacturers do not provide sufficient successors by their children, the parish work-houses again supply the deficit of hands, for those killed off in the service. Thus the scene is replenished, and the work obtains its perpetual motion, under the direction and tyranny of a proprietor, who, if he has not the disposal of the life of the slavish workmen, has at least that power over him, to give a good living, or a pittance, less than is allowed to the negro slave; or if he has not employment, leaves him and his family to starve, or go on the public charity for subsistence: a circumstance not in the power of the proprietor of the slave, who is obliged to maintain him in health and in sickness.

I aver it, and I aver it with confidence, that manufactures are a monstrous evil; that they destroy man, and subject a great portion of the human race to misery and wretchedness. Now let us see what is the proportion of benefit derived to any one from this sacrifice? Suppose that all the money which any one master manufacturer has amassed, were to be divided in equal shares between those who have been sacrificed in his employ, killed off or maimed in his service, and their imbecile children, without going further down the degenerating race, I doubt extremely that the dividend would be sufficient even to buy a small hut and a garden. Then if all were to have a share, what a pitiful employ it is at last: Whereas in America, the industrious man, can by a few years savings from common labor, purchase a small farm and even stock it, and make a permanent and improving provision for a rising and thriving generation. I ask, therefore, is it possible for any man possessing the intellect, the humanity, the integrity of a man of mind, to wish to establish manufactures in America, when he knows that to encourage the cultivation of our uncultivated soil, is to bestow a blessing on mankind, to strengthen and give dignity to our nation; and that to introduce manufactures is, to destroy our population, to render them objects of misery, and debase the character of man and of the empire. And for what purpose? Why for no purpose on earth excepting that of making disproportionally rich, some few individuals, who shall have the means to enjoy gorgeous palaces, and pampered pageantry surrounded by indigence and misery. This scene may do in England. The individual so enriched, is there on the road to an ennobled title, and enters the arcana of the oligarchy: The necessary sacrifice has been offered, and he has his pretensions: but I should blush to be the American who would wish our country to tread in the footsteps of England, either in the encouragement of manufactures or in amassing large fortunes by such means.

Let me for a moment draw your attention to the produce of the labor of a man employed in a manufactory, and that of the labor of a cultivator of the ground:

* [Swan goes to some lengths to demonstrate with highly hypothetical statistics that, dollar for dollar and day for day, agriculture produces more wealth than manufactures.—Eds.]

And here I would ask, is there any one inducement arising from the actual situation of America which demands or can even tolerate that America should adopt and follow pursuits similar to those of England? I confess I do not see one: on the contrary, I think that every principle of sound policy, every principle of humanity, every principle of justice, the love of our fellow creatures and the prosperity of our country, calls upon America to avoid the occupations that England is forced to follow, and which have thrown her into such unparalleled distress, and to turn her energy and resources to very different objects. Her large and unoccupied domains demand of her the preservation of the life of every inhabitant of the United States, by every means possible. It will be time for America to turn to those pursuits destructive of the life of man, when, as in England, when every acre of her soil is cultivated, or comparatively so; when she has a surplus of population, as in England, then manufactures may become necessary to her, to waste her surplus population, as is now done in England, but not before that time: for before that time, such pursuits by impeding the population, will destroy the first principle of duty, that America owes to man and to herself.

If it be asked: shall we not erect manufactories so as to manufacture sufficient for our home consumption?—I answer—no, with a few exceptions to be noticed hereafter; and that, because no country can profitably give into manufactures, whilst her lands are unpeopled, agriculture being the most rational, the most useful, and the most durable of all industry. We are not in need of manufactures at home, since we can draw all we want from England, the work of the enslaved poor, and we pay for them by the produce of our soil.

You gentlemen, should use all means to avoid even the introduction of the manufacturing spirit, for it is destructive; it has the allurements of gain for the masters: that being once introduced, it cannot be controlled, and there is no corrective for it: for speculators will rise at every point, and lead the fabricants by temptation, who, ignorant of the consequence will commit themselves, and who will find, that they have adopted an occupation, which enervates mind, and body, and shortens existence; but having once entered upon it, they become unfit for any other occupation.

But to go a little more minutely into this part of the inquiry, I would ask, where is the necessity to manufacture for our home consumption, when England has such a surplus of manufactures, and when we have a surplus of our natural produce which England wants, or which other countries want? We have tobacco, cottons, sugars, corn, flour, rice, and endless other articles the produce of our soil, which other countries have not, or which they want more or less; and we want in return, and can easily have the produce of their destructive labor—manufactured goods.

This reciprocity forms for us sufficient commerce, supplies all our wants, gets rid of all our surplus produce. And why is it not sufficient for us to follow a plain, simple and useful course? Nothing but vanity and folly can lead America to tread in the steps of the old European states, or to imitate them either in their occupations or their policy. It is a certain fact, that many of them make wars for no purpose on earth, but to carry off a surplus population, and others their surplus manufactured goods. The true policy of America, on the contrary, is founded on morality, on humanity and justice. It is her bounden duty to encourage and increase her population by every possible means in her power, and her means are immense. If properly applied, her dignity of character will rise in proportion, and with a surprising rapidity.

We have an immensely extensive territory, which wants hands to work it: Encourage then population by discouraging manufactures,—the untimely grave of those who labor in them. Territory and population form the real riches of nations. Encourage agriculture and we shall soon become very powerful by the cheapness of provisions which that agriculture will produce; for where the necessaries of life are cheapest, there population abounds most.

No manufactures ought to be encouraged with us but such as are of general use amongst, and necessary to, the people in the country:—for our seaport towns have but a small weight in the scale of national consideration when it is a question of what is fitting to our country at large—and these consist of leather, shoes and boots, cordage, sail-cloth, iron, and all instruments and utensils of iron for country use, earthen wares, window glass and glass bottles, and every thing in the manufactory of which wood and charcoal are used, as that clears and pays for clearing our lands; the domestic manufactures, (I mean those made by the farmers' families in the winter, when agriculture is suspended) such as coarse linen, woollen and cotton cloths, and stockings. And to encourage effectually that domestic industry, premiums might be granted to such families, as with the same number of hands shall have manufactured the greatest quantity in a given time.

These domestic manufactures are so much the more to be encouraged, as such is the havoc and destruction of man that follows in the train of large manufactures, that it is impossible for me to give you the full particulars.

The evils attending men who follow commerce, are not few, or of little weight: the anxious cares, the fatigues of mind, the failures, bankruptcies and imprisoned debtors, and the thousand other calamities which appertain to commerce, rack and distract the mind, and give an existence of wretched misery, and shorten the period of life. America, if she is wise and faithful to herself, may be exempt from all these calamities.

I am sufficiently aware that my observations will be very far from being agreeable to the maritime towns, because of their commercial habits: but I flatter myself, that there is not one evil, that I have endeavored to describe, or one fact that I have endeavored to establish, that the well informed man of commerce would venture to deny, however contrary it might be to the interest of his pursuits.

But I do not seek to disturb any of the occupations of the present men of commerce. I ask no prohibitory law; I seek no coercive enactment: I have no right to do so; every pursuit should remain as free as air. I seek only to convince my countrymen, to convince the public of what is useful and laudable to pursue. I seek to convince the Federal Legislature, to convince the government and the people at large, of facts, namely: of the destructive nature of manufactures and a too extended commerce, and I endeavor, with every humble deference to do this, in order that the public mind and the energies of the empire, may be directed to those pursuits, and to those alone, which are truly laudable, and which are to the real interest and benefit of the Republic.

My opinion is, that the policy of Great Britain will be such as to destroy all manufactures in every country, (especially in ours, who are her best customers,) that interfere with hers. Examples, if I may believe what I read in our Gazettes, are not unfrequent, in which manufactures are abandoned, others sold for half what they cost, and that almost all are fast declining: the evil then will cure itself: But that the government may express its sentiments on the subject of this address, and in order to prevent future losses, I hope some patriotic member will move the following Resolution,—namely:

" That it is the opinion of this House, that to encour-
" age the establishment of manufactories in this country,
" would be injurious to the welfare and dignity of the
" United States, and would endanger the peace and fu-
" ture happiness of the people."

I am with great respect,
  your's and the public's devoted servant,
    JAMES SWAN, *of Boston.*
Paris, September 8, 1817.

1. Very loosely quoted from Lady Sydney Morgan, *France*, 2 vols. (London 1817), vol. I, pp. 41–45.

The Philosophy of Manufactures

John Taylor of Caroline

Tyranny Unmasked (1822)

John Taylor (1753–1824) was a Jeffersonian Democrat and a defender of the agrarian order. He opposed the increasing power of the federal government and thus opposed the banking and tariff systems suggested by those supporting manufactures. His efforts to champion the farmer were both political and practical. Like several other supporters of agrarianism he promoted a system of scientific agriculture.

Taylor was the author of several books and pamphlets. His most important book was *An Inquiry into the Principles and Policy of the United States of America* (1814), a condemnation of strong central government. *Arator* (1813) included a defense of agrarian life and suggestions for scientific agriculture. The subjects of his pamphlets ranged from banking to the Missouri Compromise to the protective tariff.

The protective tariff is discussed in *Tyranny Unmasked.* This pamphlet, written in response to the 1821 Congressional Report of the Committee on Manufactures, attacks point by point that committee's recommendation of a strong protective tariff. Taylor rambles through some two hundred fifty pages, arguing that the tariff is "unconstitutional; injurious to morals, and productive of pauperism; improper to be extended; a tax on the many and a bounty to the few; a restrictive system; a destroyer of revenue; ruinous to commerce; and destructive to agriculture." Reprinted here are two of the objections. The first

---

John Taylor, *Tyranny Unmasked* (Washington City, 1822), pp. 138–141, 194–199. The report of the Committee on Manufactures can be found in *American State Papers,* Class III, Finance, volume 3, pp. 594–624. See also the dissent to this report, *Objections to an Increase of Duties on Imports,* issued by the Congressional Committee on Agriculture, volume 3, pp. 650–660.

discusses how manufactures affect morals; it is given in its entirety. The second objection deals with the economic problems agriculture will face if manufactures are protected. The excerpt of the second includes the unusual argument that it is in the joint interest of laborers and farmers to oppose the tariff.

Manufactures are injurious to morals, and produce pauperism.

This the Committee deny : and, to sustain their denial, reject the evidence of the great foreign factories, and rely on that of the Waltham factory, consisting of two hundred and sixty persons. I shall not attempt to prove that this little experiment is less to be relied on, than those made on a great scale, nor to overhaul the fact and opinions coinciding in the conclusion, that these factories degrade human nature. But leaving to the Committee all their arithmetick for estimating the thefts of the poor, it is yet necessary to remind them, that in wandering through its mazes, they have entirely overlooked political immorality, by which vices more pernicious to society are produced, and which also causes many of those peccadillos, admitted by them, and allowed by me to be bad enough. Laws for creating exclusive privileges and monopolies corrupt governments, interests, and individuals ; and substitute patronage, adulation, and favour, for industry, as the road to wealth. If it be true, as the Committee believe, that the preferences and partialities of such laws, will not produce a correspondent impoverishment, which will reach the poor and deteriorate their morals ; yet it cannot be denied that they will reach the rich, and corrupt the morals of the best informed, and of the officers of the government; in which three classes reside, the power and the influence, by which the morality and the liberty of nations are sustained or destroyed.

As to pauperism, the Committee quaintly contend, that it is not produced by hard labour. Daily wages earned by hard labour, do not prevent it. One of these general assertions balances the other, and they unite in showing how little is proved by either ; and neither can diminish the force of the fact, that pauperism and crimes are more frequently produced by hard labour for daily wages, than from any other source ; because it usually expends the wages of to-day in the subsistence of to-day, and is too improvident to lay up a defence against the occurrence of disability, or the temptations of necessity.

In a pamphlet lately published at Philadelphia, in defence of the system proposed by the Committee, we are informed that the poor list of the city of New-York has risen to fifteen thousand persons; being about an eighth of the whole population. We have also learned from State documents, that its prisons are crowded with felons and debtors. We have seen it too published in the newspapers, that one hundred and eleven persons were last year sentenced to death in four counties of England. In England the gallows groans, or ships are laden with convicts. In New-York the penitentiary overflows with them. In both, the prisons abound with debtors. And in both the proportion of paupers is about one person in eight. In England, fictitious capital, legal privileges, factories, and monopolies are abundant. At New-York they are probably more abundant, than in any other part of the United States. I have said that a partial accumulation of fictitious or legal capital in any one State, at the national expense, would not promote the general happiness or wealth of the people, even of that State.— If the proofs of the assertion in England lie too far off to be seen, that at home is visible. If a local and individual accumulation of capital united with factories, will diffuse honesty and wealth within the sphere of its influence, why do we see most crimes, most debtors, and most pauperism, wherever this policy is most prevalent? May it not therefore be possible that this policy itself generates the crimes and pauperism by which it is attended? At least we must discern, that by whatever names exclusive privileges call themselves; however earnestly they assert that they are not monopolies, and only honest encouragers of industry; that they are not chafferers for selfish acquisitions, but pleaders for general good; that far from causing crimes, they are political moralists; and that far, also, from causing pauperism, they make people work harder than they could otherwise be made to do; that yet they are constantly attended by phenomena, which very plainly contradict all these professions. Bonaparte as devoutly declared, that he was not a military despot, but a patriotick consul.

Political economists in Europe, and especially in England, have forborne to consider the effect of political immorality upon national prosperity, or its influence in begetting both individual pauperism and crimes, because they could only build their systems upon the foundation of governments so thoroughly corrupted, that they despaired of producing a reformation by a true system of political economy; and could only seek for inadequate alleviations of evils, necessarily caused by the firm establishment of the system of patronage, monopoly, and exclusive privileges. Compelled into a reverence for these abuses, they have kept at an awful distance from adversaries so dangerous and unconquerable, and contented themselves with attempting only to soften their baleful influence upon human happiness by temporary expedients. In these endeavours, though they have exhibited great ingenuity, they have been unsuccessful; and, as the causes remain, the effects follow in spite of their wisdom and philanthropy. Here, we are yet able to apply the axe to the causes themselves, which in other countries have generated bad morals and grinding poverty, in spite of fine soils and good climates.

*

Destroy Agriculture.

Neither ambition nor avarice could ever succeed in depriving nations of their liberty and property, if they did not by some artifice enlist the services of a body of men, numerically powerful. The general promises the plunder of a town to his soldiers; they take it; and he keeps most of it for himself and his officers. These are enriched, and the soldiers remain poor. A demagogue promises liberty to a rabble, and by their help makes himself their tyrant. And capitalists, by promising wealth to mechanicks, accumulate it for themselves, and become their masters. The Committee disclaim a predilection for factory capitalists, and an enmity towards agriculture. I balance this argument by disclaiming also a predilection for agriculturists, and an enmity towards mechanicks; but I avow an enmity against all modes for transferring property by exclusive privileges. As no man, however, can find the seeds from which his opinions have germinated, such protestations are frivolous, and they are also unworthy of weight; because the consequences, and not the origin of opinions, constitute their materiality. If it was important to decide, whether the policy proposed by the Committee or its competitor, could be convicted of foreign origin, the difficulty of the subject would not be increased; but I wave the unedifying enquiry, and proceed to the substantial part of the question, whether it will be most injurious to agriculturists or mechanicks. At the threshold of this enquiry, I have changed a term, by substituting mechanicks for manufacturers, to display truth more clearly. The term agriculture needs no such correction, because we have not the two conflicting classes of landlords and and tenants, as we have of capitalists and mechanicks. Where the land of a country is owned by landlords, and worked by tenants, the phrase " landed interest" refers to the landlords, who may enjoy exclusive privileges of which the tenants do not partake; and the impoverishment of one interest may contribute to the enrichment of the other. In like manner, where the factories belong to capitalists, and are worked by mechanicks, the phrase " manfacturing interest" refers to the capitalists, who may enjoy exclusive privileges of which the workmen do not partake; and their impoverishment may contribute to the enrichment of the capitalists, as the impoverishment of tenants may enrich landlords. In deciding the questions, therefore, by the test of friendship or enmity, we ought to exhibit persons, and not confound distinct interests, as the objects of these passions. A cold calculation of the profit to be made by factories, may be a vice of avarice, but a friendly sympathy for the calamities of workmen, arising from the policy of making laws to accumulate this profit, can only flow from good will towards them.

The interest of mechanicks against the factory policy, advocated by the Committee, is infinitely stronger than that of farmers, because, they may more easily be swept into factories, and the profits of their labour more completely carried into the pockets of the capitalists, than can be effected in the case of land owners. These are so powerful as to be able, when they feel a loss, to give themselves a compensation, as the English landlords have done by the corn laws; and between the capitalists and landlords in that country, the mechanicks find poverty. A keen sense of misery fraudulently inflicted, is the cause of their frequent insurrections, and fixed hatred of the government. Why are soldiers necessary to protect their masters, their work-houses and their looms, against the mechanicks themselves? The great lexicographer Johnson, in defining the condition and character of an English mechanick, has called him " mean and servile." The definition is justified by the fact, that his best resource against ending his days in a hospital or poor house, is the shortness of his life. A mechanick employed in a factory rarely acquires a competence; opulence is out of the question; and he is completely excluded from publick employments, by being doomed to a situation in which he can never acquire a capacity for them. He can hardly be considered as a citizen. A code of laws draws around him a magick circle, by making mechanical combinations punishable, lest they should check capitalist combinations; and he is re-imbursed by penalties for the loss of hope.

The condition of the mechanick in the United States has hitherto been extremely different. It neither excites insurrections, nor inculcates a hatred of the government. It does not require a regular army to cure the agonies of misery. It neither shortens life, nor devotes old age to an hospital. It never fails to acquire a competency by industry and good conduct; sometimes rises to opulence; and receives its due share of publick employments. Instead of being deemed mean and servile, it is capable of respectability, and the whole magistracy is open to it. I have heard that the son of a mechanick has been a President; and I know that a weaver, a carpenter, and a carriage-maker, (the two first from Pennsylvania, and the last from Virginia,) were at one time for a long period, worthy members of Congress. Probably there have been many other similar instances. In State legislatures mechanicks are often seen, and as magistrates and militia officers, they abound. They are real, and not nominal citizens. How often do the hirelings of a factory in England, become members of Parliament, magistrates, or militia officers?

For these enormous differences between the condition of the mechanicks in England and the United States, there must be some cause. What can it be, except that the factory and capitalist policy, deprives them of the erect attitude in society inspired by the freedom of industry, and bears hardest upon them, as the chief objects of its gripe? Has this policy bettered the condition of mechanicks, even whilst it was creating enormous fortunes for their masters? If not, the strongest motive for resisting it, is the happiness and prosperity of the mechanicks themselves; though the success of this resistance will also contribute towards the happiness and prosperity of all other useful occupations, because the freedom of talents and industry, and the absence of a system for making both subservient to the interest of avarice, is the principle which must operate beneficially to all, though most so to that occupation most immediately assailed.

To counteract facts established by a double example, the same bribe is offered to land-owners here, which has created in England, a conspiracy between landlords and capitalists against mechanicks, by which they have been reduced to perpetual labour and perpetual poverty.— The land-owners are told, that by coercing mechanicks into factories, the prices of their manufactures will be reduced, and that the land-owners will then be reimbursed for the bounties now paid to capitalists, by a future cheapness to be effected at the expense of mechanicks, thus coerced into factories. I do not deny that such would be the case, if the factory scheme could be carried to the same extent here as in England. This could not be effected, even if our populousness could furnish the materials, except by the English system of legislation to prevent mechanicks from breaking their factory chains, and compelling them to labour hard for low wages to supply the conspirators cheaply. But is not this coerced cheapness evidently imposed upon the mechanical occupation? If it could be effected in the United States, the first class of valuable and respectable citizens which would be ruined by it, would be the great body of mechanicks scattered throughout the country, who would be undersold by the factory capitalists, and compelled to relinquish their free occupations, and become hirelings at the factories. The promised consummation of the factory project, therefore, however tempting to farmers, would be a complete degradation of mechanicks from the equal and comfortable station they hold in society, to one much less desirable. Every present fraud offers a future bribe. The future cheapness offered to land-holders is too distant and uncertain, to induce them to enter into this conspiracy with the capitalists against the mechanicks; and besides, why should they get less than the English landlords for doing so? These have had their rents, and of course the value of their lands doubled or trebled into the bargain, and if without this additional bribe, cheapness would have been insufficient to compensate them for the evils of the capitalist-policy, the land-owners here may safely conclude that they will not be compensated by this promise alone, for co-operating in the conspiracy; and that to make a good bargain, they ought to have the price of their lands doubled or trebled, like the English landlords.

The solitary promise of future cheapness to farmers, to arise from the factory policy, is met by many formidable considerations: If it could be fulfilled at some distant period, the great injury to society from reducing the respectable and numerous class of mechanicks down to Johnson's definition of them; from creating a moneyed aristocracy; and from establishing the policy of exclusive privileges, in which few or no farmers can ever share, would alone suffice to prove that the bribe, if received, would bring along with it a far greater cargo of evils than of benefits.

The prices paid by farmers to the great number of free mechanicks, scattered throughout the country, and by these mechanicks to farmers, promote neighbourhood consumptions; create much domestick commerce regulated by free exchanges, and not by a fraudulent monopoly; stimulate mutual industry, and increase the value of property; but the prices paid to factory capitalists, so long as their monopoly operates, will to a great extent be employed in transferring and accumulating capital. A transfer of profit from industry to the accumulation of capital, whether the profit is agricultural or mechanical, is a mutual diminution of the fund, acting and reacting between industrious occupations, and begetting mutual prosperity. The more of his profits the agriculturist can save from the capitalist, the more employment he will give to his friend and neighbour, the mechanick; and the more of his are retained by the mechanick, the more he will consume of agricultural products, or enhance by his savings, the value of land.— In either case would domestick commerce be rendered more beneficial to the society, by diverting these funds from this intercourse, to the accumulation of pecuniary capitals?

Henry Niles

Morality of Manufactures (1823)

The anecdotal pleasantries of this short article are a relief from the ethical and rhetorical ponderousness of most writers on this subject. In its apprehensive first paragraph, though, the author acknowledges widespread suspicion about the immorality of manufactures. His assertion that factories in the United States are, indeed, morally beneficent establishments rests explicitly on an assumption that they will remain small village works like the Matteawan and Brandywine factories described.

Niles' Weekly Register was a widely influential weekly in the first half of the nineteenth century, published for most of its career in Baltimore. Although politically "nonpartisan," it was, as this article suggests, supportive of commercial and manufacturing interests.

[Henry Niles], "Morality of Manufactures," *Niles' Weekly Register,* volume 25, whole number 637 (November 29, 1823), pp. 195–196.

MORALITY OF MANUFACTURERS. Many persons apprehend that large manufacturing establishments are the great seats of vice and immorality. Whatever may be the case in Europe, they are not so in the United States, nor will they be, until our population is much more dense than it is, and our immense tracts of vacant lands are occupied: *then*, if the people shall also have lost the rights and privileges which they at present enjoy, perhaps the manufacturers may become as corrupt as the population of some of our cities is now.

An able correspondent of the New York "Statesman" gives the following account of a visit to the Matteawan Factory, near the Fish-kill mountains, when the proprietor, Mr. Schenck, gave him the following narrative:

"Before I commenced the erection of these works, said Mr. S. and established in this place the branch of cotton manufacture, the process of which you have been just examining, the man who built, and now owns that neat little tenement, had no place to shelter himself and his numerous family, but the wretched hovel which you may observe at a few rods distance from his present abode. At that time, continued my informant, his only occupation was that of fishing: or rambling in the mountains in pursuit of such game as chance might throw in his way. Of the little he obtained by this occasional and precarious mode of subsistence, a large proportion was expended in the purchase of rum; in the use of which he indulged to such an extent as to brutalize his faculties, and render him a pest to society, as well as a curse to his family; which he kept in a state of the most deplorable and squalid poverty. Of his children three or four were daughters, of various ages, from seven or eight to fourteen years; these, said Mr. on commencing my establishment, I took into the factory; where, from that period to the present time, they have always had constant and regular employment. The proceeds of their first week's labor, amounting to six or seven dollars, when paid and taken home to their parents, was an amount which, it is probable, they never before at any one time possessed. The almost immediate effect on the mind of the father appears to have been a conviction that his children, instead of being a burthen which he despaired of supporting, and, therefore, never before made an effort to accomplish, would, on the contrary, by the steady employment now provided for them, be able, by their industry, not only to sustain themselves, but also contribute to the maintenance and support of the other members of the family. From that moment, it would appear, as if he had determined to reform his vicious habits, and to emerge from that state of degradation and wretchedness into which he had plunged himself and family. He has done so, said Mr. S. and, instead of being a pest, he has become a useful member of society; instead of being a curse to his family, and occupying with them that wretched hovel yonder, fit only for swine to wallow in, he has, by his own exertions, aided by the industry and good conduct of his children, lately purchased the soil, and erected the comfortable cottage, which, said Mr. S. smiling, appears so powerfully to attract your notice."

I have also to relate a case that came under my own personal observation. Some years since, I designed to write for the REGISTER an account of the improvements on the *Brandywine*, and for that purpose leisurely walked five miles up its banks, accompanied by two gentlemen of refined minds and extensive observation. It was in the month of May or early in June, and one of the most delightful days that I ever passed in my life. The scenery is among the most lively, picturesque and romantic, I suppose, in the world, and the improvements have cost millions. I well knew the country when it all was as covered with huge rocks, with here and there only a solitary house or mill squeezed into a small space of cleared land—now there is a continual succession of mills and elegant houses and comfortable cottages, with pretty extensive fields and gardens, wrested from the late rock covered wild, and what remains of the primitive state of the country, always causes one to stop and reflect on the labor bestowed and money expended to bring the major part of it to its present improved condition. Copious notes were made of the things that I saw, but the delay of some of the manufacturers to furnish particulars, prevented the intended publication, and my notes are since lost or mislaid.

The people who inhabited the borders of this stream, and especially those on the north side of it, were as wild and as rude as the country itself. A good many of them thought that "learning was a dangerous thing" and, in general, they were poor and miserable. They were proverbial for their dissoluteness and profligacy. But they have changed in their manners and habits, almost as much as the shores of their beautiful stream have been changed —and comfortable buildings have taken place of wretched huts, that every moment seemed ready to fall down on the heads of their owners, through old age and from the want of repair.

These general remarks seemed necessary to lead to the particular thing which I wish to mention. On the Sunday morning after my walk, I visited one of the school houses which the manufacturers have built, and at which they require the attendance of the children employed by them. There were about ninety present, male and female, all dressed in clean clothes, and some of them, especially the girls, very neatly; and they looked hearty and happy. The business of the morning was opened with a prayer, and a hymn was then sung in a very agreeable stile, with much more harmony than could have been expected. The head teacher of the day then struck his desk with his hand, and the *classes* were instantly formed. The Lancasterian system was used, and the monitors were at their posts, at the head of their little squads, in a moment. Not a word was heard, unless it related to the matters of instruction going on. I passed through the school. Some little ones were learning their alphabet, others spelling, reading, writing, cyphering, &c. Several, who then read or wrote pretty well, were pointed out to me as having been utterly destitute of the knowledge of a letter some few months before—others, who read and wrote as well as children of their age generally do, had received *all* the rudiments of their education at this school! My attention was particularly directed to one boy, an overgrown rough looking youth of thirteen or fourteen. He was the monitor of a small class spelling in two syllables. I stood behind him, and thought that I had never seen one labor more in mauling rails, than he did to find out some of his letters and give their pronunciation. This boy had been twice driven from one of the factories, *or taken away by his parents*, because it had been required of him to attend the school on Sunday mornings. His father and mother thought it a great hardship that their son should be *forced* to learn to read, hear prayers, sing hymns and listen occasionally to short sermons. About five weeks before the time now alluded to, his father had brought him back, that his earnings might assist in the support of the family, and agreed that he should be compelled to attend the school. The boy suddenly fell in love with his book, and studied it at every leisure moment that he had—from being one of the worst, he became one of the best and most obedient lads in the mill; he now seemed as proud of the monitor's badge on his breast as a nobleman could be of his star, and the gentleman with whom he worked, boldly prophecied that he would become a valuable man, if he lived. After the classes had been several times changed by a stroke on the desk, without any word of command, the school was closed by prayer, and the children departed from it with as much order and propriety as ever a society of Quakers retired from a meeting. When I looked at the school and recollected what the poor people of the neighborhood were twenty five or thirty years before—though not used "to the melting mood," I could hardly refrain from shedding tears. And the gentlemen who had lived a long time in those parts, assured me, that, though the population had been so much increased, there was much less of vice and immorality than had heretofore prevailed, and that petty depredations on orchards, &c. were more rare than they had ever known them to be. In short, that the state of society had improved, as well as the face of the country—about which I could not entertain a doubt, for I read the history of it in the people's eyes.

The Philosophy of Manufactures

Thomas Cooper

The Disadvantage of Machinery (1823)

This outburst against the "curse" of machinery and the "proud and wealthy capitalists" created by the manufacturing system appears in the midst of a tract contributed by Cooper to the tariff controversy of the early 1820s. Cooper meant to scotch the notion that the tariff would benefit Americans by enabling them to take better advantage of machinery.

Thomas Cooper, M.D., *Two Tracts: On the Proposed Alteration of the Tariff; and on Weights and Measures. Submitted to the Consideration of the Members from South Carolina, in the Ensuing Congress of 1823–24* (Charleston, 1823), pp. 21–22.

To take advantage of machinery.—If capital, employed in commerce, bring 15 per cent. and capital, employed in machinery, bring 15 per cent. there is nothing gained by converting the one into the other.

Oh, but a cotton mill will perform the work of a thousand hands—Will it so? What then, if it brings me no higher than common profit? But it will bring much greater profit—Will it so? Then so many people will have cotton mills, that in a year or two, the profit will decrease to the common level, and I shall be no gainer.

The machinery of England, is, in many instances, a dreadful curse to that country; and the British manufacturing system would be so to this. The works usually go night and day, one set of boys and girls go to bed, as another set get up to work. The health, the manners, the morals, are all corrupted. They work not for themselves, but for the capitalist who employs them: they are employed on the calculation of how small a sum will subsist a human creature: they are machines, as much so as the spindles they superintend: hence they are not calculated to turn readily, from one occupation to another: they are the most discontented, the most ignorant, the most turbulent of the British population. The whole system tends to increase the wealth of a few capitalists, at the expense of the health, life, morals, and happiness of the wretches who labour for them. I would rather see treatises on the sources of national happiness, than national wealth. We want in this happy country, no increase of proud and wealthy capitalists, whose fortunes have accumulated by such means. It is not the careful, skilful superintendant of his own business, living frugally, but plentifully on reasonable profits, who expresses discontent at the present state of things—no, it is the would-be great man, anxious to acquire wealth speedily, by means of an extorted monopoly, who is most forward in petitioning, for an increase of prohibitory duties. Neither the prophecies, the promises, or the statements of these men, are to be trusted. They may pledge themselves to any thing, for they know they cannot be called on to redeem it. They calculate the imposition will last their time. Well meaning and good men, have been over persuaded by the bold assertions of those who are intrusted; and we stand now actually on the very brink of the precipice to which they have urged us. It is impossible to shut our eyes to the wonderful superiority in permanence of capital invested in agriculture, over capital invested in machinery.

Thomas Carlyle

Signs of the Times (1829)

The influence of Thomas Carlyle (1795–1881) on American men of letters and their thoughts about the power and significance of industrial technology was profound. His influence on the progress of American manufactures was negligible. As a result, he helped to define the great gulf in the nineteenth-century American imagination between literary culture and the practical arts. His critique of what he called the "Age of Mechanism," which first appeared in the *Edinburgh Review,* struck close enough to home that it elicited strong reproof in America's leading periodical, the *North American Review* (see the article by Timothy Walker, reprinted in this volume). Otherwise, Carlyle's essay appears to have been generally ignored by proponents of industrial development. Carlyle's arguments and language, however, were echoed in the work of Whittier and Thoreau (also reprinted in this volume). Emerson, who was Carlyle's most eager American disciple and promoter, had a more ambiguous relation to Carlyle's views on industry. "Signs of the Times" bears comparison to the Emerson chapters reprinted in this collection.

Thomas Carlyle, "Signs of the Times," *Edinburgh Review,* No. 98 (1829), pp. 439–459. For an extended analysis of Carlyle's essay and its influence on American men of letters see Leo Marx, *The Machine in the Garden* (New York: Oxford University Press, 1964).

ART. VII.—1. *Anticipation ; or, an Hundred Years Hence.* 8vo. London, 1829.
2. *The Rise, Progress, and Present State of Public Opinion in Great Britain.* 8vo. London, 1829.
3. *The Last Days ; or, Discourses on These Our Times, &c. &c.* By the Rev. EDWARD IRVING. 8vo. London, 1829.

IT is no very good symptom either of nations or individuals, that they deal much in vaticination. Happy men are full of the present, for its bounty suffices them; and wise men also, for its duties engage them. Our grand business undoubtedly is, not to *see* what lies dimly at a distance, but to *do* what lies clearly at hand.

> Know'st thou *Yesterday*, its aim and reason?
> Work'st thou well *To-day*, for worthy things?
> Then calmly wait the *Morrow's* hidden season,
> And fear not thou, what hap soe'er it brings!

But man's ' large discourse of reason' *will* look ' before and ' after ;' and, impatient of ' the ignorant present time,' will indulge in anticipation far more than profits him. Seldom can the unhappy be persuaded that the evil of the day is sufficient for it; and the ambitious will not be content with present splendour —but paints yet more glorious triumphs, on the cloud-curtain of the future.

The case, however, is still worse with nations. For here the prophets are not one, but many ; and each incites and confirms the other—so that the fatidical fury spreads wider and wider, till at last even a Saul must join in it. For there is still a real magic in the action and reaction of minds on one another. The casual deliration of a few becomes, by this mysterious reverberation, the frenzy of many; men lose the use, not only of their understandings, but of their bodily senses; while the most obdurate, unbelieving hearts melt, like the rest, in the furnace where all are cast, as victims and as fuel. It is grievous to think, that this noble omnipotence of Sympathy has been so rarely the Aaron's-rod of Truth and Virtue, and so often the Enchanter's-rod of Wickedness and Folly! No solitary miscreant, scarcely any solitary maniac, would venture on such actions and imaginations, as large communities of sane men have, in such circumstances, entertained as sound wisdom. Witness long scenes of the French Revolution! a whole people drunk with blood and arrogance—and then with terror and cruelty—and with desperation, and blood again! Levity is no protection against such visitations, nor the utmost earnestness of character. The New England Puritan burns witches, wrestles for months with the horrors of Satan's invisible world, and all ghastly phantasms, the daily and hourly precursors of the Last Day; then suddenly bethinks him that he is frantic, weeps bitterly, prays contritely—and the history of that gloomy season lies behind him like a frightful dream.

And Old England has had her share of such frenzies and panics; though happily, like other old maladies, they have grown milder of late: and since the days of Titus Oates, have mostly passed without loss of men's lives, or indeed without much other loss than that of reason, for the time, in the sufferers. In this mitigated form, however, the distemper is of pretty regular recurrence—and may be reckoned on at intervals, like other natural visitations; so that reasonable men deal with it, as the Londoners do with their fogs—go cautiously out into the groping crowd, and patiently carry lanterns at noon; knowing, by a well-grounded faith, that the sun is still in existence, and will one day reappear. How often have we heard, for the last fifty years, that the country was wrecked, and fast sinking; whereas, up to this date, the country is entire and afloat! The 'State in Danger' is a condition of things, which we have witnessed a hundred times; and as for the church, it has seldom been out of 'danger' since we can remember it.

All men are aware, that the present is a crisis of this sort; and why it has become so. The repeal of the Test Acts, and then of the Catholic disabilities, has struck many of their admirers with an indescribable astonishment. Those things seemed fixed and immovable—deep as the foundations of the world; and, lo! in a moment they have vanished, and their place knows them no more! Our worthy friends mistook the slumbering Leviathan for an island—often as they had been assured, that Intolerance was, and could be, nothing but a Monster; and so, mooring under the lee, they had anchored comfortably in his scaly rind, thinking to take good cheer—as for some space they did. But now their Leviathan has suddenly dived under; and they can no longer be fastened in the stream of time; but must drift forward on it, even like the rest of the world—no very appalling fate, we think, could they but understand it; which, however, they will not yet, for a season. Their little island is gone, and sunk deep amid confused eddies; and what is left worth caring for in the universe? What is it to them, that the great continents of the earth are still standing; and the polestar and all our loadstars, in the heavens, still shining and eternal? Their cherished little haven is gone, and they will not be comforted! And therefore, day after day, in all manner of periodical or perennial publications, the most lugubrious predictions are sent forth. The king has virtually abdicated; the church is a widow, without jointure; public principle is gone; private honesty is going; society, in short, is fast falling in pieces; and a time of unmixed evil is come on us. At such a period, it was to be expected that the rage of prophecy should be more than usually excited. Accordingly, the Millennarians have come forth on the right hand, and the Millites on the left. The Fifth-monarchy men prophesy from the Bible, and the Utilitarians from Bentham. The one announce that the last of the seals is to be opened, positively, in the year 1860; and the other assure us, that 'the greatest happiness principle' is to make a heaven of earth, in a still shorter time. We know these symptoms too well, to think it necessary or safe to interfere with them. Time and the hours will bring relief to all parties. The grand encourager of Delphic or other noises is—the Echo. Left to themselves, they will soon dissipate, and die away in space.

Meanwhile, we too admit that the present is an important time—as all present time necessarily is. The poorest day that passes over us is the conflux of two Eternities! and is made up of currents that issue from the remotest Past, and flow onwards into the remotest Future. We were wise indeed, could we discern truly the signs of our own time; and, by knowledge of its wants and advantages, wisely adjust our own position in it. Let us then, instead of gazing idly into the obscure distance, look calmly around us, for a little, on the perplexed scene where we stand. Perhaps, on a more serious inspection, something of its perplexity will disappear, some of its distinctive characters, and deeper tendencies, more clearly reveal themselves; whereby our own relations to it, our own true aims and endeavours in it, may also become clearer.

Were we required to characterise this age of ours by any

single epithet, we should be tempted to call it, not an Heroical, Devotional, Philosophical, or Moral Age, but, above all others, the Mechanical Age. It is the Age of Machinery, in every outward and inward sense of that word; the age which, with its whole undivided might, forwards, teaches, and practises the great art of adapting means to ends. Nothing is now done directly, or by hand; all is by rule and calculated contrivance. For the simplest operation, some helps and accompaniments, some cunning, abbreviating process is in readiness. Our old modes of exertion are all discredited, and thrown aside. On every hand, the living artisan is driven from his workshop, to make room for a speedier, inanimate one. The shuttle drops from the fingers of the weaver, and falls into iron fingers that ply it faster. The sailor furls his sail, and lays down his oar, and bids a strong, unwearied servant, on vaporous wings, bear him through the waters. Men have crossed oceans by steam; the Birmingham Fireking has visited the fabulous East; and the genius of the Cape, were there any Camoens now to sing it, has again been alarmed, and with far stranger thunders than Gama's. There is no end to machinery. Even the horse is stripped of his harness, and finds a fleet fire-horse yoked in his stead. Nay, we have an artist that hatches chickens by steam—the very brood-hen is to be superseded! For all earthly, and for some unearthly purposes, we have machines and mechanic furtherances; for mincing our cabbages; for casting us into magnetic sleep. We remove mountains, and make seas our smooth highway; nothing can resist us. We war with rude nature; and, by our resistless engines, come off always victorious, and loaded with spoils.

What wonderful accessions have thus been made, and are still making, to the physical power of mankind; how much better fed, clothed, lodged, and, in all outward respects, accommodated, men now are, or might be, by a given quantity of labour, is a grateful reflection which forces itself on every one. What changes, too, this addition of power is introducing into the social system; how wealth has more and more increased, and at the same time gathered itself more and more into masses, strangely altering the old relations, and increasing the distance between the rich and the poor, will be a question for Political Economists—and a much more complex and important one than any they have yet engaged with. But leaving these matters for the present, let us observe how the mechanical genius of our time has diffused itself into quite other provinces. Not the external and physical alone is now managed by machinery, but the internal and spiritual also. Here, too, nothing follows its spontaneous course, nothing is left to be accomplished by old, natural methods. Every thing has its cunningly devised implements, its pre-established apparatus; it is not done by hand, but by machinery. Thus we have machines for Education: Lancastrian machines; Hamiltonian machines—Monitors, maps, and emblems. Instruction, that mysterious communing of Wisdom with Ignorance, is no longer an indefinable tentative process, requiring a study of individual aptitudes, and a perpetual variation of means and methods, to attain the same end; but a secure, universal, straight-forward business, to be conducted in the gross, by proper mechanism, with such intellect as comes to hand. Then, we have Religious machines, of all imaginable varieties—the Bible Society, professing a far higher and heavenly structure, is found, on enquiry, to be altogether an earthly contrivance, supported by collection of monies, by fomenting of vanities, by puffing, intrigue, and chicane—and yet, in effect, a very excellent machine for converting the heathen. It is the same in all other departments. Has any man, or any society of men, a truth to speak, a piece of spiritual work to do, they can nowise proceed at once, and with the mere natural organs, but must first call a public meeting, appoint committees, issue prospectuses, eat a public dinner; in a word, construct or borrow machinery, wherewith to speak it and do it. Without machinery they were hopeless, helpless—a colony of Hindoo weavers squatting in the heart of Lancashire. Then every machine must have its moving power, in some of the great currents of society: Every little sect among us, Unitarians, Utilitarians, Anabaptists, Phrenologists, must each have its periodical, its monthly or quarterly magazine—hanging out, like its windmill, into the *popularis aura*, to grind meal for the society.

With individuals, in like manner, natural strength avails little. No individual now hopes to accomplish the poorest enterprise single-handed, and without mechanical aids; he must make interest with some existing corporation, and till his field with their oxen. In these days, more emphatically than ever, ' to live, sig-' nifies to unite with a party, or to make one.' Philosophy, Science, Art, Literature, all depend on machinery. No Newton, by silent meditation, now discovers the system of the world from the falling of an apple; but some quite other than Newton stands in his Museum, his Scientific Institution, and behind whole batteries of retorts, digesters, and galvanic piles, imperatively ' inter-' rogates Nature,'—who, however, shows no haste to answer. In defect of Raphaels, and Angelos, and Mozarts, we have Royal Academies of Painting, Sculpture, Music; whereby the languishing spirit of Art may be strengthened by the more generous diet of a Public Kitchen. Literature, too, has its Paternoster-row

mechanism, its Trade dinners, its Editorial conclaves, and huge subterranean, puffing bellows; so that books are not only printed, but, in a great measure, written and sold, by machinery. National culture, spiritual benefit of all sorts, is under the same management. No Queen Christina, in these times, needs to send for her Descartes: no King Frederick for his Voltaire, and painfully nourish him with pensions and flattery : But any sovereign of taste, who wishes to enlighten his people, has only to impose a new tax, and with the proceeds establish Philosophic Institutes. Hence the Royal and Imperial Societies, the Bibliothèques, Glypeothèques, Sechnothèques, which front us in all capital cities, like so many well-finished hives, to which it is expected the stray agencies of Wisdom will swarm of their own accord, and hive and make honey. In like manner, among ourselves, when it is thought that religion is declining, we have only to vote half a million's worth of bricks and mortar, and build new churches. In Ireland, it seems they have gone still farther—having actually established a ' Penny-a-week Purga-
' tory Society!' Thus does the Genius of Mechanism stand by to help us in all difficulties and emergencies; and, with his iron back, bears all our burdens.

These things, which we state lightly enough here, are yet of deep import, and indicate a mighty change in our whole manner of existence. For the same habit regulates, not our modes of action alone, but our modes of thought and feeling. Men are grown mechanical in head and in heart, as well as in hand. They have lost faith in individual endeavour, and in natural force, of any kind. Not for internal perfection, but for external combinations and arrangements, for institutions, constitutions—for Mechanism of one sort or other, do they hope and struggle. Their whole efforts, attachments, opinions, turn on mechanism, and are of a mechanical character.

We may trace this tendency, we think, very distinctly, in all the great manifestations of our time; in its intellectual aspect, the studies it most favours, and its manner of conducting them; in its practical aspects, its politics, arts, religion, morals; in the whole sources, and throughout the whole currents, of its spiritual, no less than its material activity.

Consider, for example, the state of Science generally, in Europe, at this period. It is admitted, on all sides, that the Metaphysical and Moral Sciences are falling into decay, while the Physical are engrossing, every day, more respect and attention. In most of the European nations, there is now no such thing as a Science of Mind; only more or less advancement in the general science, or the special sciences, of matter. The French were the first to desert this school of Metaphysics; and though they have lately affected to revive it, it has yet no signs of vitality. The land of Malebranche, Pascal, Descartes, and Fenelon, has now only its Cousins and Villemains; while, in the department of Physics, it reckons far other names. Among ourselves, the Philosophy of Mind, after a rickety infancy, which never reached the vigour of manhood, fell suddenly into decay, languished, and finally died out, with its last amiable cultivator, Professor Stewart. In no nation but Germany has any decisive effort been made in psychological science; not to speak of any decisive result. The science of the age, in short, is physical, chemical, physiological, and, in all shapes, mechanical. Our favourite Mathematics, the highly prized exponent of all these other sciences, has also become more and more mechanical. Excellence, in what is called its higher departments, depends less on natural genius, than on acquired expertness in wielding its machinery. Without undervaluing the wonderful results which a Lagrange, or Laplace, educes by means of it, we may remark, that its calculus, differential and integral, is little else than a more cunningly-constructed arithmetical mill, where the factors being put in, are, as it were, ground into the true product, under cover, and without other effort on our part, than steady turning of the handle. We have more Mathematics certainly than ever; but less Mathesis. Archimedes and Plato could not have read the *Méchanique Céleste;* but neither would the whole French Institute see aught in that saying, ' God geometrises !' but a sentimental rodomontade.

From Locke's time downwards, our whole Metaphysics have been physical; not a spiritual Philosophy, but a material one. The singular estimation in which his Essay was so long held as a scientific work, (for the character of the man entitled all he said to veneration,) will one day be thought a curious indication of the spirit of these times. His whole doctrine is mechanical, in its aim and origin, in its method and its results. It is a mere discussion concerning the origin of our consciousness, or ideas, or whatever else they are called; a genetic history of what we see *in* the mind. But the grand secrets of Necessity and Freewill, of the mind's vital or non-vital dependence on matter, of our mysterious relations to Time and Space, to God, to the universe, are not, in the faintest degree, touched on in their enquiries; and seem not to have the smallest connexion with them.

The last class of our Scotch Metaphysicians had a dim notion that much of this was wrong; but they knew not how to right it. The school of Reid had also from the first taken a mechanical course, not seeing any other. The singular conclusions at which Hume, setting out from their admitted premises, was arriving, brought this school into being; they let loose Instinct, as an undiscriminating bandog, to guard them against these conclusions—they tugged lustily at the logical chain by which Hume was so coldly towing them and the world into bottomless abysses of Atheism and Fatalism. But the chain somehow snapped between them; and the issue has been that nobody now cares about either—any more than about Hartley's, Darwin's, or Priestley's contemporaneous doings in England. Hartley's vibrations and vibratiuncles one would think were material and mechanical enough; but our continental neighbours have gone still farther. One of their philosophers has lately discovered, that 'as the liver secretes bile, so does the brain secrete thought;' which astonishing discovery Dr Cabanis, more lately still, in his *Rapports du Physique et du Morale de l'Homme*, has pushed into its minutest developements. The metaphysical philosophy of this last enquirer is certainly no shadowy or unsubstantial one. He fairly lays open our moral structure with his dissecting-knives and real metal probes; and exhibits it to the inspection of mankind, by Leuwenhoeck microscopes and inflation with the anatomical blowpipe. Thought, he is inclined to hold, is still secreted by the brain; but then Poetry and Religion (and it is really worth knowing) are 'a product of the smaller intestines!' We have the greatest admiration for this learned doctor: with what scientific stoicism he walks through the land of wonders, unwondering—like a wise man through some huge, gaudy, imposing Vauxhall, whose fire-works, cascades, and symphonies, the vulgar may enjoy and believe in—but where he finds nothing real but the saltpetre, pasteboard, and catgut. His book may be regarded as the ultimatum of mechanical metaphysics in our time; a remarkable realization of what in Martinus Scriblerus was still only an idea, that 'as the jack had a meat-roast-'ing quality, so had the body a thinking quality,'—upon the strength of which the Nurembergers were to build a wood and leather man, 'who should reason as well as most country par-'sons.' Vaucasson did indeed make a wooden duck, that seemed to eat and digest; but that bold scheme of the Nurembergers remained for a more modern virtuoso.

This condition of the two great departments of knowledge; the outward, cultivated exclusively on mechanical principles—the inward finally abandoned, because, cultivated on such principles, it is found to yield no result—sufficiently indicates the intellectual bias of our time, its all-pervading disposition towards that line of enquiry. In fact, an inward persuasion has long been diffusing itself, and now and then even comes to utterance, that except the external, there are no true sciences; that to the inward world (if there be any) our only conceivable road is through the outward; that, in short, what cannot be investigated and understood mechanically, cannot be investigated and understood at all. We advert the more particularly to these intellectual propensities, as to prominent symptoms of our age; because Opinion is at all times doubly related to Action, first as cause, then as effect; and the speculative tendency of any age, will therefore give us, on the whole, the best indications of its practical tendency.

Nowhere, for example, is the deep, almost exclusive faith, we have in Mechanism, more visible than in the Politics of this time. Civil government does, by its nature, include much that is mechanical, and must be treated accordingly. We term it, indeed, in ordinary language, the Machine of Society, and talk of it as the grand working wheel from which all private machines must derive, or to which they must adapt, their movements. Considered merely as a metaphor, all this is well enough; but here, as in so many other cases, the 'foam hardens itself into a shell,' and the shadow we have wantonly evoked stands terrible before us, and will not depart at our bidding. Government includes much also that is not mechanical, and cannot be treated mechanically; of which latter truth, as appears to us, the political speculations and exertions of our time are taking less and less cognisance.

Nay, in the very outset, we might note the mighty interest taken in *mere political arrangements*, as itself the sign of a mechanical age. The whole discontent of Europe takes this direction. The deep, strong cry of all civilized nations—a cry which, every one now sees, must and will be answered, is, Give us a reform of Government! A good structure of legislation—a proper check upon the executive—a wise arrangement of the judiciary, is *all* that is wanting for human happiness. The Philosopher of this age is not a Socrates, a Plato, a Hooker, or Taylor, who inculcates on men the necessity and infinite worth of moral goodness, the great truth that our happiness depends on the mind which is within us, and not on the circumstances which are without us; but a Smith, a De Lolme, a Bentham, who chiefly inculcates the reverse of this—that our happiness depends entirely on external circumstances; nay that the strength and dignity of the mind within us is itself the creature and consequence of these. Were the laws, the government, in good order, all were well with us; the rest would care for itself! Dissentients from this opinion, expressed or implied, are now rarely to be met with; widely and angrily as men differ in its application, the principle is admitted by all.

Equally mechanical, and of equal simplicity, are the methods proposed by both parties for completing or securing this all-sufficient perfection of arrangement. It is no longer the moral, religious, spiritual condition of the people that is our concern, but their physical, practical, economical condition, as regulated by public laws. Thus is the Body-politic more than ever worshipped and tended: But the Soul-politic less than ever. Love of country, in any high or generous sense, in any other than an almost animal sense, or mere habit, has little importance attached to it in such reforms, or in the opposition shown them. Men are to be guided only by their self-interests. Good government is a good balancing of these; and, except a keen eye and appetite for self-interest, requires no virtue in any quarter. To both parties it is emphatically a machine: to the discontented, a 'taxing-machine;' to the contented, a 'machine for securing pro-'perty.' Its duties and its faults are not those of a father, but of an active parish constable.

Thus it is by the mere condition of the machine; by preserving it untouched, or else by re-constructing it, and oiling it anew, that man's salvation as a social being is to be insured and indefinitely promoted. Contrive the fabric of law aright, and without farther effort on your part, that divine spirit of freedom which all hearts venerate and long for, will of herself come to inhabit it; and under her healing wings every noxious influence will wither, every good and salutary one more and more expand. Nay, so devoted are we to this principle, and at the same time so curiously mechanical, that a new trade, specially grounded on it, has arisen among us, under the name of 'Codification,' or code-making in the abstract; whereby any people, for a reasonable consideration, may be accommodated with a patent code—more easily than curious individuals with patent breeches, for the people does *not* need to be measured first.

To us who live in the midst of all this, and see continually the faith, hope, and practice of every one founded on Mechanism of one kind or other, it is apt to seem quite natural, and as if it could never have been otherwise. Nevertheless, if we recollect or reflect a little, we shall find both that it has been, and might again be, otherwise. The domain of Mechanism,—meaning thereby political, ecclesiastical, or other outward establishments,—was once considered as embracing, and we are persuaded can at any time embrace, but a limited portion of man's interests, and by no means the highest portion.

To speak a little pedantically, there is a science of *Dynamics* in man's fortunes and nature, as well as of *Mechanics*. There is a science which treats of, and practically addresses, the primary, unmodified forces and energies of man, the mysterious springs of Love, and Fear, and Wonder, of Enthusiasm, Poetry, Religion, all which have a truly vital and *infinite* character; as well as a science which practically addresses the finite, modified developements of these, when they take the shape of immediate 'motives,' as hope of reward, or as fear of punishment.

Now it is certain, that in former times the wise men, the enlightened lovers of their kind, who appeared generally as Moralists, Poets, or Priests, did, without neglecting the Mechanical province, deal chiefly with the Dynamical; applying themselves chiefly to regulate, increase, and purify the inward primary powers of man; and fancying that herein lay the main difficulty, and the best service they could undertake. But a wide difference is manifest in our age. For the wise men, who now appear as Political Philosophers, deal exclusively with the Mechanical province; and occupying themselves in counting up and estimating men's motives, strive by curious checking and balancing, and other adjustments of Profit and Loss, to guide them to their true advantage: while, unfortunately, those same 'motives' are so innumerable, and so variable in every individual, that no really useful conclusion can ever be drawn from their enumeration. But though Mechanism, wisely contrived, has done much for man, in a social and moral point of view, we cannot be persuaded that it has ever been the chief source of his worth or happiness. Consider the great elements of human enjoyment, the attainments and possessions that exalt man's life to its present height, and see what part of these he owes to institutions, to Mechanism of any kind; and what to the instinctive, unbounded force, which Nature herself lent him, and still continues to him. Shall we say, for example, that Science and Art are indebted principally to the founders of Schools and Universities? Did not Science originate rather, and gain advancement, in the obscure closets of the Roger Bacons, Keplers, Newtons; in the workshops of the Fausts and the Watts—whereever, and in what guise soever Nature, from the first times downwards, had sent a gifted spirit upon the earth? Again, were Homer and Shakspeare members of any beneficed guild, or made Poets by means of it? Was Painting and Sculpture created by forethought, brought into the world by institutions for that end? No; Science and Art have, from first to last, been the free gift of Nature; an unsolicited, unexpected gift—often even a fatal one. These things rose up, as it were, by spontaneous growth, in the free soil and sunshine of Nature. They were not planted or grafted, nor even greatly multiplied or improved by the culture or manuring of institutions. Generally speaking, they have derived only partial help from these; often enough have suffered damage. They made constitutions for themselves. They originated in the Dynamical nature of man, not in his Mechanical nature.

Or, to take an infinitely higher instance, that of the Christian Religion, which, under every theory of it, in the believing or the unbelieving mind, must ever be regarded as the crowning glory, or rather the life and soul, of our whole modern culture: How did Christianity arise and spread abroad among men? Was it by institutions and establishments, and well-arranged systems of mechanism? Not so; on the contrary, in all past and existing institutions for those ends, its divine spirit has invariably been found to languish and decay. It arose in the mystic deeps of man's soul; and was spread abroad by the 'preaching of the 'word,' by simple, altogether natural and individual efforts; and flew, like hallowed fire, from heart to heart, till all were purified and illuminated by it; and its heavenly light shone, as it still shines, and as sun or star will ever shine, through the whole dark destinies of man. Here again was no Mechanism; man's highest attainment was accomplished, Dynamically, not Mechanically. Nay, we will venture to say that no high attainment, not even any far-extending movement among men, was ever accomplished otherwise. Strange as it may seem, if we read History with any degree of thoughtfulness, we shall find, that the checks and balances of Profit and Loss have never been the grand agents with men; that they have never been roused into deep, thorough, all-pervading efforts by any computable prospect of Profit and Loss, for any visible, finite object; but always for some invisible and infinite one. The Crusades took their rise in Religion; their visible object was, commercially speaking, worth nothing. It was the boundless, Invisible world that was laid bare in the imaginations of those men; and in its burning light, the visible shrunk as a scroll. Not mechanical, nor produced by mechanical means, was this vast movement. No dining at Freemasons' Tavern, with the other long train of modern machinery; no cunning reconciliation of 'vested interests,' was required here: only the passionate voice of one man, the rapt soul looking through the eyes of one man; and rugged, steel-clad Europe trembled beneath his words, and followed him whither he listed. In later ages, it was still the same. The Reformation had an invisible, mystic, and ideal aim: the result was indeed to be embodied in external things; but its spirit, its worth, was internal, invisible, infinite. Our English Revolution, too, originated in Religion. Men did battle, even in those days, not for Purse sake, but for Conscience sake. Nay, in our own days, it is no way different. The French Revolution itself had something higher in it than cheap bread and a Habeascorpus act. Here, too, was an Idea; a Dynamic, not a Mechanic force. It was a struggle, though a blind and at last an insane one, for the infinite, divine nature of Right, of Freedom, of Country.

Thus does man, in every age, vindicate, consciously or unconsciously, his celestial birthright. Thus does nature hold on her wondrous, unquestionable course; and all our systems and theories are but so many froth-eddies or sand-banks, which from time to time she casts up and washes away. When we can drain the Ocean into our mill-ponds, and bottle up the Force of Gravity, to be sold by retail, in our gas-jars; then may we hope to comprehend the infinitudes of man's soul under formulas of Profit and Loss; and rule over this too, as over a patent engine, by checks, and valves, and balances.

Nay, even with regard to Government itself, can it be necessary to remind any one that Freedom, without which indeed all spiritual life is impossible, depends on infinitely more complex influences than either the extension or the curtailment of the ' democratic interest?' Who is there that ' taking the high ' *priori* road,' shall point out what these influences are; what deep, subtle, inextricably entangled influences they have been, and may be? For man is not the creature and product of Mechanism; but, in a far truer sense, its creator and producer: it is the noble people that makes the noble Government; rather than conversely. On the whole, Institutions are much; but they are not all. The freest and highest spirits of the world have often been found under strange outward circumstances: Saint Paul and his brother Apostles were politically slaves; Epictetus was personally one. Again, forget the influences of Chivalry and Religion, and ask,—what countries produced Columbus and Las Casas? Or, descending from virtue and heroism, to mere energy and spiritual talent: Cortes, Pizarro, Alba, Ximenes? The Spaniards of the sixteenth century were indisputably the noblest nation of Europe; yet they had the Inquisition, and Philip II. They have the same government at this day; and are the lowest nation. The Dutch, too, have retained their old constitution; but no Siege of Leyden, no William the Silent, not even an Egmont or De Witt, any longer appear among them. With ourselves, also, where much has changed, effect has nowise followed cause, as it should have done: two centuries ago, the Commons' Speaker addressed Queen Elizabeth on bended knees, happy that the virago's foot did not even smite him; yet the people were then governed, not by a Castlereagh, but by a Burghley; they had their Shakspeare and Philip Sidney, where we have our Sheridan Knowles and Beau Brummel.

These and the like facts are so familiar, the truths which they preach so obvious, and have in all past times been so universally believed and acted on, that we should almost feel ashamed for repeating them; were it not that, on every hand, the memory of them seems to have passed away, or at best died into a faint tradition, of no value as a practical principle. To judge by the loud clamour of our Constitution-builders, Statists, Economists, directors, creators, reformers of Public Societies; in a word, all manner of Mechanists, from the Cartwright up to the Code-maker; and by the nearly total silence of all Preachers and Teachers who should give a voice to Poetry, Religion, and Morality, we might fancy either that man's Dynamical nature was, to all spiritual intents, extinct—or else so perfected, that nothing more was to be made of it by the old means; and henceforth only in his Mechanical contrivances did any hope exist for him.

To define the limits of these two departments of man's activity, which work into one another, and by means of one another, so intricately and inseparably, were by its nature an impossible attempt. Their relative importance, even to the wisest mind, will vary in different times, according to the special wants and dispositions of these times. Meanwhile, it seems clear enough that only in the right co-ordination of the two, and the vigorous forwarding of *both*, does our true line of action lie. Undue cultivation of the inward or Dynamical province leads to idle, visionary, impracticable courses, and especially in rude eras, to Superstition and Fanaticism, with their long train of baleful and well-known evils. Undue cultivation of the outward, again, though less immediately prejudicial, and even for the time productive of many palpable benefits, must, in the long run, by destroying Moral Force, which is the parent of all other Force, prove not less certainly, and perhaps still more hopelessly, pernicious. This, we take it, is the grand characteristic of our age. By our skill in Mechanism, it has come to pass that, in the management of external things, we excel all other ages; while in whatever respects the pure moral nature, in true dignity of soul and character, we are perhaps inferior to most civilized ages.

In fact, if we look deeper, we shall find that this faith in Mechanism has now struck its roots deep into men's most intimate, primary sources of conviction; and is thence sending up, over his whole life and activity, innumerable stems—fruit-bearing and poison-bearing. The truth is, men have lost their belief in the Invisible, and believe, and hope, and work only in the Visible; or, to speak it in other words, This is not a Religious age. Only the material, the immediately practical, not the divine and spiritual, is important to us. The infinite, absolute character of Virtue has passed into a finite, conditional one; it is no longer a worship of the Beautiful and Good; but a calculation of the Profitable. Worship, indeed, in any sense, is not recognised among us, or is mechanically explained into Fear of pain, or Hope of pleasure. Our true Deity is Mechanism. It has subdued external Nature for us, and, we think, it will do all other things. We are Giants in physical power: in a deeper than a metaphorical sense, we are Titans, that strive, by heaping mountain on mountain, to conquer Heaven also.

The strong mechanical character, so visible in the spiritual pursuits and methods of this age, may be traced much farther into the condition and prevailing disposition of our spiritual nature itself. Consider, for example, the general fashion of Intellect in this era. Intellect, the power man has of knowing and believing, is now nearly synonymous with Logic, or the mere power of arranging and communicating. Its implement is not Meditation, but Argument. 'Cause and effect' is almost the only category under which we look at, and work with, all Nature. Our first question with regard to any object is not, What is it? but, How is it? We are no longer instinctively driven to apprehend, and lay to heart, what is Good and Lovely, but rather to enquire, as onlookers, how it is produced, whence it comes, whither it goes? Our favourite Philosophers have no love and no hatred; they stand among us not to do, or to create any thing, but as a sort of Logic-mills to grind out the true causes and effects of all that is done and created. To the eye of a Smith, a Hume, or a Constant, all is well that works quietly. An Order of Ignatius Loyola, a Presbyterianism of John Knox, a Wickliffe, or a Henry the Eighth, are simply so many mechanical phenomena, caused or causing.

The *Euphuist* of our day differs much from his pleasant predecessors. An intellectual dapperling of these times boasts chiefly of his irresistible perspicacity, his 'dwelling in the daylight 'of truth,' and so forth; which, on examination, turns out to be a dwelling in the *rush*-light of 'closet-logic,' and a deep unconsciousness that there is any other light to dwell in; or any other objects to survey with it. Wonder, indeed, is, on all hands, dying out: it is the sign of uncultivation to wonder. Speak to any small man of a high, majestic Reformation, of a high, majestic Luther to lead it, and forthwith he sets about 'accounting' for it! how the 'circumstances of the time' called for such a character, and found him, we suppose, standing girt and road-ready, to do its errand; how the 'circumstances 'of the time' created, fashioned, floated him quietly along into the result; how, in short, this small man, had he been there, could have performed the like himself! For it is the 'force of 'circumstances' that does every thing; the force of one man can do nothing. Now all this is grounded on little more than a metaphor. We figure Society as a 'Machine,' and that mind is opposed to mind, as body is to body; whereby two, or at most ten, little minds must be stronger than one great mind. Notable absurdity! For the plain truth, very plain, we think, is, that minds are opposed to minds in quite a different way; and *one* man that has a higher Wisdom, a hitherto unknown spiritual Truth in him, is stronger, not than ten men that have it not, or than ten thousand, but than *all* men, that have it not; and stands among them with a quite ethereal, angelic power, as with a sword out of Heaven's own armoury, sky-tempered, which no buckler, and no tower of brass, will finally withstand.

But to us, in these times, such considerations rarely occur. We enjoy, we see nothing by direct vision; but only by reflexion, and in anatomical dismemberment. Like Sir Hudibras, for every Why we must have a Wherefore. We have our little *theory* on all human and divine things. Poetry, the workings of genius itself, which in all times, with one or another meaning, has been called Inspiration, and held to be mysterious and inscrutable, is no longer without its scientific exposition. The building of the lofty rhyme is like any other masonry or bricklaying: we have theories of its rise, height, decline, and fall— which latter, it would seem, is now near, among all people. Of our 'Theories of Taste,' as they are called, wherein the deep, infinite, unspeakable Love of Wisdom and Beauty, which dwells in all men, is 'explained,' made mechanically visible, from 'Associ- 'ation,' and the like, why should we say any thing? Hume has written us a 'Natural History of Religion;' in which one Natural History, all the rest are included. Strangely, too, does the general feeling coincide with Hume's in this wonderful problem; for whether his 'Natural History' be the right one or not, that Religion must have a Natural History, all of us, cleric and laic, seem to be agreed. He indeed regards it as a Disease, we again as Health; so far there is a difference; but in our first principle we are at one.

To what extent theological Unbelief, we mean intellectual dissent from the Church, in its view of Holy Writ, prevails at this day, would be a highly important, were it not, under any circumstances, an almost impossible enquiry. But the Unbelief, which is of a still more fundamental character, every man may see prevailing, with scarcely any but the faintest contradiction, all around him; even in the Pulpit itself. Religion, in most countries, more or less in every country, is no longer what it was, and should be—a thousand-voiced psalm from the heart of Man to his invisible Father, the fountain of all Goodness, Beauty, Truth, and revealed in every revelation of these; but for the most part, a wise prudential feeling grounded on mere calculation; a matter, as all others now are, of Expediency and Utility; whereby some smaller quantum of earthly enjoyment may be exchanged for a far larger quantum of celestial enjoyment. Thus Religion, too, is Profit; a working for wages; not Reverence, but vulgar Hope or Fear. Many, we know, very many, we hope, are still religious in a far different sense; were it not so, our case were too desperate: But to witness that such is the temper of the times, we take any calm observant man, who agrees or disagrees in our feeling on the matter, and ask him whether our *view* of it is not in general well-founded.

Literature, too, if we consider it, gives similar testimony. At no former era has Literature, the printed communication of Thought, been of such importance as it is now. We often hear that the Church is in danger; and truly so it is—in a danger it seems not to know of: For, with its tithes in the most perfect safety, its functions are becoming more and more superseded. The true Church of England, at this moment, lies in the Editors of its Newspapers. These preach to the people daily, weekly; admonishing kings themselves; advising peace or war, with an authority which only the first Reformers, and a long-past class of Popes, were possessed of; inflicting moral censure; imparting moral encouragement, consolation, edification; in all ways, diligently 'administering the Discipline of the Church.' It may be said, too, that in private disposition, the new Preachers somewhat resemble the Mendicant Friars of old times: outwardly full of holy zeal; inwardly not without stratagem, and hunger for terrestrial things. But omitting this class, and the boundless host of watery personages who pipe, as they are able, on so many scrannel straws, let us look at the higher regions of Literature, where, if anywhere, the pure melodies of Poesy and Wisdom should be heard. Of natural talent there is no deficiency: one or two richly-endowed individuals even give us a superiority in this respect. But what is the song they sing? Is it a tone of the Memnon Statue, breathing music as the *light* first touches it? a 'liquid wisdom,' disclosing to our sense the deep, infinite harmonies of Nature and man's soul? Alas, no! It is not a matin or vesper hymn to the Spirit of all Beauty, but a fierce clashing of cymbals, and shouting of multitudes, as children pass through the fire to Molech! Poetry itself has no eye for the Invisible. Beauty is no longer the god it worships, but some brute image of Strength; which we may well call an idol, for true Strength is one and the same with Beauty, and its worship also is a hymn. The meek, silent Light can mould, create, and purify all nature; but the loud Whirlwind, the sign and product of Disunion, of Weakness, passes on, and is forgotten. How widely this veneration for the physically Strongest has spread itself through Literature, any one may judge, who reads either criticism or poem. We praise a work, not as ' true,' but as ' strong;' our highest praise is that it has ' affected' us, has ' terrified' us. All this, it has been well observed, is the ' maximum of the Barbarous,' the symptom, not of vigorous refinement, but of luxurious corruption. It speaks much, too, for men's indestructible love of truth, that nothing of this kind will abide with them; that even the talent of a Byron cannot permanently seduce us into idol-worship; but that he, too, with all his wild syren charming, already begins to be disregarded and forgotten.

Again, with respect to our Moral condition: here also, he who runs may read that the same physical, mechanical influences are everywhere busy. For the ' superior morality,' of which we hear so much, we, too, would desire to be thankful: at the same time, it were but blindness to deny that this ' su-' perior morality' is properly rather an ' inferior criminality,' produced not by greater love of Virtue, but by greater perfection of Police; and of that far subtler and stronger Police, called Public Opinion. This last watches over us with its Argus eyes more keenly than ever; but the ' inward eye' seems heavy with sleep. Of any belief in invisible, divine things, we find as few traces in our Morality as elsewhere. It is by tangible, material considerations that we are guided, not by inward and spiritual. Self-denial, the parent of all virtue, in any true sense of that word, has perhaps seldom been rarer: so rare is it, that the most, even in their abstract speculations, regard its existence as a chimera. Virtue is Pleasure, is Profit; no celestial, but an earthly thing. Virtuous men, Philanthropists, Martyrs, are happy accidents; their ' taste' lies the right way! In all senses, we worship and follow after Power; which may be called a physical pursuit. No man now loves Truth, as Truth must be loved, with an infinite love; but only with a finite love, and as it were *par amours*. Nay, properly speaking, he does not *believe* and know it, but only ' *thinks*' it, and that ' there is every probability!' He preaches it aloud, and rushes courageously forth with it—if there is a multitude huzzaing at his back! yet ever keeps looking over his shoulder, and the instant the huzzaing languishes, he too stops short. In fact, what morality we have takes the shape of Ambition, of Honour; beyond money and money's worth, our only rational blessedness is Popularity. It were but a fool's trick to die for conscience. Only for 'character,' by duel, or, in case of extremity, by suicide, is the wise man bound to die. By arguing on the 'force 'of circumstances,' we have argued away all force from ourselves; and stand leashed together, uniform in dress and movement, like the rowers of some boundless galley. This and that may be right and true; *but* we must not do it. Wonderful 'Force of 'Public Opinion!' We must act and walk in all points as it prescribes; follow the traffic it bids us, realize the sum of money, the degree of 'influence' it expects of us, *or* we shall be lightly esteemed; certain mouthfuls of articulate wind will be blown at us, and this what mortal courage can front? Thus, while civil Liberty is more and more secured to us, our moral Liberty is all but lost. Practically considered, our creed is Fatalism; and, free in hand and foot, we are shackled in heart and soul, with far straiter than feudal chains. Truly may we say, with the Philosopher, 'the deep meaning of the Laws of Me'chanism lies heavy on us;' and in the closet, in the marketplace, in the temple, by the social hearth, encumbers the whole movements of our mind, and over our noblest faculties is spreading a nightmare sleep.

These dark features, we are aware, belong more or less to other ages, as well as to ours. This faith in Mechanism, in the all-importance of physical things, is in every age the common refuge of Weakness and blind Discontent; of all who believe, as many will ever do, that man's true good lies without him, not within. We are aware also, that, as applied to ourselves in all their aggravation, they form but half a picture; that in the whole picture there are bright lights as well as gloomy shadows. If we here dwell chiefly on the latter, let us not be blamed: it is in general more profitable to reckon up our defects than to boast of our attainments.

Neither, with all these evils more or less clearly before us, have we at any time despaired of the fortunes of society. Despair, or even despondency, in that respect, appears to us, in all cases, a groundless feeling. We have a faith in the imperishable dignity of man; in the high vocation to which, throughout this his earthly history, he has been appointed. However it may be with individual nations, whatever melancholic speculators may assert, it seems a well-ascertained fact that, in all times, reckoning even from those of the Heraclides and Pelasgi, the happiness and greatness of mankind at large has been continually progressive. Doubtless this age also is advancing. Its very unrest, its ceaseless activity, its discontent, contains matter of promise. Knowledge, education, are opening the eyes of the humblest—are increasing the number of thinking minds without limit. This is as it should be; for, not in turning back, not in resting, but only in resolutely struggling forward, does our life consist. Nay, after all, our spiritual maladies are but of Opinion; we are but fettered by chains of our own forging, and which ourselves also can rend asunder. This deep, paralysed subjection to physical objects comes not from nature, but from our own unwise mode of *viewing* Nature. Neither can we understand that man wants, at this hour, any faculty of heart, soul, or body, that ever belonged to him. ' He who has been born, has been a ' First Man;' has had lying before his young eyes, and as yet unhardened into scientific shapes, a world as plastic, infinite, divine, as lay before the eyes of Adam himself. If Mechanism, like some glass bell, encircles and imprisons us, if the soul looks forth on a fair heavenly country which it cannot reach, and pines, and in its scanty atmosphere is ready to perish—yet the bell is but of glass; ' one bold stroke to break the bell in pieces, ' and thou art delivered !' Not the invisible world is wanting, for it dwells in man's soul, and this last is still here. Are the solemn temples, in which the Divinity was once visibly revealed among us, crumbling away? We can repair them, we can rebuild them. The wisdom, the heroic worth of our forefathers, which we have lost, we can recover. That admiration of old nobleness, which now so often shows itself as a faint *dilettantism*, will one day become a generous emulation, and man may again be all that he has been, and more than he has been. Nor are these the mere daydreams of fancy—they are clear possibilities; nay, in this time they are even assuming the character of hopes. Indications we do see, in other countries and in our own, signs infinitely cheering to us, that Mechanism is not always to be our hard taskmaster, but one day to be our pliant, all-ministering servant; that a new and brighter spiritual era is slowly evolving itself for all men. But on these things our present course forbids us to enter.

Meanwhile that great outward changes are in progress can be doubtful to no one. The time is sick and out of joint. Many things have reached their height; and it is a wise adage that tells us, ' the darkest hour is nearest the dawn.' Whenever we can gather any indication of the public thought, whether from printed books, as in France or Germany, or from Carbonari rebellions and other political tumults, as in Spain, Portugal, Italy, and Greece, the voice it utters is the same. The thinking minds of all nations call for change. There is a deep-lying struggle in the whole fabric of society; a boundless, grinding collision of the New with the Old. The French Revolution, as is now visible enough, was not the parent of this mighty movement, but its offspring. Those two hostile influences, which always exist in human things, and on the constant intercommunion of which depends their health and safety, had lain in separate masses accumulating through generations, and France was the scene of their fiercest explosion; but the final issue was not unfolded in that country; nay, it is not yet anywhere unfolded. Political freedom is hitherto the object of these efforts; but they will not and cannot stop there. It is towards a higher freedom than mere freedom from oppression by his fellow-mortal, that man dimly aims. Of this higher, heavenly freedom, which is 'man's 'reasonable service,' all his noble institutions, his faithful endeavours and loftiest attainments, are but the body, and more and more approximated emblem.

On the whole, as this wondrous planet, Earth, is journeying with its fellows through infinite space, so are the wondrous destinies embarked on it journeying through infinite time, under a higher guidance than ours. For the present, as our Astronomy informs us, its path lies towards *Hercules*, the constellation of *Physical Power*: But that is not our most pressing concern. Go where it will, the deep HEAVEN will be around it. Therein let us have hope and sure faith. To reform a world, to reform a nation, no wise man will undertake; and all but foolish men know that the only solid, though a far slower reformation, is what each begins and perfects on *himself*.

Edward Everett

Fourth of July at Lowell (1830)

Edward Everett (1794–1865) was the most precocious and eloquent voice of the New England Whig establishment during the decades of industrial ascendance in the region. He was congressman, senator, governor of Massachusetts, ambassador to England, secretary of state, president of Harvard, as well as the first American to earn a Ph.D. When he spoke in behalf of American manufactures on Independence Day 1830, he was congressman from a district (Middlesex) that included both the cultural stillwaters of Cambridge and the industrial headwaters of Lowell. Everett's rhetoric in this excerpt is an example of the "technological sublime" at its most orotund and influential. His unqualified encouragement of manufactures in 1830 represented a shift in his position. In the controversies preceding the Tariff of 1824, he had taken the side of Boston merchants who were not yet involved in manufacturing and who opposed high tariffs. The arguments Everett adopted in this 1830 address presumably indicated a general shift in the views of commercial Boston as well.

Edward Everett, "Fourth of July at Lowell," *Orations* (Boston, 1850), volume 3, pp. 52–65. See also Everett's address to the American Institute, "American Manufactures" (1831), volume 3, pp. 69–99.

It is the spirit of a free country which animates and gives energy to its labor; which puts the mass in action, gives it motive and intensity, makes it inventive, sends it off in new directions, subdues to its command all the powers of nature, and enlists in its service an army of machines, that do all but think and talk. Compare a hand loom with a power loom; a barge, poled up against the current of a river, with a steamer breasting its force. The difference is not greater between them than between the efficiency of labor under a free or despotic government; in an independent state or a colony. I am disposed to think that the history of the world would concur with our own history, in proving that, in proportion as a community is under the full operation of the encouraging prospects and generous motives which exist in a free country, precisely in that proportion will its labor be efficient, enterprising, inventive, and productive of all the blessings of life.

This is a general operation of the establishment of an independent government in the United States of America, which has not perhaps been enough considered among us. We have looked too exclusively to the mere political change, and the substitution of a domestic for a foreign rule, as an historical fact, flattering to the national vanity. There was also another consequence of very great practical importance, which, in celebrating the declaration of independence at Lowell, ought not to pass unnoticed. While we were colonies of Great Britain, we were dependent on a government in which we were not represented. The laws passed by the Imperial Parliament were not passed for the benefit of the colony as their immediate object, but only so far as the interest of the colony was supposed to be consistent with that of the mother country. It was the principle of the colonial system of Europe, as it was administered before the revolution, to make the colonies subserve the growth and wealth of the parent state. The industry of the former was accordingly encouraged where it contributed to this object; it was discouraged and restrained where it was believed to have an opposite tendency. Hence the navigation law, by which the colonies were forbidden to trade directly with any but British ports. It is not easy to form a distinct conception of the paralyzing effect of such a restraint upon the industry of a population like ours, seated upon a coast which nature has indented with capacious harbors, and with a characteristic aptitude, from the earliest periods of our existence as a community, for maritime adventure.

The case was still worse in reference to manufactures. The climate of the northern and middle colonies is such as to make the manufacture of clothing one of the great concerns of civilized life. Apart from all views to the accumulation of wealth, the manufactures of wool, (and of late years of cotton,) of iron, leather, and wood, are connected with the comfortable subsistence of every family. And yet to all these branches of industry, except so far as they were carried on for household consumption, not only was no legislative favor extended by the home government, but they were from time to time made the subject of severe penal statutes

In this respect there was an important difference in the condition of the northern and southern colonies. The industry of the latter was encouraged; bounties were offered for the cultivation of some of their staples; the growth of a rival article, in the case of tobacco, was prohibited in the mother country; and a free market opened to all the agricultural produce, and the raw materials of manufacture, which the colony could export. The Northern States were hardly able to feed themselves. They had no agricultural produce to give in exchange for foreign manufactures; and that species of industry which was so peculiarly necessary, not so much to their prosperity and growth as to their subsistence, was inhibited by act of Parliament.

Accordingly, when the country entered upon the condition of independent political existence, of the three great branches of human industry, its agriculture had been fostered and patronized; its navigation, though subject to restraint, still vigorous within the permitted limits; and in the department of the fisheries, as we have seen, carried on with such boldness and success as to attract the admiration of the world: but its manufacturing interests were suffering under the effect of a century of actual warfare, and the loss of all the skill which would have been acquired in a century's experience.

The establishment, therefore, of a prosperous manufacturing town like Lowell, regarded in itself, and as a specimen of other similar seats of American art and industry, may with propriety be considered as a peculiar triumph of our political independence They are, if I may so express it, the complement of the revolution. They redress the peculiar hardships of the colonial system. They not only do that which was not done, but which was not permitted to be done before the establishment of an independent government.

It is no part of my present purpose — in fact, I conceive it would be out of place on an occasion like this — to discuss the protective policy which has been extended to the manufactures of the country as far as it has been made a party question. It will, however, I think, strike every one that the view I am now taking of the subject is peculiarly appropriate to the fourth of July; and it is only in this connection that I propose to treat the subject. It is well known that the sagacious and intelligent persons, who have been principally concerned in establishing the manufactures of Massachusetts, have never been friends of what has been called a high tariff policy; and that all they have desired, in this respect, was, that after a very large amount of capital had, under the operation of the restrictive system, as it was called, and the war of 1812, been led to take this direction, and had grown up into one of the most important interests of the country, it should not be deprived of that moderate protection which might be accorded to it under the general revenue laws of the Union, and which was necessary to shield the American manufacturer against the fluctuations of the foreign market, and the effect of a condition of the laborer in foreign countries, to which no one can desire to see the labor of this country reduced.

Without, therefore, going at all into the merits of such a system, as a matter of political controversy, I have thought it appropriate to the occasion to point out, in a summary way, the connection of the growth of our manufactures with the independence of the country; and I believe it would not be difficult to show that no event, consequent upon the establishment of our independence, has been of greater public benefit.

Let us consider, first, the addition made to the capital of the country, by bringing into action the immense mechanical power which exists at the falls and rapids of our streams

Could the choice have been given to us, for the abode of our population, of a dead alluvial plain of twice the extent, every one feels that it would have been bad policy to accept the offer. Every one perceives that this natural water power is a vast accession to the wealth and capital of a state. The colonial system annihilated it, or, what was the same thing, prevented its application. To all practical purposes, it reduced the beautiful diversity of the surface — nature's grand and lovely landscape gardening of vale and mountain — to that dull alluvial level. The rivers broke over the rapids; but the voice of nature and Providence, which cried from them, " Let these be the seats of your creative industry," was uttered in vain. It was an element of prosperity which we held in unconscious possession. It is scarcely credible how completely the thoughts of men had been turned in a different direction. There is probably no country on the surface of the globe, of the same extent, on which a greater amount of this natural capital has been bestowed by Providence; but a century and a half passed by, not merely before it began to be profitably applied on a large scale, but before its existence even began to be suspected, and this in places where some of its greatest accumulations are found. If a very current impression in this community is not destitute of foundation, the site of Lowell itself was examined, no very long time before the commencement of the first factories here, and the report brought back was, that it presented no available water power. Does it not strike every one who hears me, that, in calling this water power into action, the country has gained just as much as it would by the gratuitous donation of the same amount of steam power; with the additional advantage in favor of the former, that it is, from the necessity of the case, far more widely distributed, stationed at salubrious spots, and unaccompanied with most of the disadvantages and evils incident to manufacturing establishments moved by steam in the crowded streets and unhealthy suburbs of large cities?

Of all this vast wealth bestowed upon the land by Providence, — brought into the common stock by the great partner

Nature, — the colonial system, as I have observed, deprived us; and it is only since the establishment of our own manufactures that we have begun to turn it to account. Even now, the smallest part of it has been rendered available; and what has thus far been done is not so much important for its own sake as for pointing the way and creating an inducement for further achievements in the same direction. There is water power enough in the United States, as yet unapplied, to sustain the industry of a population a hundred fold as large as that now in existence.

I do not wish to overstate this point, or to imply that it was owing to the restrictions of the colonial system that such a town as Lowell had not grown up in America in the middle of the last century, or at some still earlier period. There were not only no adequate accumulations of capital at that time, but those inventions and improvements in machinery had not been made, which have contributed so much to the growth of manufactures within the last fifty years. There is something, however, quite remarkable in the eagerness with which our forefathers, at a very early period, turned their attention to manufactures. Our colonial history contains very curious facts in reference to this subject; and it is not to be doubted that, if no legislative obstacles to the pursuit of this branch of industry had existed, and it had received the same kind of encouragement which was extended to the staple agriculture of the plantation colonies, a very different state of things would have existed at the revolution. Such certainly was the opinion entertained in England; for such was the principle of the whole legislation of the mother country. The great and sagacious statesmen who ruled her councils for a century would not, under all administrations, have persevered in a course of policy towards the colonies, manifestly arbitrary in its character and extremely vexatious in its operation, if they had not been persuaded that, but for this legislation, successful attempts would be made for the development of manufacturing industry.

Connected with this is another benefit of the utmost importance, and not wholly dissimilar in kind. The population gathered at a manufacturing establishment is to be fed, and this gives an enhanced value to the land in all the neighboring region. In this new country the land often acquires a value in this way for the first time. A large number of persons in this assembly are well able to contrast the condition of the villages in the neighborhood of Lowell with what it was ten or twelve years ago, when Lowell itself consisted of two or three quite unproductive farms. It is the contrast of production with barrenness; of cultivation with waste; of plenty with an absence of every thing but the bare necessaries of life. The effect, of course, in one locality is of no great account in the sum of national production throughout the extent of the land. But wherever a factory is established this effect is produced; and every individual to whom they give employment ceases to be a producer, and becomes a consumer of agricultural produce. The aggregate effect is, of course, of the highest importance.

This circumstance constitutes that superiority of a domestic over a foreign market, which is acknowledged by the most distinguished writers on political economy.

"The capital which is employed," says Adam Smith, "in purchasing in one part of the country, in order to sell in another the produce of the industry of that country, generally replaces by every such operation two distinct capitals, that had both been employed in the agriculture or manufactures of that country, and thereby enables them to continue that employment. When it sends out from the residence of the merchant a certain value of commodities, it generally brings back in return at least an equal value of other commodities. When both are the produce of domestic industry, it necessarily replaces, by every such operation, two distinct capitals, which had both been employed in supporting productive labor, and thereby enables them to continue that support. * * * The capital employed in purchasing foreign goods for home consumption, when this purchase is made with the produce of domestic industry, replaces too, by every such operation, two distinct capitals; but one of them only is employed in supporting domestic industry. * * * Though the returns, therefore, of the foreign trade of consumption should be as quick as those of the home trade, the capital employed in it will give but one half of the encouragement to the industry or productive labor of the country.

"But the returns of the foreign trade of consumption are very seldom so quick as those of the home trade. The returns of the home trade gen erally come in before the end of the year, and sometimes three or four times in the year. The returns of the foreign trade of consumption seldom come in before the end of the year, and sometimes not till after two or three years. A capital, therefore, employed in the home trade, will sometimes make twelve operations, or be sent out and returned twelve times, before a capital employed in the foreign trade of consumption has made one. If the capitals are equal, therefore, the one will give four and twenty times more encouragement and support to the industry of the country than the other." *

It is a familiar remark, of which all, I believe, admit the justice, that a variety of pursuits is a great advantage to a community. It affords scope to the exercise of the boundless variety of talent and capacity which are bestowed by nature, and which are sure to be developed by an intelligent population, if encouragement and opportunity are presented. In this point of view, the establishment of manufacturing industry, in all its departments, is greatly to be desired in every country, and has had an influence in ours of a peculiar character. I have already alluded to the fact that, with the erection of an independent government, a vast domain in the west was for the first time thrown freely open to settlement. As soon as the Indian frontier was pacified by the treaty of Greenville, a tide of emigration began to flow into the territory north-west of the Ohio; and from no part of the country more rapidly than from Massachusetts. In many respects this was a circumstance by no means to be regretted. It laid the foundation of the settlement of this most important and interesting region by a kindred race; and it opened to the mass of enterprising adventurers from the older states a short road to competence. But it was a serious drain upon the population of good old Massachusetts. The temptation of the fee simple of some of the best land in the world for two dollars an acre, and that on credit, (for such, till a few years ago, was the land system of the United States,) was too powerful to be resisted by the energetic and industrious young men of the New England States, in which there is but a limited quantity of fertile land, and that little of course

---

* Smith's Wealth of Nations, Vol. II. pp. 135, 136, Edinburgh ed. 1817

to be had only at a high price. The consequence was, that although the causes of an increase of population existed in New England to as great a degree, with this exception, as in any other part of the world, the actual increase was far from rapid ; scarcely amounting to one half of the average rate of the country.* The singular spectacle was exhibited of a community abounding in almost all the elements of prosperity, possessing every thing calculated to engage the affections of her children, annually deserted by the flower of her population. These remarks apply with equal force to all the other New England States, with the exception of Maine, where an abundance of unoccupied fertile land counterbalances the attractions of the west.

But this process of emigration has already received a check, and is likely to be hereafter adequately regulated by the new demand for labor of every kind and degree, consequent upon the introduction of manufactures. This new branch of industry, introduced into the circle of occupations, is creating a demand for a portion of that energy and spirit of acquisition which have heretofore carried our young men beyond the Ohio, and beyond the Wabash. Obvious and powerful causes will continue to direct considerable numbers in the same path of adventure; but it will not be, as it was at the commencement of this century, almost the only outlet for the population of the older states. In short, a new alternative of career is now presented to the rising generation.

There is another point of importance, in reference to manufactures, which ought not to be omitted in this connection, and it is this — that in addition to what may be called their direct operation and influence, manufactures are a great school for all the practical arts. As they are aided themselves, in the progress of inventive sagacity, by hints and materials from every art and every science, and every kingdom of nature, so, in their turn, they create the skill and furnish the instruments for carrying on almost all the other pursuits. Whatever per-

---

* From 1820 to 1830, although some check had been given to emigration from this state, the rate of increase of the population of Massachusetts was sixteen and one half per centum; that of the whole United States thirty-two and four fifths.

tains to machinery, in all the great branches of industry, will probably be found to have its origin, directly or indirectly, in that skill which can be acquired only in connection with manufactures. Let me mention two striking instances, the one connected with navigation, and the other with agriculture. The greatest improvement in navigation, since the invention of the mariner's compass, is the application of steam for propelling vessels. Now, by whom was this improvement made? Not by the merchant, or the mariner, fatigued by adverse winds and weary calms. The steam engine was the production of the machine shops of Birmingham, where a breath of the sea breeze never penetrated; and its application as a motive power on the water, was a result wrought out by the sagacity of Fulton from the science and skill of the millwright and the machinist. The first elements of such a mechanical system as the steam engine, in any of its applications, must be wanting in a purely commercial or agricultural community. Again, the great improvement in the agriculture of our Southern States, and in its results one of the greatest additions to the agricultural produce of the world, dates from the invention of the machine for separating the seed from the staple. This invention was not the growth of the region which enjoys its first benefits. The peculiar faculty of the mind to which these wonderful mechanical contrivances of modern art owe their origin, is not likely to be developed in the routine of agricultural operations. These operations have their effects on the intellectual character, — salutary effects, — but they do not cultivate the principle of mechanical contrivance, which peoples your factories with their lifeless but almost reasoning tenants.*

I cannot but think that the loss and injury unavoidably accruing to a people, among whom a long-continued exclu-

---

* At the time this Oration was delivered, a few miles only of railroad had been built in the United States, and the locomotive engine was hardly known. It need not be said that this application of the steam engine furnishes a still more striking illustration of the benefits conferred on every other interest by the mechanical skill which is not likely to be acquired except in the service of manufactures.

sive pursuit of other occupations has prevented the cultivation of the inventive faculty and the acquisition of mechanical skill, is greater in reference to the general affairs and business of life than in reference to the direct products of manufactures. The latter is a great economical loss, the nature and extent of which are described in the remarks which I have quoted from the great teacher of political economy; but a community in which the inventive and constructive principle is faintly developed is deprived of one of the highest capacities of reasoning mind. Experience has shown that the natural germ of this principle — the inborn aptitude — is possessed by our countrymen in an eminent degree; but, like other natural endowments, it cannot attain a high degree of improvement without cultivation. In proportion as a person, endowed with an inventive mechanical capacity, is acquainted with what has been already achieved, his command is extended over the resources of art, and he is more likely to enlarge its domain by new discoveries. Place a man, however intelligent, but destitute of all knowledge in this department, before one of the complicated machines in your factories, and he would gaze upon it with despairing admiration. It is much if he can be brought, by careful inspection and patient explanation, to comprehend its construction. A skilful artist, at the first sight of a new machine, comprehends, in a general way, the principles of its construction. It is only, therefore, in a community where this skill is widely diffused, and where a strong interest is constantly pressing for every practicable improvement, that new inventions are likely to be made, and more of those wonderful contrivances may be expected to be brought to light, which have changed the face of modern industry.

These important practical truths have been fully confirmed by the experience of Lowell, where the most valuable improvements have been made in almost every part of the machinery by which its multifarious industry is carried on. But however interesting this result may be, in an economical point of view, another lesson has been taught at Lowell, and our other well-conducted manufacturing establishments, which I deem vastly more important. It is well known that the degraded condition of the operatives in the old world had created a strong prejudice against the introduction of manufactures into this country. We were made acquainted, by sanitary and parliamentary reports, detailing the condition of the great manufacturing cities abroad, with a state of things revolting to humanity. It would seem that the industrial system of Europe required for its administration an amount of suffering, depravity, and brutalism, which formed one of the great scandals of the age. No form of serfdom or slavery could be worse. Reflecting persons, on this side of the ocean, contemplated with uneasiness the introduction, into this country, of a system which had disclosed such hideous features in Europe; but it must be frankly owned that these apprehensions have proved wholly unfounded. Were I addressing an audience in any other place, I could with truth say more to this effect than I will say on this occasion. But you will all bear me witness, that I do not speak the words of adulation when I say, that for physical comfort, moral conduct, general intelligence, and all the qualities of social character which make up an enlightened New England community, Lowell might safely enter into a comparison with any town or city in the land. Nowhere, I believe, for the same population, is there a greater number of schools and churches, and nowhere a greater number of persons whose habits and mode of life bear witness that they are influenced by a sense of character.

In demonstrating to the world that such a state of things is consistent with the profitable pursuit of manufacturing industry, you have made a discovery more important to humanity than all the wonderful machinery for weaving and spinning, — than all the miracles of water or steam. You have rolled off from the sacred cause of labor the mountain reproach of ignorance, vice, and suffering under which it lay crushed. You have gained, for the skilled industry required to carry on these mighty establishments, a place of honor in the great dispensation by which Providence governs the world. You have shown that the home-bred virtues of the parental roof are not required to be left behind by those who resort for a few years to these crowded marts of social industry; and, in the fruits of your honest and successful labor, you are daily carrying gladness to the firesides where you were reared.

The alliance which you have thus established between labor and capital (which is nothing but labor saved) may truly be called a *holy alliance*. It realizes, in a practical way, that vision of social life and action which has been started abroad, in forms, as it appears to me, inconsistent with the primary instincts of our nature, and wholly incapable of being ingrafted upon our modern civilization. That no farther progress can be made in this direction, I certainly would not say. It would be contrary to the great laws of human progress to suppose that, at one effort, this hard problem in social affairs had reached its perfect solution. But I think it may be truly said, that in no other way has so much been done, as in these establishments, to mingle up the interests of society; to confer upon labor, in all its degrees of cultivation, (from mere handiwork and strength up to inventive skill and adorning taste,) the advantages which result from previous accumulations. Without shaking that great principle by which a man calls what he has *his own*, whether it is little or much, (the corner stone of civilized life,) these establishments form a mutually beneficial connection between those who have nothing but their muscular power and those who are able to bring into the partnership the masses of property requisite to carry on an extensive concern, — property which was itself, originally, the work of men's hands, but has been converted, by accumulation and thrift, from labor into capital. This I regard as one of the greatest triumphs of humanity, morals, and, I will add, religion. The labor of a community is its great wealth — its most vital concern. To elevate it in the social scale, to increase its rewards, to give it cultivation and self-respect, should be the constant aim of an enlightened patriotism. There can be no other basis of a progressive Christian civilization. Woe to the land where labor and intelligence are at war! Happy the land whose various interests are united together by the bonds of mutual benefit and kind feeling!

Timothy Walker

Defence of Mechanical Philosophy
(1831)

This defense is an attack on Thomas Carlyle's essay "Signs of the Times." Timothy Walker (1802–1856), a young Harvard-trained lawyer, undertook to promote "the mechanical philosophy" in opposition to Carlyle's "mysticism." In this brief on behalf of what he called "modernism," Walker defended not only the technology of the "Age of Machinery" but also its science, philosophy, and religion. He connected mechanical progress with the progress of knowledge and the progress of the human race: "Mechanism," he argued, would emancipate the mind and allow people leisure to create in the image of the "Omnipotent Mind." Like many other promoters of American industry, Walker concluded that technological progress is the work of Providence.

Timothy Walker, "Defence of Mechanical Philosophy," *North American Review*, volume 31 (July 1831), pp. 122–136. For a discussion of Walker's controversy with Carlyle, see Leo Marx, *The Machine in the Garden*, pp. 174–191.

Art. V.—*Defence of Mechanical Philosophy.*
Signs of the Times.—Article VII. in the Ninety-eighth Number of the Edinburgh Review, for June, 1829.

The article which we have just named raises the grave and solemn question, whether mankind are advancing or not, in moral and intellectual attainments? The writer expresses his opinion, with sufficient distinctness, in the following words; 'In whatever respects the pure moral nature, in true dignity of soul and character, we are, *perhaps,* inferior to most civilized ages.' If this be true, it is a truth of deep and melancholy import. But is it true? Well may we pause, and ponder the matter carefully. What are the petty controversies which agitate sects, parties, or nations, compared with one which concerns the destinies of the whole human race? When we essay to cast the world's horoscope, and interpret auguries for universal man, it becomes us to approach the task with diffidence. And we do approach it with unfeigned diffidence. We despair of being able to rise to the height of the theme, on which we are to speak. Yet we feel that good may come even from the attempt.

Are we, then, in fact, degenerating? Has the hand been moved backward on the dial-plate of Time? Has the human race, comet-like, after centuries of advancement, swept suddenly round its perihelion of intelligence, and commenced its retrogradation? The author of the article before us, as we have seen, expresses, though with a *perhaps,* his belief of the affirmative. Throughout the whole article, with the exception of the last paragraph or two, of which the complexion is somewhat more encouraging, he draws most cheerless conclusions from the course which human affairs are taking. If the writer do not, as he humanely assures us in the end, ultimately despair of the destinies of our ill-starred race, he does, nevertheless, perceive baleful influences hanging over us. Noxious ingredients are working in the caldron. He has detected the 'midnight hag' that threw them in, and her name is Mechanism. A more malevolent spirit, in his estimation, does not come from the hateful abodes. The fated inhabitants of this planet are now under her pernicious sway, and she is most industriously plotting against their weal. To countervail her malignant efforts, the author invokes a spirit of a character most unlike the first. Her real name, as we shall see, is Mysticism, though this is not pronounced in the incantations.

Now we cannot help thinking, that this brilliant writer has conjured up phantoms for the sake of laying them again. At all events, we can see nothing but phantoms in what he opposes. In plain words, we deny the evil tendencies of Mechanism, and we doubt the good influences of his Mysticism. We cannot perceive that Mechanism, as such, has yet been the occasion of any injury to man. Some liberties, it is true, have been taken with Nature by this same presumptuous intermeddler. Where she denied us rivers, Mechanism has supplied them. Where she left our planet uncomfortably rough, Mechanism has applied the roller. Where her mountains have been found in the way, Mechanism has boldly levelled or cut through them. Even the ocean, by which she thought to have parted her quarrelsome children, Mechanism has encouraged them to step across. As if her earth were not good enough for wheels, Mechanism travels it upon iron pathways. Her ores, which she locked up in her secret vaults, Mechanism has dared to rifle and distribute. Still further encroachments are threatened. The terms uphill and downhill are to become obsolete. The horse is to be unharnessed, because he is too slow; and the ox is to be unyoked, because he is too weak. Machines are to perform all the drudgery of man, while he is to look on in self-complacent ease.

But where is the harm and danger of this? Why is every lover of the human race called on to plant himself in the path, and oppose these giant strides of Mechanism? Does this writer fear, that Nature will be dethroned, and Art set up in her place? Not exactly this. But he fears, if we rightly apprehend his meaning, that mind will become subjected to the laws of matter; that physical science will be built up on the ruins of our spiritual nature; that in our rage for machinery, we shall ourselves become machines. This we take to be the import of the following unusually plain passages; 'Not the external and physical alone is now managed by machinery, but the internal and spiritual also.'—'Philosophy, Science, Art, Literature, all depend upon Machinery.'—'Men are grown mechanical in head and in heart, as well as in hand.'—'Their whole efforts, attachments, opinions, turn on Mechanism, and are of a mechanical character.' These are pretty broad and sweeping assertions, and we might quote many equally positive, and of the same style and meaning. In fact, the whole article is a series of repetitions of this leading idea, under various shapes; and this idea we propose to examine and controvert.

And, on the face of the matter, is it likely that mechanical ingenuity is suicidal in its efforts? Is it probable that the achievements of mind are fettering and enthralling mind? Must the proud creator of Mechanism stoop to its laws? By covering our earth with unnumbered comforts, accommodations, and delights, are we, in the words of this writer, descending from our 'true dignity of soul and character?' Setting existing facts aside, and reasoning in the abstract, what is the fair conclusion? To our view, directly the contrary. We maintain, that the more work we can compel inert matter to do for us, the better will it be for our minds, because the more time shall we have to attend to them. So long as our souls are doomed to inhabit bodies, these bodies, however gross and unworthy they may be deemed, must be taken care of. Men have animal wants, which must and will be gratified at all events; and their demands upon time are imperious and peremptory. A certain portion of labor, then, must be performed, expressly for the support of our bodies. But at the same time, as we have a higher and nobler nature, which must also be cared for, the necessary labor spent upon our bodies should be as much abridged as possible, in order to give us leisure for the concerns of this better nature. The smaller the number of human beings, and the less the time it requires, to supply the physical wants of the whole, the larger will be the number, and the more the time left free for nobler things. Accordingly, in the absolute perfection of machinery, were that attainable, we might realize the absolute perfection of mind. In other words, if machines could be so improved and multiplied, that all our corporeal necessities could be entirely gratified, without the intervention of human labor, there would be nothing to hinder all mankind from becoming philosophers, poets, and votaries of art. The whole time and thought of the whole human race could be given to inward culture, to spiritual advancement. But let us not be understood as intimating a belief, that such a state of things will ever exist. This we do not believe, nor is it necessary to our argument. It is enough, if there be an approach thereto. And this we do believe is constantly making. Every sober view of the past confirms us in this belief.

In the first ages of the world, when Mechanism was not yet

known, and human hands were the only instruments, the mind scarcely exhibited even the feeblest manifestations of its power. And the reason is obvious. As physical wants could only be supplied by the slow and tedious processes of hand-work, every one's attention was thereby completely absorbed. By degrees, however, the first rudiments of Mechanism made their appearance, and effected some simple abbreviations. A portion of leisure was the necessary result. One could now supply the wants of two, or each could supply his own in half the time previously required. And now it was, that mind began to develope its energies, and assert its empire over all other things. Leisure gave rise to thought, reflection, investigation; and these, in turn, produced new inventions and facilities. Mechanism grew by exercise. Machines became more numerous and more complete. The result was a still greater abridgment of labor. One could now do the work of ten, or each could do his own, in one tenth of the time before required. It is needless to follow the deduction farther. Every one knows that now, in many of the departments of labor, one can perform, by the help of machinery, the work of hundreds; or, supposing no division, each could perform his own in a hundredth part of the time before required. The consequence is, that there has never been a period, when so large a number of minds, in proportion to the whole, were left free to pursue the cultivation of the intellect. This is altogether the result of Mechanism, forcing inert matter to toil for man. And had it been reached gradually, commencing at the Creation, and continuing until now, the blessing would have been without alloy. But unhappily the progress has not been gradual. Of late, Mechanism has advanced *per saltum*, and the world has felt a temporary inconvenience from large numbers being thrown suddenly out of employment, while unprepared to embark in any thing else. But this evil must be from its nature temporary, while the advantage resulting from a release of so large a proportion of mankind from the thraldom of physical labor, will be as lasting as the mind. And hence it is, that we look with unmixed delight at the triumphant march of Mechanism. So far from enslaving, it has emancipated the mind, in the most glorious sense. From a ministering servant to matter, mind has become the powerful lord of matter. Having put myriads of wheels in motion by laws of its own discovering, it rests, like the Omnipotent Mind, of which it is the image, from its work of creation, and pronounces it good.

When we attempt to convey an idea of the infinite attributes of the Supreme Being, we point to the stupendous machinery of the universe. From the ineffable harmony and regularity, which pervade the whole vast system, we deduce the infinite power and intelligence of the Creating Mind. Now we can perceive no reason, why a similar course should not be pursued, if we would form correct conceptions of the dignity and glory of man. Look at the changes he has effected on the earth; so great, that could the first men revisit their mortal abodes, they could scarcely recognize the planet they once inhabited. Fitted up as it now is, with all the splendid furniture of civilization, it no more resembles the bleak, naked, incommodious earth, upon which our race commenced their improvements, than the magnificent palace resembles the low, mud-walled cottage. From the effect, turn your attention to the cause. Examine the endless varieties of machinery which man has created. Mark how all the complicated movements co-operate, in beautiful concert, to produce the desired result. Before we conclude that man's dignity is depreciated in the contrivance and use of this machinery, let us remember, that a precisely analogous course of reasoning must conduct us to the conclusion, that the act of Creation subtracted from the glory of the Creator; that the Infinite Mind, as it brooded from eternity over chaos, was more transcendently glorious, than when it returned from its six days' work, to contemplate a majestic world. We accordingly believe there is nothing irreverent in the assertion, that the finite mind in no respect approximates so nearly to a resemblance of the Infinite Mind, as in the subjugation of matter, through the aid of Mechanism, to fixed and beneficial laws,—to laws ordained by God, but discovered and applied by man.

If the views now presented be correct, it follows that the mechanical enterprise, with which our age is so alive, far from being unfavorable to our spiritual growth, is the one thing needful to furnish the freedom and leisure necessary for intellectual exercises,—to establish mankind in the *otium cum dignitate,* in a higher sense than even Cicero conceived it. But we may be referred, by way of refutation, to some of the renowned nations of antiquity, for which Mechanism effected little or nothing, but which, nevertheless, ' in true dignity of soul and character,' would be pronounced by the writer, whose views we are examining, superior to any of the present day.

Greece may be selected as the most prominent illustration. To Greece, then, let us look. But let it be borne in mind, that we are speaking of society in the mass, and that our doctrine is, that men must be released from the bondage of perpetual bodily toil, before they can make great spiritual attainments. And now the question is, how came Greece to achieve her high intellectual supremacy, when all her work was performed by hand? The answer, so far as it respects this discussion, is ready. The Greeks themselves did not toil. Every reader of their history knows, that labor, physical labor, was stigmatized as a disgrace. Their wants were supplied by levying tribute upon all other nations, and keeping slaves to perform their drudgery at home. Hence their leisure. Force did for them, what machinery does for us. But what was the condition of the surrounding world? It is explained in a word. All other men had to labor for them; and as these derived no helps from Mechanism, manual labor consumed their whole lives. And hence their spiritual acquisitions have left no trace in history.

Now if we are willing to recur to that barbarous principle, that one nation may purchase itself leisure, as the Greeks did, by aggressions upon the rest, and if all other nations can be persuaded to submit to the experiment, we may no doubt behold a people, spurning all mechanical improvements, and yet attaining to a surprising 'dignity of soul and character.' But so long as it continues to be settled by compact among the nations, that each shall produce the means of subsistence within itself, or else an equivalent to exchange with others; and so long as the balance of power continues to be so adjusted, as to prevent any one from living upon the rest through the force of superior numbers; we see not how we can avoid the conclusion, that that nation will make the greatest intellectual progress, in which the greatest number of labor-saving machines has been devised. It may not produce a Newton, Milton, or Shakspeare, but it will have a mass of thought, reflection, study, and contemplation perpetually at work all over its surface, and producing all the fruits of mental activity.

But this writer has not confined his warfare to the world as a whole. He has divided mankind into classes, and attacked them in detail. We shall try to follow him through his campaign. One remark, however, upon the name which he has given to the age. 'It is not an Heroical, Devotional, Philosophical, or Moral Age, but, above all others, the Mechanical Age. It is the Age of Machinery, in every outward and *inward* sense of that word.' It may puzzle our readers as much as it does ourselves, to understand what is meant by the 'inward sense of machinery.' We are still more perplexed to understand how the following charge, which seems intended as unusually severe, can be construed by thinking men into any thing else than substantial eulogy. 'With its whole, undivided might, it [this age] forwards, teaches, and practises, *the great art of adapting means to ends.* Nothing is now done directly, or by hand; all is by rule, and calculated contrivance. For the simplest operation, some helps and accompaniments, some cunning, abbreviating process is in readiness.' Now take away the lurking sneer with which this is said, and we see not how it would be possible to crowd more praise into a smaller compass. It is no small part of wisdom, to possess 'the capacity of adapting means to ends.' What would the writer have us do? Pursue ends without regard to means?

*
On the whole, we have no wish to disguise the feeling of strong dissatisfaction, excited in us, by the article under consideration. We consider its tendency injurious, and its reasoning unsound. That it has some eloquent passages must be admitted, but when we hear distinguished philosophers spoken of as 'logic-mills,'—the religion of the age as 'a working for wages,'—our Bible societies as 'supported by fomenting of vanities, by puffing, intrigue, and chicane,'—and all descriptions of men 'from the cartwright up to the code-maker,' as mere 'mechanists;' when we further hear 'the grand secrets of necessity and free-will,'—' our mysterious relations to time and space,'—and 'the deep, infinite harmonies of nature and man's soul,'—brought repeatedly forward under the most varied forms of statement, as the legitimate objects of philosophical inquiry, and the most illustrious of the living and the dead, men whom we never think of but as benefactors of our race, made the objects of satire and ridicule, because they have preferred the *terra firma* of mechanical philosophy to the unstable quagmire of mystic conjecture;—we find it difficult not to regard the Essay rather as an effort of paradoxical ingenuity,—the sporting of an adventurous imagination with settled opinions,—than as a serious inquiry after truth.

Indeed the writer himself seems to think, towards the end, that he has gone too far; and deems it prudent, in contradiction, as it seems to us, to the assertion first quoted, as well as to the whole tenor of the article, to insert the following *saving clause*:—' It seems a well ascertained fact, that in all times, reckoning even from those of the Heraclides and Pelasgi, *the happiness and greatness of mankind at large, have been continually progressive.*' This is one of the few assertions in the article, in which we altogether agree with the author. We do entertain an unfaltering belief in the permanent and continued improvement of the human race, and we consider no small portion of it, whether in relation to the body or the mind, as the result of mechanical invention. It is true, that the progress has not always been regular and constant. In happy times it has been so rapid, as to fill the benevolent with inexpressible joy. But anon, clouds have gathered over the delightful prospect,—evil influences, but not mechanical, have operated,—evil times have succeeded,—and human nature has undergone a disastrous eclipse. But it has been only an eclipse, not an extinguishment of light. And frequent as these alternations have been, mankind are found to have been constant gainers. The flood has always been greater than the ebb. Each great billow of time has left men further onward than its predecessor. This could be proved, if necessary, by a thousand references. Darkness has indeed given a name to some ages, but light on the whole has immensely preponderated; and it is this conviction which nerves the heart and invigorates the arm of philanthropy. They who feel this divine impulse, know that the labors of kindred spirits in past ages have not been in vain. They see Atlantis, Utopia, and the Isles of the Blest, nearer than those who first descried them. These imaginary abodes of pure and happy beings, which have been conceived by the most ardent lovers of their kind, we delight to contemplate; for we regard them as types and shadows of a higher and better condition of human nature, towards which we are surely though slowly tending.

But let us not be misunderstood. The condition we speak of, is not one of perfection. This we neither believe in, nor hope for. Supposing it possible in the nature of things, it would be any thing but desirable. For with nothing left to achieve nor gain, existence would become empty and vapid. But if, with this explanation, our views should pass for visionary, we cannot help it. We cannot go back to the origin of mankind and trace them down to the present time, without believing it to be a part of the providence of God, that his creatures should be perpetually advancing. The first men must have been profoundly ignorant, except so far as the Supreme Being communicated with them directly. But with them commenced a series of inventions and discoveries, which have been going on, up to the present moment. Every day has beheld some addition to the general stock of information. When the exigency of the times has required a new truth to be revealed, it has been revealed. Men gifted beyond the ordinary lot, have been raised up for the purpose; witness Cadmus, Socrates, and the other sages of Greece, Cicero and the other sages of Rome, Columbus, Galileo, Bacon, Newton, and the other giant spirits of modern times. We cannot regard it as an abuse of language to call such men inspired, that is, pre-eminently endowed beyond all their contemporaries, and moved by the invisible agency of God, to enlighten the world on subjects, which had never till they spoke, occupied the minds of men. In other words, we believe that the appearance of such men, at the exact times when all things were ready for the disclosures they were to make, was not the result of accident, but the work of an overruling Providence. And if such has been the beneficent operation of Providence upon the minds of men in all past times,—if whenever a revelation was needed, He has communicated it, and in the exact measure in which it was needed,—how can we, without irreverence, adopt any other conclusion, than that He, who changeth not, will still continue, through all future time, to make known through gifted men, as fast as the world is prepared to receive them, new truths from His exhaustless store?

Alexander Everett

Memorial of the New York Convention of the Friends of Domestic Industry (1831)

The New York Convention of the Friends of Domestic Industry met in October 1831 to urge continued high protective tariffs at a time when the 1828 "Tariff of Abominations" was under wide attack. The immediate cause of this convention was an antitariff convention held in Philadelphia a few months earlier. Nearly five hundred manufacturers, most of them from New England and the Middle Atlantic states, attended the New York meeting. They heard speeches on the need for a high tariff and appointed committees to prepare reports on the condition of various industries and memorials to the Senate and House of Representatives.

This memorial to the House of Representatives was written by Alexander Everett (1790–1847), brother of Edward Everett, diplomat, and editor of the *North American Review*. Much of his discussion of the advantages of industry had appeared in his work of social commentary, *America* (1827).

Reprinted here is the section of the memorial on the "economic, moral, and political" advantages of manufactures. The remainder of the piece deals with the manner in which protection for manufactures should be provided.

[Alexander Everett], "Memorial of the New York Convention, to the Congress of the United States," in *Journal of the Proceedings of the Friends of Domestic Industry, in General Convention Met at the City of New York, October 26, 1831* (Baltimore, 1831), pp. 134–136, 144–152. The entire *Journal* has been reprinted: *Journal of the Proceedings of the Friends of Domestic Industry*, with introduction by Michael Hudson (New York: Garland Publishers, 1975). See also Michael Hudson's *Economics and Technology in Nineteenth-Century American Thought* (New York: Garland Publishers, 1975), pp. 158–165.

The time has been, and it is not a very remote one, when the question, whether the operation of domestic manufactures is or is not injurious, would have been readily answered by many persons in this country in the affirmative. They were habitually viewed with distrust, as likely to exercise a pernicious influence on the morals of the persons employed in them. In the earlier discussions of the protecting policy in Congress, and up to the period immediately preceding the war, this was the argument principally urged in opposition to it, and it was urged with peculiar zeal by the representatives of the eastern states. Experience has since corrected this error, and has shown that, in well-managed manufacturing establishments, the only ones which can long thrive, and which, of course, determine the average condition of the whole, the standard of morals is as high, if not higher, than it is in those belonging to any other branch of industry, not excepting agriculture in its best estate. The same experience has amply demonstrated the great positive advantages resulting from the possession of domestic manufactures, the most important of which your memorialists will now very briefly recapitulate.

The first and principal one is an accession of wealth, population, and political importance, exactly proportioned to the whole amount of capital, and the whole number of persons which they employ. This proposition may appear at first so trite and obvious as to be nearly or quite self-evident. That an agricultural village, town, and country, which obtains its supplies of manufactured articles within its own limits, is, in the same proportion more wealthy, populous, and flourishing, than one, in other respects similarly situated, which sends for them to a distant city, is a truth too familiar to every one from actual observation, to be made the subject of question. It is also apparent that the remark is equally true of communities politically independent. But, as this truth, however obvious, is constantly denied by the opponents of the protecting policy, and as the denial of it forms the first step in most of the reasoning by which they endeavour to support their opinions, it may be proper to dwell upon it for a few moments, for the purpose of stating it in a more precise form, and examining the objections that are alleged against it.

It may be remarked then, that, in every community, the wants of the people are regularly supplied by the co-operation of the three great branches of industry, agriculture, commerce and manufactures, in proportions determined by the degree of civilization. The three classes of laborers who are respectively engaged in these three different employments, all derive their means of subsistence, and the manufacturers their materials, from the products of agriculture, and must receive their share alike, whether they dwell within or without the country. If they live abroad, they still consume the same amount as before, of the products of the community for which they work, and the wealth and population of the latter are of coure regularly smaller in the same proportion. If it be supposed, for example—and such is nearly the case in the wealthy and civilized parts of the Christian world—that these three classes of laborers are regularly equal in number, then a community which receives its manufactures from abroad will regularly export one-third part of its agricultural produce in exchange for them, and will be one-third less wealthy and populous, than it would be if they were all supplied at home. Besides this, the labor required for exchanging the products of the two classes of workmen now belonging to distinct communities, regularly divides itself between those communities, and the agricultural country will thus sustain the loss of half its commercial, in addition to the whole of its manufacturing population. The distance at which the exchanges are made being now greater, it requires a greater amount of labor than before to effect them; or, in other words, to carry on the necessary commerce; so that, if this branch of industry before occupied a third part of the laborers, it will now occupy more. On this first and simplest view of the effect on a community of the absence of domestic manufactures, there is, therefore, a loss of more than half the wealth and population that naturally belong to it. But the wealth and population of every country form the true measures of its general prosperity and political importance; and a community which receives its manufactures from abroad, sustains, therefore, in each of these respects, a positive loss of half its natural advantages. But this is not all. The wealth and population thus lost, go to swell the wealth and population of some other country, of necessity, one with which the losing people have a close relation. If the elements of wealth and power belonging to the countries thus situated, be, in other respects, naturally equal, one now gains, and the other loses more than half the amount, and the relative forces of the two become as three to one. Thus, the want of domestic manufactures deprives a country of half its positive, and two-thirds of its relative importance; degrades it of course from its rank among the nations, and places it at the mercy of the powers with which it has the closest connexion, and to which it is naturally equal.

*

The *second* great advantage, resulting from the establishment of manufactures, is the superior steadiness of the home market. All speculations, founded on the supposed situation of distant markets, are, in their nature, doubtful; and, when they occupy capital and labor to any great extent, introduce a continual and disastrous fluctuation into commerce, and, indirectly, into the whole industry of the community. The home market, on the other hand, is in general fixed and certain. Its extent may be calculated, and the probable increase or diminution of demand foretold with sufficient exactness. Foreign commerce is a sort of game, in which fortune exercises as much influence at least, as prudence and skill. All calculations connected with it are not only more or less uncertain at the time when they are made, but are continually defeated by occurrences that intervene before their results can be realized. The whole capital embarked in this pursuit, and with it the happiness of its owners and their families, lie at the mercy of political events, or, in other words, of the caprice and violence of foreign powers. We cast our bread upon the waters, but whether, in this case, it return to us again after many days, is a matter of mere chance. When we have covered the sea with our products, a wanton belligerent (and some war is continually going on in one quarter or another,) issues a decree, and sweeps the whole into his own coffers. The United States, during the short period of their history, have experienced but too severely the truth of this remark in the injuries of this kind which they have sustained, not merely from the great powers, such as England, France and Spain, but even from the secondary States of Denmark, Naples, Holland, and the republics of Spanish America. Under such circumstances, we remonstrate —negotiate—go to war, perhaps—possibly, after the lapse of twenty or thirty years, obtain some partial satisfaction. In the mean time, the unfortunate individuals who were the victims of this legalized piracy, have seen their prospects for life blasted, and gone down, with their wives and children, in sorrow to the grave. No association can stand the force of these fatal shocks. Insurance companies sink under them like private fortunes. The only effectual remedy is the one employed by England, of maintaining a public navy sufficiently strong to command the ocean, and defy attack from any quarter; but the burden which such an establishment imposes upon private industry, makes the remedy nearly as bad as the disease.

Independently of the violent attacks to which the capital employed in these distant exchanges is exposed, the mere circumstance of dealing with foreign markets at a great distance, creates a dangerous uncertainty in the whole business. In time of war, the most extensive colonial and European markets are thrown open to our flour and other provisions, and our cultivators extend their enterprise in all directions, perhaps for years in succession. Peace comes at length, and all these markets are at once hermetrically sealed. Flour falls from ten or fifteen dollars a barrel to three or four, and ruin stalks at large through the fair fields of the middle states. Again: a panic terror is felt in England on account of a supposed deficiency in the supply of cotton actually on hand, and the value of the article takes a sudden rise. Our speculating merchants, incapable of estimating the correctness of these apprehensions, go on buying for exportation at extravagant prices. Immense supplies arrive in Europe. In the mean time, the imagined deficiency is found to be of little or no importance. The market is overstocked, and the merchants ruined. Finally, we are forced ourselves into war—as happened in the year 1812, by the mere effect of the embarrassments resulting from the defenceless state of the capital employed in foreign trade—the usual supply of foreign manufactures is checked. Immediately, large amounts of capital, following the direction which they would naturally have taken in time of peace had it not been for the very peculiar circumstances in which the United States have been placed, and to which your memorialists have already briefly alluded, are invested in domestic establishments, for the purpose of making up the deficiency. Every thing goes on prosperously until the war comes to a close. Within a few months after, our markets are inundated with British goods, cheaper than we can make them of equal quality, and the manufacturers are, in their turn, involved in one common ruin. It is in this way that the fluctuations incident to these distant and uncertain exchanges reach successively all the great branches of industry. The results which your memorialists have thus described, are not accidental, but the regular consequences of the state of things to which they are attributed, and must continue to recur, from time to time, wherever such a state of things exists. No foresight, prudence, or probity, furnishes the means of avoiding them; and so extensive have been the disasters which they have brought upon the United States since the close of the late war, that there are probably very few individuals in the country who have not felt them within the circle of their own immediate connexions. It is true, that where there is hazard, there is also occasionally gain as well as loss; but one result is hardly less pernicious than the other, though in a different way. Large and rapid fortunes, whether considered in their effect on the persons obtaining them, or on the community, are highly injurious to good morals and regular habits of industry. These, on the contrary, are promoted by a course of trade, which, when carried on with honesty and prudence, produces slow and moderate but certain profits; and such is that which naturally takes place in a community where the three principal branches of industry furnish each other with a reciprocal home market for the greater part of their respective products.

In pointing out the uncertainty and fluctuations which are naturally incident to foreign commerce, your memorialists would not be understood to intimate that this branch of industry is not to be pursued at all, but merely that it cannot, with advantage, be made the basis of all the business operations of the community. This is almost necessarily the case in countries of limited extent and dense population, which are thus condemned by the necessity of their position, to go through a continual course of ruinous revulsions. In more extensive countries, possessing within themselves the materials for a large internal trade, to which foreign commerce is in its nature subsidiary; the latter, instead of creating fluctuation and uncertainty, becomes itself a sort of regulator of the home market, and serves as a wholesome check upon the occasional fluctuations to which even the internal trade is exposed. Thus the trade with this country is habitually employed by the British manufacturers, (as far as our policy will permit them to use it,) to relieve the encumbered state of their own home market; and the indirect advantage which they obtain at home, by exporting to this country a part of their superfluous produce, enables them to sell it here, without eventual loss, at prices very often considerably below the actual cost, and much below the remunerating price of the labor employed upon them in either country. This particular result, while it lasts, is no doubt advantageous to us as well as to them; but it is not advantageous to us to be brought within the vortex of these tremendous revulsions, and to find the prices of the necessaries of life in our own markets rising and falling with the political revolutions of nations on the other side of the globe. By fixing our industry on the basis of agriculture, manufactures, and the home trade, and making foreign commerce a subsidiary branch, we convert it from a cause of disturbance into a useful regulator and ally of all the rest.

In the remarks which they have thus far made upon the advantages resulting from the establishment of domestic manufactures, your memorialists have supposed, throughout, that the quantity consumed in the country is the same, whether they are imported from abroad, or made at home. But this is very far from being the case; and a *third* beneficial result of their establishment is a greatly increased consumption of the necessaries and comforts of life, which is attended of course with a corresponding increase in the civilization and general welfare of the people.

The necessity of this result is easily seen. The exchange which naturally takes place between the two classes of agricultural and manufacturing laborers, is that of the means of subsistence for the products of art. The cultivator supplies the manufacturer with food and materials, and receives from him in return the articles of use, comfort, and luxury, into which these materials have been fashioned. But this exchange can never take place to any great extent, excepting when the two classes are situated in the neighborhood of each other, and belong to the same political society. Provisions are too bulky, and, in many cases, too perishable, to bear transportation from one quarter of the globe to another. If not consumed on the spot where they are raised, they cannot be consumed at all. Or, were it even possible to surmount this difficulty, it is, and always will be, and ought to be, the standing policy of other nations, to interdict their importation, excepting in extraordinary cases of actual necessity. We find accordingly that our provisions are excluded, in ordinary cases, from almost all the markets of Europe. They constitute, nevertheless, the staple product of at least two-thirds, according to some, of four-fifths, of our whole population. What then follows? Our cultivators, in most parts of the country, and in the usual state of commerce, have nothing to offer in exchange for foreign manufactures, and, of course, no means whatever of obtaining them. The manufacturing population of the old world is represented in every precinct of our territory by a few shopkeepers, and the amount of agricultural products consumed by their families, is the only reciprocal demand upon a county or township of our cultivators, created by their whole consumption of European manufactures, which, under these circumstances, must, of course, reduce itself to nothing. In particular sections of the union, the inconveniences resulting from this state of things are partially relieved by an extensive cultivation of the materials employed in the European manufactures, particularly cotton, and some other articles, such as rice and tobacco, to which our climate and soil are more favorable than those of Europe. Our exports of these pay for the foreign manufactures which we in fact consume. The transportation is effected by the navigation of the eastern and middle states, which, in this way, obtain a share of the returns. But the cultivators throughout the interior of these vast and populous regions, and throughout the whole western country, have nothing to offer in exchange for foreign manufactures, excepting the provisions of the ordinary kinds of which the caprice, or necessities of foreign powers, from time to time, permit the introduction into their markets. Their consumption of foreign manufactures must, under these circumstances, as your memorialists have already remarked, be extremely small; and if they have no domestic manufactures in their neighborhood, they are compelled to live without a knowledge of the arts, or an enjoyment of the comforts, of life. But it is the extent of this knowledge and enjoyment, that forms the distinction between the civilized and uncivilized states of society. A community thus destitute of home manufactures, and excluded from an intercourse with foreign markets, has a constant tendency to decline into rudeness and barbarism. This tendency has, in the particular case of the United States, been thus far counteracted by strong political and moral causes, but the only effectual and permanent remedy is to remove the principle of evil. It was, therefore, with great reason that Mr. Jefferson, in one of his private letters written in 1816, declares his strong conviction of the expediency of bringing the producer into the neighborhood of the consumer. When, in any township, county, or section of country, there is a fair proportion between the number of families employed, respectively, in agriculture, manufactures, commerce and the liberal professions, there is, in consequence, a sure and steady market for the products of all; and all are supplied with the articles of comfort and luxury which are essential to civilization, and to the enjoyment of life. Prices must, of course, be paid, whatever they may be, and it is of little or no importance whether these are or are not the same on the other side of the globe. If a cultivator in the interior obtain from his neighbors in exchange for a part of his grain, good clothes and furniture, and a good education for his children, of what consequence is it to him whether he gives, for these comforts and blessings, more or less grain than they cost in Europe? He cannot send his grain to buy clothes and furniture, nor his children to be educated. His wants must be supplied by his neighbours, who will consume his provisions in exchange for what they give him, or not at all. If they be not supplied, he loses the sense of moral dignity that results from a civilized mode of life, ceases to produce any more grain than what is necessary to furnish him with bread and whiskey, sinks into idleness, and dies of intemperance; while his children, growing up without education, of course follow his example. Every article of use and comfort, which he can get at home in exchange for his surplus products, is, therefore, so much clear profit to him, although it should cost him twice as much as a similar article is worth in England, France, or China.

A *fourth* important advantage resulting from the establishment of domestic manufactures, is, that they render a people *independent* of all others in regard to the supplies of the necessaries and comforts of life. The intercourse between different countries is liable, from political causes, and particularly war, to frequent interruptions, which sometimes, endure for many years in succession. The effect of the return of peace, and of the renewal of intercourse after one of these interruptions, in prostrating the manufactures that have grown up during its continuance, has been already noticed; but it is also worth remark, that the first effects of such a suspension of intercourse, is to deprive a merely agricultural country of its ordinary supply of the necessaries and comforts of life, including the very arms and ammunition with which the war, that occasions it, is to be carried on. The impolicy of encountering this result, as far as the means of military defence are concerned, seems to be universally admitted. The most decided opponents of the protecting policy have recognized the expediency of providing, within our limits, and under the direction of the General Government, for the manufacture of the weapons and munitions of war. But it is quite obvious, that the very same principle goes the full length of justifying, or rather of imperiously dictating, the policy of protecting, by legislative aid, the manufacture of all the ordinary comforts and necessaries of life. It is no doubt important for the successful defence of the country against the attacks of a foreign enemy, that our troops should be furnished with arms and ammunition, but it is also absolutely necessary for this purpose, as well as for the comfort and well-being of the citizens at large, that the usual supplies of clothing, furniture, and other manufactured articles, should not be cut off, as, if received from abroad, they, of course, must be, on the occasion of a suspension of intercourse with the manufacturing country. The force of this consideration is apparent, and especially in its application to the United States, in as much as the nation from which we receive most of our supplies of manufactures, is also the one with which we are most likely to be involved in political disputes. It is sometimes vaguely said, in answer to this suggestion, that the danger of non-intercourse, or war with the manufacturing nations of Europe, is imaginary; and that it is impolitic to legislate in anticipation of any such improbable contingencies. Unfortunately, however, there is too much reason to believe that the occasional occurrence of war, even among the most civilized nations, must be calculated on as inevitable. The most recent experience shows that, between such nations, and at this enlightened day, it is quite within the compass of possibility, not only that wars should occur, but that they should last, with little interruption, for five and twenty years in succession. During the short period of their independent existence, the United States have been involved in war, or in political relations which had the same effect on our foreign trade, for about fifteen years, or nearly a third part of the time. It has been calculated that, for the period of about two centuries, which has elapsed since the conclusion of the peace of Westphalia, and during which Europe claims to have exhibited a higher degree of civilization than was ever known in any other part of the world, every alternate year has been, on an average, a year of war. Is it then supposed that the United States, and the foreign governments with which they have intercourse, are to be forever exempt from the operation of the evil passions, and conflicting temporary interests, that drive the nations to these terrific extremities? Does theory or experience justify any such belief? Do the tone and character of our negotiations, since the close of the late war with Great Britain, render it probable that it is to be followed by a halcyon age of perpetual peace? Your memorialists fear that the necessary answer to these questions is far too obviously in the negative to admit the possibility of giving any other. Much as they deprecate the occurrence of future hostilities with any foreign power; anxiously as they desire that the good understanding which now so happily exists between the United States and most of the governments with which they maintain habitual relations, may be perpetual; sincerely as they have rejoiced at the recent conclusions of several arrangements which seem to authorize the expectation of an improvement in the future character of these relations, they must still regard it as the strict and bounden duty of an American statesman to consider the occurrence of war with any foreign power as a thing within the compass of ordinary probability, and to act upon the supposition. It is not our policy, nor yet the policy of foreign countries, but rather the naturally imperfect condition of human nature, which will occasian these hostilities whenever they may happen, and which renders the supposition of their possibility necessary. Should the international relations of the great powers of the Christian world be on no worse a footing for the next two centuries than they have been for the two last—and it would surely be rash, whatever we may wish and hope, to reason and act on the hypothesis that the next following age will be better than the best in the history of our race—we must still calculate, as your memorialists have already remarked, that on an average, every alternate year will be one of war. So far indeed are recent and present occurrences from warranting the expectation of any immediate change in this respect in the habits of the world, that, although the great Christian powers have been at peace among themselves since the conclusion of the treaty of Paris in 1815, there has not been a moment during that time when the sword has been sheathed in all parts of Christendom, and hardly any two or three years in succession, when there has not been a strong probability of the occurrence of another general war, from which it would require the exercise of great discretion and ability in the government of the United States to keep them clear. At this moment, the danger of such an occurrence appears to be again very imminent; but whatever may be the issue of the present crisis, it is, at all events, clearly the duty of an American statesman to suppose, and to act upon the supposition, that the country is continually liable to be drawn into war with any of the foreign powers with which we have relations. If then we look to these foreign powers for our regular supply of the ordinary comforts of life, what is to become of us during these periods of occasional hostilities, which may last three years or thirty as the quarrel happens to turn? By what miracle are we to find, at a moment's warning, resources, before unemployed, which will furnish us with substitutes for this supply? Are we to extemporise at the commencement of every war, as we did at that of the last, a set of manufactures sufficient to meet the demand of twelve, fifteen, or, looking forward only to the end of the next five and twenty years, twenty million consumers, only to see them all shaken to their foundations by the return of peace, and sinking in one general ruin as they did before? Your memorialists can hardly imagine how any intelligent person can be so totally blind to the most obvious considerations of expediency as to counsel such a policy. They leave it with confidence to your honorable body to decide, whether it be not the duty of a prudent government to provide, by every imaginable means, against the recurrence of such widely spreading disasters; whether, were it even true, as it is not, that domestic manufactures would be, in the end, dearer than foreign ones, an annual pecuniary sacrifice of considerable extent made in this form, would not be decidedly preferable, both on the score of interest and feeling, to supporting the incalculable losses and miseries of every kind produced by these convulsions. For themselves, your memorialists have no hesitation in saying, that were there no other argument for the expediency of a protecting policy, except the single consideration to which they have last adverted, they should still regard it as established beyond the possibility of doubt.

The last advantage which your memorialists propose to mention, as resulting from the establishment of domestic manufactures, is their effect in restraining emigration from the settled to the unsettled parts of the country. It is true, as a general principle, that manufactures add to the wealth and population of a country, the whole amount of the capital and labor to which they give employment; but, in the particular case of the United States, where large tracts of good unoccupied land are continually for sale at low prices, it is probable, as your memorialists have already remarked, that some of the persons who, under the influence of the protecting policy, invest their capital and labor in manufactures, would, if this field of employment had not been opened to them at home, have emigrated to some of the unsettled parts of the country, and been occupied in clearing land. But when an individual can obtain a profitable market for his labor at his own door, in the midst of his friends and kindred, and of objects that are connected with the agreeable associations of his early years, he will hardly be tempted to go in search of it to a distant unexplored wilderness. The increase of population which thus takes place in the manufacturing states, by creating an encreased demand for provisions and materials, renders it in turn more advantageous for the agricultural states to extend their industry at home, than to send off continually new colonies. In this way, the tide of emigration, without being wholly dammed up, is considerably checked throughout all the settled parts of the union, and the population of all begins to put on a more consolidated shape. This result, although it amounts in fact, as has been intimated, to a change in the direction of a part of the agricultural labor of the country, and a transfer of some of it to manufactures, not only furnishes no objection to the encouragement of this branch of industry, but is itself a strong argument in favor of such a policy. These remarks are not made under impressions in any way unfavorable to the character or interests of the younger members of the union. Your memorialists, in common with all their fellow-citizens, feel a just pride in the flourishing condition of the new states. They consider the rapid progress of these states in wealth, population, and general prosperity, as a spectacle unparalleled in moral magnificence by any thing to be met with in the annals of the world. Your memorialists are fully of opinion, that the sudden expansion of our population over the unsettled territories of the union, has been thus far productive of good. It has thrown open a broad and ample field for the national industry, and has brought into action a new political element, which serves as a sort of mediator between sectional interests, which might otherwise have proved to be irreconcilably hostile. But, admitting the reality of these great benefits, it is also certain that if, in a region like the interior of the United States, which cannot be supplied with manufactures from abroad, the whole population devote themselves exclusively to agriculture, and as fast as they increase, continue to spread themselves more and more widely over the unlimited regions that are accessible to them, they must live, in a considerable degree, without the knowledge or enjoyment of the arts of life, and be in continual danger of sinking to a lower degree of civilization. The singularly excellent character of the settlers, their industrious habits, and the high tone of patriotic sentiment which has always pervaded the whole population of the new states, have hitherto maintained them at a point of civilization which, considering their circumstances, is hardly less wonderful than the rapidity of their progress in wealth and greatness. But the only way in which the advances they have made can be secured, and a solid foundation laid for the fabric of social improvement, is by naturalizing, on the spot, the cultivation of the useful arts. As far as the protecting policy may have the effect of diverting, into this channel, a portion of the labor and capital of the country, which would otherwise be employed in clearing land on the borders of the union, it will work, undoubtedly, a material change for the better. It is almost superfluous to add, that no one section of the more anciently settled parts of the union is more particularly interested in this result than the others. It is well known that the emigration from the southern Atlantic states has been of late even more considerable than from any other quarter. In this respect, there is a complete identity of interest among all the different sections of the union.

It has been said, in reference to this particular branch of the subject, that "a comparison might be drawn with advantage between the respective situations, at the end of thirty years, of the working men who have availed themselves of the natural advantages offered by the facility with which the rich unimproved lands of America may be acquired, and of those who have attached themselves to a manufacturing establishment." Your memorialists are inclined to doubt whether, on a fair comparison of all the circumstances of their respective conditions through life, that of an emigrant who goes off with his axe into the western wilderness, would be considered as decidedly preferable to that of the citizen who finds employment as a mechanical or manufacturing laborer in the settled parts of our country. They are well aware that industry, probity, and perseverance, will ensure success under almost any circumstances; but examples are certainly not wanting of mechanics who have acquired, in shorter periods of time than the one just mentioned, independent fortunes, and the most respectable stations in the society of our wealthiest and most populous cities. But it is not the intention of your memorialists to institute comparisons between different occupations, or to encourage any at the expense of the rest. They desire that every citizen should be left at liberty to select that pursuit which, in his opinion, will most conduce to his happiness; and it is for the precise purpose of affording to those citizens who may think it will contribute to their happiness to invest their capital in manufactures, an opportunity of so doing, which they could not have if they were exposed to a ruinous foreign competition, that your memorialists are anxious for the continuance of the system of protection.

Such are the advantages, economical, moral. and political, which, as your memorialists conceive, result from the establishment of domestic manufactures. If real, they certainly afford sufficient evidence of the expediency of maintaining the protecting policy.

Reproduced courtesy of Merrimack Valley Textile Museum.

Review of The Results of Machinery (1832)

The Results of Machinery, attributed to Charles Knight, was a response to the machinery-breaking riots in England. Knight argued that improved technology benefited society, especially the working classes, and suggested that improved knowledge among those classes of the effects of machinery would result in its acceptance. Addressing workers, Knight announced his purpose:

> We shall attempt to make you feel . . . that although your individual labour may be partially displaced, or unsettled for a time, by the use of a cheaper and a better power, which power is machinery, you are great gainers by the general use of that power. We shall strive to shew you, that through this power you possess however poor you may be, many of the comforts which make the difference between man in a civilized and man in a savage state; and further that, in consequence of machinery having rendered productions of all sorts cheaper, and therefore caused them to be more universally purchased, it has really increased the demand for that manual labour, which it appears to some of you, reasoning only from a few instances, it has a tendency to diminish.

Knight's book attracted wide attention in this country. It was frequently reviewed, and four American editions were published before 1840. American reviewers strongly agreed with Knight. Further they blamed

---

"The Working Man's Companion.—The Results of Machinery—namely, Cheap Production and Increased Employment, exhibited: being an address to the Working Men of the United Kingdom," *American Quarterly Review,* volume 12 (December 1832), pp. 299–315. Other reviews: Anon., "Effects of Machinery," *North American Review,* volume 34 (January 1832), pp. 220–246; reprinted in Thomas Parke Hughes, ed., *Changing Attitudes Toward American Technology* (New York: Harper and Row, 1975), pp. 120–141. Anon., "Review and Extracts of Chas. Knight's: 'The Result of Machinery,'" *Mechanics Magazine and Register of Inventions and Improvements,* volume 1 (May 1833), pp. 247–250.

the antimachinery feeling in England on the English political situation. In America, they agreed, there was no strong opposition to machinery, and what opposition there was came from "Superficial observers" who saw only the temporary problems and not the long-term benefits.

The author of the review excerpted here begins with a plea for "universal education," without which the "practical men" of the United States would not be able to act on the ideas of the "enlightened and scientific" few. The reviewer sees Charles Knight's book as a step toward the education of the masses in the truths of political economy. The body of the review is a summary of Knight's arguments. The reviewer concludes with examples of the "results of machinery" selected from Knight's book. Excerpted here is the central portion of the review.

The "practical man," if he happens to live near a manufactory, upon the introduction of an improvement in machinery, whereby the work formerly performed by six men can now be done by two, sees a number of poor labourers thrown out of employment, and a number of families reduced to want. He is induced to suppose that labour-saving machinery is an evil, and productive of poverty and wretchedness—and if he is a passionate man as well as a practical one, he thinks the workmen would serve their employers right by destroying the machines. The scientific political economist, on the contrary, from the examination and comparison of many facts, and from a train of comprehensive and accurate reasoning, is convinced, that notwithstanding the partial and transient evil caused by their introduction, every improvement in machinery by which the cost of production is diminished, is a permanent advantage to *all* classes of society.

But it is not by the prejudices of the ignorant, and the efforts of misguided and uneducated reason alone, that the advance of accurate knowledge on the subject of national economy is retarded. Where reason is exerted at all, sound argument and convincing proof, if they can only obtain a hearing, will in the end be triumphant. But the passions of the multitude are sometimes aroused; deceived by appearances which they cannot comprehend, goaded by distress which is the result partly of their own ignorance and imprudence, and partly of the necessary organization of society, it is not wonderful that being unable to understand the true cause of their hardships, they should believe that to be the real which is the apparent one, and that feeling themselves miserable and believing themselves oppressed, want and indignation united should sometimes drive them to desperation.

This has not unfrequently occurred in England, where the distress generally immediately consequent upon the introduction of improvements in machinery, has sometimes excited the labourers to such a degree that social order and the rights of private property were disregarded, and the restraints of the law trampled upon by the ignorant fury of a mob.

It is upon occasions such as these, when the effects of ignorance are to destroy the blessings of security and order, and to overturn the whole fabric of society, that we feel the advantages of knowledge. With the benevolent view of affording those advantages, the little volume whose title we have placed at the head of this article, was published. It is generally ascribed to Lord Brougham, and is not unworthy of the greatness of his intellect, or the variety and extent of his learning. Its object is to demonstrate the real effects of improvements in machinery, and to show how entirely all the blessings and comforts of civilized life are dependant upon it. It is addressed to "the working men of the united kingdom," and is written in a clear, plain, and familiar style, adapted to the minds which it is intended to enlighten —but it contains so much valuable, and to the generality of readers, curious information upon a subject of universal interest— so much clear, sound, and accurate reasoning, and opens such important views of the prospects of society and the sources of its improvement, that it cannot be read by any one without much pleasure and much instruction.

When we survey the complicated organization of civilized society, the busy multitudes of a populous nation pursuing their various avocations in harmony and order—their vast undertakings, their great achievements—their numerous transactions and various interests, and the regularity of their operation—their wealth, their power, their luxury; and compare the situation of such a community with that of a tribe of savages, few in number, thinly scattered over a vast tract of desolate and uncultivated territory, dwelling in rude huts unfurnished with the conveniences of life, scantily clothed with the skins, and half fed with the flesh of the wild beasts which their whole time is occupied in pursuing,—exposed, unprotected, to the inclemencies of the weather, without government, or social order, or knowledge— toiling only to supply the lowest and most obvious animal wants —living only to gratify the coarsest animal appetites—debased, degraded, miserable—the reflection does not instantly occur to every mind, that the difference between the civilized community and the savage tribe, is caused by machinery alone, and that deprived of it, the wealthiest and most flourishing nation would speedily be reduced to the situation of a wretched and barbarous horde.

And yet this is strictly true. What constitutes the difference between the member of a civilized community, and the wandering child of the forest? The possession of a cultivated mind, and of the conveniences and comforts which minister in the greatest degree to his physical enjoyment. The civilized man has knowledge—knowledge of the nature and properties of the material objects which surround him, and of the means by which they may be rendered subservient to his use—knowledge of the past—of the actions of his species for many centuries, from which he draws conclusions which form rules of conduct for the future—knowledge of the mechanism of his own body, of its faculties and its diseases, by which he is enabled to preserve his health or to alleviate the pangs of sickness—knowledge of the powers and operations of his own mind, from which he learns what he can attempt with prospects of success, and the means calculated to increase his individual happiness. From this knowledge springs his power over the material world, and the brute creation, and his superiority over those of his own species who possess it not. Hence come also, the pleasures of taste, the aspirations of ambition, the exalted enjoyments of intellectual superiority and exertion, and all the refined delights of civilized and social life. And how was this knowledge, which forms the best and richest treasure of the human race, because it is the source of every other blessing, obtained? By the patient investigation and diligent study of a small portion of mankind, devoted in successive ages to its acquisition. But how were these men enabled thus to devote their time to such labours? For it is evident that they must have devoted nearly all their time, and that if they had been obliged to toil in procuring for themselves the necessaries of life, in supplying their animal wants by their own labour, they would have had little left for other occupations. The answer is, by machinery, which, by increasing the productive powers of industry, supplies the wants of the *whole* community by means of the labour of a portion of it, and supplies those wants infinitely better and more abundantly, than the labour of the whole community could do without its aid. It is thus that leisure is afforded to some, for the cultivation of their minds, and for the acquisition and increase of that knowledge upon which the prosperity of all depends. If in addition to this, we add the reflection, that the present improved state of knowledge, and its diffusion among all classes, are caused solely by the art of printing, we shall need nothing more to convince us, that the possession of knowledge, of a cultivated mind, which distinguishes the civilized man from the savage, in as marked a manners as the possession of reason at all does the human being from the brute, results from machinery alone. But to this knowledge thus resulting from the labours of a portion of the community, enabled by the productive powers of machinery to devote their time to its acquisition—machinery itself—simple, feeble, and inartificial in its first advances, owes its innumerable applications and improvements.

The first steps being conquered, improvement was rapid, and as the increase of wealth and population necessarily kept pace with augmented productive powers of industry—a greater number of the community were constantly relieved from the necessity of labour, and enabled to devote their minds to the pursuits of science. Accurate observation, diligent and extensive research, and sound induction, became, at length, the characteristics of philosophy, which being directed to the attainment of a knowledge of the properties and laws of matter, has already discovered and performed so much, that the commonest necessaries of life are now the production of the most complicated and wonderful inventions, the condition of the humble peasant in point of solid comfort and even luxury superior to that of the wealthiest noble three centuries ago, the conveniences and splendour of the rich, such as the monarchs of old never imagined even in their wildest dreams, and the common and daily spectacles of life, of a character that would have startled our ancestors as the work of supernatural agency. Let any one who enjoys, even in a moderate degree, the advantages of fortune, look around him and see which of the luxuries or conveniences which constitute the difference between himself and the poorest and most wretched of his species, he could obtain without the assistance of machinery. He lives in a spacious and commodious house, itself an elaborate machine, filled with innumerable contrivances to promote ease and save labour. His table is covered with wholesome and dainty food, which is either produced at home by the assistance of machines, or brought from foreign lands by other machines the most complicated and wonderful. This food he conveys to his mouth by means of various implements produced by machinery, which render the process of eating cleanly, agreeable, and refined. He clothes himself with garments of various material and texture, which are soft, pliable, and beautiful—warm or cool—all the products of complicated machinery. Does he want knowledge? He goes to his shelves where stand in many a glittering row, numerous ingenious little machines, filled with the wisdom of experience and the lore of centuries, ever ready to inform, to delight, to improve his mind. Does he wish to travel? The steam-engine propels a floating edifice, filled with every comfort and convenience for his accommodation, with ease and rapidity against wind and tide, or whirls with winged speed its long and ponderous train over the iron highway. In all his avocations, in all his pursuits of business or pleasure, he uses machines, and deprived of them, he would be helpless, degraded, and miserable.

But it is said by some, that notwithstanding the improvements in machinery, poverty still exists, and the poor are still wretched. True—but this wretchedness is not caused by machinery, it is chiefly the effect of ignorance, vice, and imprudence, and would exist in a much greater degree, were it not that machinery, by producing commodities of all kinds in infinitely greater abundance, and with much less labour than they could be produced without it, and by producing many commodities which could not be produced without it at all, places within the reach even of the poorest, a thousand comforts which were unknown to the rich in less civilized ages, and furnishes the humble cottage, if industry, neatness, and sobriety preside over it, with every necessary for substantial enjoyment. It is this effect of improvements in machinery which is dearest to the philanthropist. Poverty, destitution of the comforts of life, experience and reason teach us, must be generally accompanied by ignorance, degradation and vice. Self-respect, intelligence, sobriety, and virtue, are produced by the possession of these comforts. The lowest orders of society ordinarily mean the poorest—and the highest, the richest. Sensual excess, want of intelligence, and moral debasement, distinguish the former—knowledge, intellectual superiority, and refined, social, and domestic affections, the latter. The different classes of society rank in general estimation according to the means which they severally possess of supplying themselves most easily with the comforts and luxuries of life; and we find it every where true, that in proportion as mankind are relieved from the necessity of exerting bodily labour for the supply of their wants, they will exert mental labour,—that the possession of comfort, convenience, and leisure, is generally followed by intellectual development and moral exaltation. If this be true, whatever tends to add to the comforts of the poor man, to enable him to supply his wants, and to procure the innocent enjoyments of life with a smaller amount of labour, tends also to raise his condition in the scale of society, to improve him intellectually and morally, and to make him a happier and more useful member of the community. Now, this is the precise tendency of every improvement in machinery. It lessens the cost of production; it makes that cheap which was formerly dear, and thereby either places within reach of the poor, commodities which before could be purchased only by the rich, or gives to them at a cheaper rate those comforts which they already possess—thus leaving them the means of extending the sphere of their enjoyments. It is thus that machinery has already improved the condition of all who possess it. If it gives to the rich luxury, it gives to the poor comfort, and those things which are now called the *necessaries* of life, are thus called, because machinery has made them so abundant and so cheap, that they have become universal, and are enjoyed both by rich and poor; they were once considered luxuries. Our author, addressing the working classes, thus sums up the advantages of machinery to them.

"This increase of comfort, some of you may say, is a question that more affects the rich than it affects us. This again is a mistake. The whole tendency of the improvements of the last four hundred years, has not only been to lift the meanest of you, in regard to a great many comforts, far above the condition of the rich four hundred years ago, but absolutely to place you, in many things, upon a level with the rich of your own day. You are surrounded, as we have constantly shown you throughout this book, with an infinite number of comforts and conveniences which had no existence two or three centuries ago; and those comforts and conveniences are not used only by a few, but are within the reach of almost all men. Every day is adding something to your comfort. Your houses are better built—your clothes are cheaper—you have an infinite number of domestic utensils, whose use even was unknown to your ancestors—you can travel cheaply from place to place, and not only travel at less expense, but travel ten times quicker than the richest man could travel two hundred years ago. Above all, you are not only advancing steadily to the same level in point of many comforts with the rich, but you are gaining that knowledge, which was formerly their exclusive possession. Keep fast hold of that last and best power; and you will learn what your true individual interest is, in every situation in which you can be placed: you will learn now, that it is useless in any way to struggle against that progress of society, whose tendencies are to make all of us more comfortable, more instructed, more virtuous, and therefore more happy."

That the situation of all classes of society is infinitely improved in every respect since the introduction of the modern improvements in machinery, may be clearly illustrated by comparing some of the conveniences and sources of comfort which every one possesses at the present day, with those possessed by our ancestors. What industrious poor man need be without a tight, warm, well-ventilated house, a good bed, convenient furniture, a variety of comfortable clothing—woollen, linen, and cotton, and plenty of wholesome food? At least in this happy land, where the causes are not operating which produce so much poverty and wretchedness in other countries, these blessings may be obtained by every one. Let us now look at the condition of the mass of the people in respect to the advantages we have enumerated, some centuries ago. An old writer gives the following account of the improvements in the building and furniture of houses in his time.

"Neither do I speak this reproach of any man, as God is my judge, but rather I do rejoice to see how God has blest us with his good gifts, and to behold how that, in a time where all things are grown to such excessive prices, we do yet find the means to obtain and achieve such furniture as heretofore has been found impossible. There are old men yet dwelling in the village where I remain, who have noted three things to be marvellously altered in England within their sound remembrance. One is the multitude of chimneys lately erected, whereas in their young days there were but two or three, if so many, in most uplandish towns of the realm, the religious houses and manor houses of their lords always excepted, and peradventure some great personage. But each made his fire against a rere dosse in the hall where he dined, and dressed his meat. The second is the great amendment in lodging: for, said they, our fathers and ourselves have lain full oft on straw pallets, covered only with a sheet, under coverlets of dog's waine and hop harlots (I use their own terms) and a good round log under their head as a bolster. If it were so that the father or good man of the house had a mattress or flock bed, and thereon a sack of chaff to rest his head upon, he thought himself as well lodged as the lord of the town. Pillows, said they, are thought fit only for sick women; as for servants, if they had any sheet above them it was well, for seldom had they any under their bodies to keep them from the pricking straws that ran oft through the canvass and rased their hardened hides. The third thing they tell of is the exchange of trene platters (so called from tree, wood) into pewter, and wooden spoons into silver or tin; for so common were all sorts of trene vessels in those times, that a man could hardly find four pieces of pewter (of which one was peradventure a salt) in a good farm-house."—*Hollingshed's Chron.*

It is hardly necessary to say how differently every labouring man is lodged at the present day, or to remark upon the numerous and convenient household utensils with which every cottage, however humble, is filled. The clothing of both poor and rich, upon which so much of our comfort and refinement depends, is now produced entirely by machinery, and that it is of various texture and material, and so cheap as to be within the reach, in great variety, even of the poorest, is owing to some modern inventions which lessen the labour and cost of producing it. In 1750, before Arkwright had invented his machine for spinning cotton, a cotton dress was a luxury which the wealthy could alone enjoy. Cotton was then spun and wove by the hand, and therefore with much labour; it was thus very dear; it is now spun and wove by machinery which produces a great quantity with little labour; it is therefore cheap, and forms a cleanly, healthy, and agreeable article of dress for all classes, particularly for women, thereby elevating and improving their condition. The same observations apply to wool, which is a material universally used in clothing, the cheapness and general use of which depend altogether upon late improvements in machinery. In 1589, the stocking machine was invented by William Lea, a clergyman; before that time, stockings, and those very bad ones, were a luxury confined to the rich—the poorest now consider them a necessary part of dress.

One valuable quality of improvements in machinery is, that their benefits must be diffusive and pervading. They owe their existence to extensive demand. The price men are willing to pay for the gratification of their desires, forms the motive and the reward of the invention. Every improvement in machinery which diminishes the cost of production, increases the demand for the article produced—by giving many the power of purchasing, who before had the desire to possess it, but were not able to pay the price demanded. Demand for any commodity does not mean merely the desire to possess it, for that is universal and indefinite; but this desire united to the power of paying an equivalent in exchange. Every poor man, as he walks the streets, desires to possess the various rich and elegant articles which he sees in the shops; but that is not demand until the wish to possess be accompanied by the power of purchasing. The tendency of improvements in machinery is, by making commodities cheaper, to make this power of purchasing more and more general, or in other words, to increase demand. This is all very clear—there is a certain demand for cloth, for example, which costs six dollars a yard:—that is, there are a certain number of persons in the community who desire to consume a certain quantity of cloth, and are willing and able to pay six dollars per yard for it. There is also a certain demand for cloth which costs three dollars per yard, which means that there are a certain number of persons willing and able to pay that price. It is evident that every one would be glad to get cloth at the former price, who cannot afford to buy it—and that those who purchase at the latter price, do so, not because they do not *desire* cloth of a better quality, but because they are not able to pay for it. Suppose an improvement in machinery to be introduced, by which the cloth which formerly cost six, can now be sold for three dollars. It is evident that all those who formerly purchased at six, will now purchase in larger quantities at three, and that those who formerly purchased cloth of an inferior quality at three dollars, will now be glad to get it twice as good at the same price. The demand is thus more than doubled, and the comforts of the poorer classes increased.

Such are the effects of machinery—it increases the comfort and convenience of all; it tends to elevate the condition of the poor, and to exalt and ennoble the character of man. It does all the drudgery, all that requires mere brute force, leaving to man the higher task of exerting chiefly his mind, and by that exertion surrounding himself with the sources of ease and refinement. How grand would be the spectacle of a nation whose inhabitants were all abundantly supplied with every article of comfort, luxury, and taste, by machinery alone, and whose whole time should be occupied in the pursuit and enjoyment of that happiness which springs from the exercise and improvement of the mind, the enjoyment of the social and domestic affections, and the refined pleasures of taste! Such a state of society is indeed impossible, but the nearer we can approach to it the better: the direct tendency of every improvement in machinery is to bring us nearer to it than before, and by producing abundantly with little labour, to require from man the exertion of his mind which ennobles him, rather than the corporeal drudgery which degrades him.

"And who can doubt," says our author, "whether instead of a state of society where the labourers were few and wretched, wasting human strength, unaided by art, in labours which could be better performed by wind, and water, and steam—by the screw and the lever—it would not be better to approach as nearly as we can to a state of society where the labourers would be many and lightly tasked, exerting human power in its noblest occupation, that of giving a direction by its intelligence to the mere physical power which it had conquered? Surely, a nation so advanced as to apply the labour of its people to occupations where a certain degree of intelligence was required, leaving all that was purely mechanical to machines and to inferior animals, would produce for itself the greatest number of articles of necessity and convenience, of luxury and taste, at the cheapest cost. But it would do more. It would have its population increasing with the increase of those productions: and that population employed in those labours alone which could not be carried on without that great power of man, by which he subdues all other power to his use—his reason."

The general improvement in the condition of man, the advancement of society in civilization, knowledge, numbers, and happiness, which has been caused by machinery, would be sufficient to convince most men of its importance to mankind, and to make them wish for its still greater improvement. They would look to the past and compare it with the present, and seeing that the numerous population, the pervading comfort and knowledge and happiness, the superior accommodation, and the intellectual and moral advancement of the present day are the results of improvements in machinery—that the direct and obvious tendency of these improvements is to advance still further the condition of man, by surrounding him with new conveniences, and by relieving him more and more from the necessity of exerting bodily labour for the supply of his wants—the philanthropist of enlarged mind would need no further argument. He would judge that the causes which had already produced so much general good, would, if they continued to operate, produce still more. Even if it were proven to him that this cause of general and lasting benefit was also frequently the cause of particular and transient evil, he would say that the happiness of the mass is to be secured even at the expense of hardship to a few, and that the advancement of society and the interests of posterity are not to be abandoned, because the means of their promotion produce in their operation some short-lived distress. There are some, however, who do not take so liberal a view of the subject; who, reasoning only from a few facts occurring in a short space of time, and neglecting, for the most part, a careful analysis of the facts which they do observe, think that they see in the wretchedness generally existing among the labourers who are thrown out of employment by the introduction of an improvement in machinery, a triumphant argument against any invention by which the necessity for human labour is diminished. These superficial observers say, that when the demand for any commodity is fully supplied, any improvement in machinery by which the commodity is produced at a cheaper rate, only supplies this demand at a lower price, and benefits the consumers indeed so far, but at the expense of the labourers, who are reduced to the greatest distress by being thrown out of employment. This argument is founded on a very obvious fallacy, for the discovery of which it is only necessary to understand the true nature of demand, which means, as we before observed, the desire to possess accompanied by the power of purchasing. The amount which a man will give for any article is the evidence and measure of his demand; an invention, therefore, which furnishes a commodity at a cheaper rate, not only supplies the old demand for it at a lower price, but creates a new demand by placing the commodity produced within the power of purchasing of many who before had the desire to possess it but were without that power. It thus not only benefits those who formerly consumed the commodity at a high price, but also all those who desired to consume before, but were not able to pay the high, and are able to purchase at the reduced price. It is perfectly evident, therefore, that the only way of supplying the wants of the people, is by increasing the demand for all sorts of commodities, which can only be done by producing them with so little labour that they can be sold at a cheap rate, and yet afford to the producer the ordinary profits of capital. Now the wants of mankind are unbounded; those of the poorest equal those of the richest; the great object of government and a system of society is to supply those wants as plentifully as possible; if machinery were to go on improving for a thousand years as rapidly as it has done for the last hundred, it would never supply them: there is no fear of its improving too fast, and it would be a happy thing for mankind if all their wants could be converted into demands, for then the condition of all, in point of comfort, ease, and convenience, would be equal to that of the richest now.

The objection usually urged against improvements in machinery, is, that the poor are deprived of employment. It is true, that at the introduction of an invention which produces the same quantity with less labour than was before required, some of the labourers are thrown out of employment—but this though a serious evil is a transient one, and not for a moment to be weighed against the permanent advantages which result from the improvement to the community generally, and particularly to the labourers themselves. The commodity is not only furnished to them in common with others at a cheaper rate, but the lasting effect of every improvement in machinery is, increased employment. This can be proved by innumerable facts—and is a conclusion which might be arrived at by *à priori* reasoning. It has been shown that by the cost of production being diminished the price is diminished; the price being diminished, the demand is increased; if the demand is increased, in order to supply that demand, a proportionably greater quantity of the commodity must be produced, and to produce this augmented supply, a greater number of labourers is required. It has generally been found in practice that the increased demand consequent upon diminished price has been so great, that many more labourers were required to supply it even with the improved machines, than were required to supply the old demand with the old machines, although they required more labourers to work them.

It was to combat and refute the objection that improvements in machinery deprive the poor of employment, and to show that their permanent effect is to give them increased employment, that "The Results of Machinery" was written. The truth of this position the author shows by proof so clear as to be intelligible to the commonest mind, and perfectly unassailable by the brightest. The proof consists of the history of the most important of those inventions in every branch of art and industry which have supplied civilized man with the means of comfort and happiness. In all, the facts clearly show that every improvement by which the cost of production has been diminished, has, besides adding to the comfort of all classes, increased the demand for the labour of the working classes.

Zachariah Allen

The Practical Tourist (1832)

Zachariah Allen (1795–1882) was a notable mechanic, entrepreneur, mill proprietor, and propagandist for the "Rhode Island system" of textile manufacture. Allen's account of his visit to Manchester, England, in *The Practical Tourist* capriciously mixes mundane detail and sublime rhetoric. His memoir, however, affords the clearest insight we have into the consciousness of an American manufacturer applying his cultural and ethical preconceptions to British industrial realities. Allen's work as a mechanic and businessman are considered more amply in *The New England Mill Village*, volume 2 of this series, in which other of his writings are reprinted.

Zachariah Allen, *The Practical Tourist* (Providence, 1832), 2 volumes, volume 1, pp. 121–161. There is no scholarly study of Allen's life or works. His voluminous papers are at the Rhode Island Historical Society.

MANCHESTER.

In a humid state of the atmosphere, the traveller is apprised of his approach to Manchester, when from the summit of some hill over which the road may wind, he first beholds at a distance the dark mass of smoke, which hovers like a sooty diadem over this queen of manufacturing cities. On approaching nearer, he views the numerous tall chimneys with smoky tops rising high above the roofs of the houses. A remarkable elevation is given to the vents of the furnaces, for the purpose of increasing the draught to render the combustion of the fuel more complete, and also to discharge the smoke into the air far above the windows of the houses. Notwithstanding these precautions, the inhabitants of the region below live amid sulphureous vapors, and the very walls of the houses are stained to a sombre hue by the coal smoke. During the summer, and also in dry and windy weather, Manchester might be deemed a pleasant place for a residence. But, at other times, and particularly on calm mornings in the early part of the spring, whilst a bright sun cheers the adjacent country, it displays to the inhabitants of Manchester its broad red disk, scarcely affecting the feeblest eye which gazes upon it through the dusky vapors, by which it is obscured. During the frequent foggy days in winter, an artificial twilight so completely shrouds the place, that at times the use of the gas lights becomes necessary, even at mid-day, for certain nice operations in manufactures. For the same reason, the lights in the large cotton mills are not extinguished until nine o'clock in the morning, and are rekindled to form a brilliant illumination, as early as about half past three in the afternoon. Most of the labor at such periods is performed by the aid of artificial light. Nearly one half of the surface of the exterior walls of the manufactories is composed of spacious glazed sashes, which are arranged in profusion to admit all the scanty light which a naturally hazy atmosphere, rendered still more obscure by smoke, will transmit. When a slight breeze arises, this dark cloud is put in motion, and is borne away over the country in an unbroken murky volume, perceptible at the distance of twenty or thirty miles, like the long train of smoke which streams from the chimney of a steamboat, and leaves a dusky line extended far over the waters and shores. It is only when a fresh supply of fuel is added to the furnace fires that the palpable black smoke spouts upwards. On a calm morning it affords amusement for a few minutes to watch these columns ascending perpendicularly several hundred feet, like a gigantic tree sprouting upward and expanding its dimensions, until the rolling masses, representing spreading foliage, meet each other from adjacent chimney tops and become intermingled. When the kindled fuel burns clear on the grates, the trunk of this ideal tree appears to be cut off from its apparent resting place on the chimney top, and thus detached to float off into the air.*

Having a letter of introduction to the proprietor of one of the most extensive cotton-spinning mills in Manchester, containing nearly 90,000 spindles,* he very civilly accom-

*At a court lately holden in Manchester, a fine of £50 ($250) was imposed on a manufacturer "for having suffered too large quantities of smoke to be emitted from the chimney of his factory." This was a cure for a smoking chimney with a vengeance. It had, however, the desired effect, as it was alledged, for the fuel was afterwards constantly supplied regularly in small quantities to the furnace, by the man whose duty it was to attend it, and the cause of the smoke, in the too abundant supply of coal, was thus obviated.

*Each of these spindles produces actually more yarn than the most skilful spinner could have made 60 or 70 years ago. Thus at this manufactory as much labor is accomplished by the aid of machinery as would have required 90,000 laborers to have furnished by the old mode of hand spinning on the single wheel. In Bolton, an adjacent town, it is stated that

panied me in a ramble over his vast works. The buildings are all of brick, in the form of a hollow square, the principal front of which towers to the height of eight stories, and the four outer fronts of the building measure more than 800 feet. The entrance is by a great gate, at which a porter is always in attendance to refuse admission to intruders from without, and to watch lest property should be conveyed furtively from within.

After passing the gate and beneath an arch formed under a side of the building, I entered the open court-yard or square, inclosed within the four interior walls of the manufactory. In the centre of this square is a sheet of navigable water bordered by a quay, on which canal boats may be seen discharging their freights of raw cotton and coals in the heart of the works, and receiving the packages of yarns. A tunnel or arched passage is made beneath the mill, to connect this interior basin with one of the principal canals which traverses a considerable part of England. Every possible facility is thus afforded for transporting the raw material to the very centre of the mill, and for shipping the manufactured goods in return to London or Liverpool.

In the preparatory process of picking over and assorting the sea island cotton, before it enters the machinery, there were more than 60 persons at work in one apartment, beating the flakes of cotton with sticks, in order to open them for more minute inspection. On suddenly entering this apartment, and viewing so many men and women, all simultaneously brandishing rods and beating the cotton, the loose locks of which flutter in every direction from beneath the strokes of the rods, descending with a deafen-

---

there are two manufacturers, who have each 100,000 spindles employed on their account. These four spinning mills, including the two referred to in Manchester, yield as much as all the spindles in a considerable nation could have once accomplished.

ing clatter, you may readily suppose that you are witnessing the disorderly scene of a mad house. The dust and small particles of cotton, floating in the air in this room, are almost suffocating, and must prove most pernicious to the health of the workmen.

When the doors of the various long apartments are successively thrown open, you view the wheels revolving on long lines of shafts, and ranges of machines with the metallic brass bright and glittering, as if polished by some careful housewife. The heads of the numerous busy attendants are visible above the machinery as they move to and fro at their tasks. In going from one apartment to another the spectacle almost produces the bewildering sensations which are sometimes excited by the strange visions of a dream.

The apartments are all warmed by steam from the boilers of the engine, conducted through cast iron pipes, in some cases arranged near the floor, with the design of distributing the heat more uniformly. Threads almost as fine as those of the web of the spider, and almost as silently spun, are drawn out upon the spinning mules. The finest yarns are always spun upon mules, and the process is slow. The labor of three persons at a mule of 300 spindles is required for a week, to spin four pounds of sea-island cotton into yarn of the fineness of 300 hanks* to the pound, at an expense for labor alone of about two dollars and a quarter for each pound of yarn produced. A respectable manufacturer in Manchester stated to me, that a single pound of sea-island cotton wrought into lace, had been sold for fifty-four guineas (about $270.)

From these fine threads the delicate tissues are fabri-

---

*The relative fineness of cotton yarn is calculated by the number of hanks required to weigh one pound. A hank contains 840 yards, and No. 300 yarn, therefore, has a thread sufficiently fine to measure, $840 \times 300 = 252,000$ yards, or 143 miles in length to one pound of the yarn!

cated, which are prized by the ladies, as being rather ornamental than useful. Unlike the texture of the wedding gown, which the good wife of the Vicar of Wakefield bought for the wear rather than for the looks, these light fabrics are lifted by the breath, or leave the form of beauty half revealed, enveloped in folds of transparent drapery. One bale of sea-island cotton, manufactured in this way, would produce nearly $100,000—a sufficient sum to purchase two or three cargoes of the raw material.

This manufactory which gives employment, directly and indirectly, to nearly thirteen hundred persons, and rivals in magnitude and importance many national works, was erected by Mr. Murray, who removed to Manchester about forty years ago, and commenced his career as a common mule-spinner. The cotton manufacture was at first, as he stated, "almost all profit." As competition gradually reduced these profits, he continued to enlarge his works; and the result in the aggregate, on a greater amount of production, he observed, has continued nearly the same. Separated from Mr. Murray's mill only by a narrow street, is another cotton mill, of equal, or even of greater magnitude. I was informed that one firm, engaged in spinning coarse yarn, has during the last year manufactured upwards of six thousand bales of cotton. It appears from a published statement, that the number of large cotton factories in the immediate parish or town of Manchester, was in 1820, fifty-four—in 1823, fifty-six—in 1826, seventy-two—in 1828, seventy-three. But the whole neighboring country abounds in them.

The vast business of the production and manufacture of cotton originated from the most humble enterprises and inventions. A brief sketch of the successive improvements, made during a long course of years, will afford to the reader a more perfect conception of the present complete mechanism of a cotton mill, than a labored description of the several machines which it contains. Few spectacles present a more impressive evidence of the successful exertion of human skill than the interior of a modern cotton manufactory, where all the varieties of complex wheel-work and machinery are admirably combined, and directed with suitable moving forces to so many different processes, each performing its assigned functions as if instinctively.

The rapid increase of the population of Manchester establishes a further proof of the prosperous diffusion of useful employment and consequent facilities for human subsistence. Within the last ten years, the population of this town has been increased by an accession of about 40,000 inhabitants, being a ratio of increase of above 35 per cent. Although the population within the circumscribed limits of the immediate town of Manchester is rated by the census of 1831, at only 142,026 inhabitants, yet including the population of the suburbs within the circuit of two miles, it contains 233,380 inhabitants ; and within nine miles of this centre of manufacturing industry, a million of people have concentrated their habitations. The county of Lancashire contained, in 1821, 1,052,200 inhabitants, and in 1831, the returns are given 1,335,000 ; showing a rate of increase of about 33 per cent in 10 years—which is fully equal to that of some of the most prosperous districts of the United States.

When all the machinery of the cotton mills is simultaneously stopped at the usual hours of intermission, to allow the laborers to withdraw to their meals, the streets of Manchester exhibit a very bustling scene; the side walks at such times being crowded by the population which is poured forth from them, as from the expanded doors of the churches at the termination of services on the Sabbath in the large cities of the United States. On first beholding these multitudes of laborers issuing from the mill doors, I paused to examine their personal appearance, expecting to behold in them the sickly crowd of miserable beings, so vividly described in Espriella's letters, as " keeping up the *laus perennis* of the devil, before furnaces which are never suffered to cool, and breathing in vapors which inevitably produce disease and death." In this respect my anticipations were disappointed; for the females were in general well dressed, and the men in particular displayed countenances which were red and florid from the effects of beer, or of " John Barleycorn," as Robert Burns figuratively called his favorite potation, rather than pale and emaciated by excessive toil in unwholesome employments in "hot task houses." Every branch of business being in a prosperous state when I had an opportunity of noticing them, they may have appeared, perhaps, under favorable circumstances, and in the possession of more than their usual share of comforts and enjoyments. The children employed in the cotton mills appeared also to be healthy, although not so robust as those employed as farmer's boys in the pure air of the open country.

*

The abundance of wild and unappropriated lands in the United States, forms the certain resource of the mechanics, and indeed of all other classes of workmen, when thrown out of employment by any of the vicissitudes of business, and serves as a sort of balance wheel to regulate fluctuations in the prices of labor.

There appears to be a greater difference between the quantity of the necessaries of life which a laborer obtains for his day's work in England, and what a similar laborer obtains in the United States, than there is between the nominal pecuniary standard of value. With an equal amount of wages, the mechanic in the United States may purchase nearly double the quantity of bread and other provisions necessary for himself and family that the English mechanic can purchase in England. Beef now sells here at from 13 to 17 cents per pound; bread 5 cents per pound. The day laborer for his three shillings earns 16 pounds of bread and 5 pounds of beef. In the New-England and Middle States, the best pieces of beef sell for about 8 cents to 9, and bread $3\frac{1}{4}$ cents per pound. Such turkeys as may often be seen in the hand of the American mechanic, on his return from market, would cost here three dollars or more; and of course are beyond the means of most of the laboring classes in England. Provisions of the coarser sort are also much dearer in Manchester than in the United States, from being more generally consumed by the poor. A sheep's head and offal, which may be bought in the United States for 8 or 10 cents, and are there frequently thrown away for the want of a purchaser, will sell here for 30 to 50 cents.

On account of these high prices, the laborers are under the necessity of living here much more economically on a stinted and inferior fare. In consequence of the heavy taxes and tithes, and other exactions, many of the very manufactures of England are retailed in her own shops for home consumption, at higher prices than they are sold for in the shops of the United States. Various fruits, such as apples, melons, &c. which in England constitute the luxury of the rich, and are cultivated in green houses and under the shelter of lofty brick walls constructed expressly for this object, are so abundant in the United States as to be found almost equally on the tables of the poor and of the rich. The sum of the enjoyment within the reach of the mechanics of the two countries appears, therefore, to be greatly in favor of those of the United States.

The most highly colored sketches of the moral depravity and vices of many of the laboring classes of Manchester, fall short of the reality. A stranger, if he walk leisurely through some of the streets during pleasant evenings, is frequently addressed by abandoned females, who press their solicitations with earnestness, and even take uninvited possession of the arm, should the position of the stranger's bended elbow happen to offer a loop favorable to their design. Unless, indeed, peremptorily repulsed at once, they acquire assurance, and press their importunities with a shamelessness, that can only be the result of long practised habits of vice.

The manufacturing operations of the United States are carried on in little villages or hamlets, which often appear to spring up as if by magic in the bosom of some forest, around the water-fall which serves to turn the mill wheel. These manufacturing villages are scattered over a vast extent of country from Indiana to the Atlantic, and from

Maine to Georgia. A stranger, travelling in the United States, commonly forms but an imperfect estimate of the extent of manufacturing operations carried on in the country. Where steam engines are in use instead of water power, the laboring classes are collected together, to form that crowded state of population, which is always favorable, in commercial as well as in manufacturing cities, to the bold practices of vice and immorality, by screening offenders from marked ignominy. In the narrow circle of a small community or country village, the finger of public scorn and disapprobation is pointed at the vicious and forms a repulsive circle around them. Intercourse with their former companions becomes, in a degree, cut off, and they find not, in a village, a sufficient number of new ones of vicious character, to countenance their indulgence in a course of depravity. But in a place like Manchester, in addition to the vast manufacturing population, there is a great influx of strangers, of boatmen from the numerous canals which centre here, and also of the various workmen from the machine shops, founderies, and other subordinate manufactories, forming a population as before stated, of nearly 150,000 persons.

In most of the manufactories in the United States, sprinkled along the glens and meadows of solitary watercourses, the sons and daughters of respectable farmers, who live in the neighborhood of the works, find for a time a profitable employment. The character of each individual of these rural manufacturing villages, is commonly well scanned, and becomes known to the proprietor, personally; who finds it for his interest to discharge the dissolute and vicious.

The proprietor of a manufactory in Manchester has many hundred persons daily entering his gates to labor, of most of whom he does not even know the names. He rarely troubles himself with investigations of their conduct whilst they are without the walls of his premises, provided they are reported to be regular at their labor whilst within them. The virtuous and vicious females are thus brought into communion without inquiry and without reproach. The contamination spreads, and the passing traveller is induced to pause at the sight, like Southey, in his letters of Espriella, to denounce the sources of present wealth, however overflowing and abundant, whilst the enriching stream is contaminating, and undermining the best interests of man. Whilst he sees plenty scattered over a smiling land, and every prospect pleases, he may sigh on finding that "only man is vile." God forbid, however fondly the patriot may cherish the hope of increasing the resources of his country by opening and enlarging the channels of national industry, that there ever may arise a counterpart of Manchester in the New World.

It may be intended as a blessing that an all-wise Providence has denied to the barren hills of New-England the mines of coal, which would allow the inhabitants to congregate in manufacturing cities, by enabling them to have recourse to artificial power, instead of the natural water power so profusely furnished by the innumerable streams, that in their course to the ocean descend over beds furrowed in the rocks of an iron bound country.— Whilst a cold climate and an ungrateful soil render the inhabitants from necessity industrious, thus distributed in small communties around the waterfalls, their industry is not likely to be the means of rendering them licentious; and of impairing the purity of those moral principles, without which neither nations nor individuals can become truly great and happy.

*
1   The dread of incendiary attempts of mechanics to destroy newly introduced labor-saving machinery as well as of riotous mobs, and the frequent threats of personal violence, all conspire to prevent proprietors of mills from

adopting the use of such machinery. Even farmers in the agricultural counties have been compelled to relinquish the advantages of using threshing machines for fear of having their barns burnt. This great obstacle has retarded the march of improvement in England, and has often been the cause of ruin even to the very inventors of useful machines. Whenever an important labor-saving machine is introduced, a town or whole district is put in commotion, and the result commonly is decided by the presence of glittering sabres. Individual suffering and wretchedness are there carried home to the fire-side of the poor laborer, who becomes destitute of all the worldly comforts which can render life desirable, and even of the very bread to nourish those dependant on him. Excited by feverish desperation, he is almost ready to rush back upon the sabres which enforce the laws, to throw away a life that seems not worth preserving. He knows that when his occupation is gone, to which he has served his apprenticeship, he will find every other branch of business around him fully occupied; and that without the opportunity of obtaining skill afforded by long practice, he can hardly compete in a new business with well trained rivals. Indeed it is already a subject of regret with the humane portion of the English nation, that new inventions are daily lessening the demand for the labor of a crowded, and yet an increasing people.

For a very contrary reason, every new invention to supersede manual labor is hailed with pleasure in the United States, as cheapening the cost of living to the agriculturists, who compose the great and leading interest of every country. Broad and fertile lands, at a low price are attainable by almost every unemployed man in the United States; and always afford a ready resort, whatever may be the fluctuations in commerce or manufactures, agriculture having been the first employment of the early Pilgrim Emigrants, and still remaining the never failing resource of their descendants when unfortunate.

The mechanics of the Eastern States, from their education and habits, (sprightly and enterprising) frequently turn their hands to two or three different trades before they become finally settled in one, quitting each with alacrity as soon as it becomes profitless or unpleasant; and leaving the service of their employers with equal indifference. Without general information on the subject of any other branch of business, the English laborer, when deprived of his accustomed profession by any of the vicissitudes which must at times attend the affairs of a manufacturing as well as of a commercial people, is left helpless and destitute, often without either the disposition or the ability to turn his hands to other avocations. In New-England, if a man does not succeed in one branch of business, he may be found readily essaying another; even in some instances officiating in the profession of law or of medicine, after commencing his career with the labors of the plane or of the anvil. It is undoubtedly true that, in most instances, this versatility is attended with a profitless result, as in the present state of the arts and sciences, a long period of arduous labor is required to attain skill and experience in any branch of business. For this reason, although the American artist, whilst thus shifting from one business to another, may display more of a spirit of independence and enterprise, yet he must thus often fail of attaining that practical skill in the use of machinery acquired by the long-trained artist in the workshops of England.

1. [in Manchester, England]

John P. Kennedy

Address to the American Institute (1833)

John Pendleton Kennedy (1795–1870) was a Baltimore gentleman lawyer who devoted his leisure to writing novels and promoting Whig politics. A congressman (1838–1839, 1841–1845) and supporter of the Bell-Everett ticket in 1860, his eloquence in behalf of protection and manufactures bears comparison with Edward Everett's. Kennedy opened this address with some genial remarks in praise of his hosts, the gentleman of the American Institute, and the exhibit of mechanical improvements they had mounted in their headquarters. He then proceeded to a retrospective view of economics in the decade since the passage of the tariff of 1824; this portion of the address is excerpted here.

John P. Kennedy, "An Address Delivered before the American Institute, in the city of New-York, October 17, 1833," *Mechanics Magazine and Register of Inventions and Improvements,* volume 2 (1834), pp. 228–240.

It is well to pause at intervals and look back upon our career, that we may compare fact with philosophy,—performance with promise. He who does so upon the last ten years, will find much to occupy his thoughts and instruct his mind. I will not pretend to draw even an outline of this survey as it strikes my view: time would not serve me to array the vicissitudes of opinion and the developements of history that belong to such a labor; but I will ask you to note the most prominent feature in the whole picture, and almost the only one which, throughout that period, has been without variation,—I mean the steady, onward march of the nation from one stage of good fortune to another; its career upon a plane of continued elevation. I would ask you to mark, too, the enchanting prospect from its present height. You will look over a landscape gilded with the purest sunshine, in an atmosphere redolent with fragrance: you will see how content has shed its balm upon the people; and how healthfully labor has walked to its toil. You will hear the frequent stroke of the woodman's axe, sending its dull echo through the frontier forest; and perceive the rich uncovered earth turned up to the sun, over many a former waste and distant wild. You will find huts grown into comfortable houses, hamlets into villages, villages into cities, and cities into great and gorgeous marts. Canals and roads may be seen stretching forth their serpent lines into the bosom of the remote vallies: fossils more rich than gold, will be found to have been dug up in abundance from the dark chambers of a thousand mines: the smoke of unnumbered furnaces will be discerned rising above the screen of the great wilderness; flocks infinite will be seen whitening the summits of the interior hills; and, on the Atlantic, commerce redoubling her busy fleets. The sound of the hammer, the din of the shuttle, and the clamor of the mill, have made the universal air vocal; and every where the incessant murmur and gush of business tells of a generation intent upon aggrandizing a vast and scattered empire, which now, like a strong man, " walks on its way rejoicing."

This is the picture afforded by the retrospect of ten years, and its hues are the more brilliant because they are warmed (to use the painter's phrase,) by their contrast with the scene presented in the previous interval of the same duration. Of that interval, embracing the space from the conclusion of the war until the era of the tariff, but a melancholy account can be rendered. Its unhappy prominent points may be shortly enumerated in a concise story of disappointed hopes and fruitless endeavors. It began with a hollow and unreal show of vigor in trade; an unnatural animation pervaded the departments of enterprise, more like the quickly exhausting fire of a fever, than the wholesome glow of health; —and the end was marked by deep disaster and pervading bankruptcy. Between these extremes we successively saw the evils of a depreciated currency, a sated commerce, and an overthrown industry. Our sturdiest population mourned their fate in sackcloth and ashes, and our best and most active citizens were whelmed in all the horrors of poverty.

A philosophical statesman would dwell with intense interest upon these pictures, and he would ask what wrought that marvellous change which made the first so beautiful? The reply would be,—that necessity is the parent of wisdom, and national instinct is not less strong than individual: want and privation are not the categories in which man is likely to repose; the restless desire to attain to good will make him astute in his perception, active and incessant in his toil; and the pressure of difficulty, better than all other masters, will teach him the true philosophy. These were the influences which produced the change, and the infallibility of their action was signally manifested in the sagacity with which the American people betook themselves to the most certain, and perhaps the only cure for the evils that encompassed them,—the adoption, namely, of the AMERICAN SYSTEM.

At the period of the adoption of this system, the reflecting portion of the citizens of the United States were divided by two theories in regard to the promotion and preservation of domestic prosperity; and time, although it may have softened the asperity of the collisions of opinion, has even yet failed to produce unanimity: many acute and learned minds are still to be found in the ranks of both. I allude to the advocates of the commercial system, or, as it is more familiarly known, the free trade theory, on the one hand, and to those, on the other, who defend the policy which supposes it wise to encourage and promote domestic industry by restraints upon importation. The majority of the nation coincided with the latter; and we may indulge the hope, that, as experience grows apace, and passion subsides,—as the fell spirit of party is lulled asleep, and good men, on either side, cultivate a conciliatory temper towards each other, the day is not far off when we shall

" In mutual, well-beseeming ranks,
March all one way."—

So far, indeed, is that happy anticipation now realized, that we may discuss the topic with reciprocal good will, and express our several opinions, free from the dread of personal exasperation or unkind surmise.

The free trade theory is of modern origin. It dates no further back than the middle of the last century, and from that time until the present it has been, in the land of its birth, a mere speculation. It is profitable to study its history and character amongst those to whom it owes its existence. I hope, by such an examination, to show that it is misunderstood in our country, and is quite inapplicable to our circumstances.

In Great Britain and in France, where the discussion of this doctrine has been most animated, it owes its popularity to a condition of things of which we have no parallel. In the first of those nations especially, (and in a not much inferior degree in the latter,) the whole machinery of municipal organization is curiously artificial. Government is complicated by an elaborate division of ranks and orders, which hold antagonist positions to each other, rendering the lower portions painfully subservient to the interests of the upper. Wealth is there distributed rather in lakes than rivers, and these large reservoirs are perpetually attracting to themselves the smaller accumulations, gathering

"their sum of more
To that which had too much."

Taxes without stint, the price of all their power, bear with a grievous weight upon the body of the community, and the constant strife has been each man to shift them for himself upon his neighbor, like an uneasy burden, which, in this world, the crafty ever compel the weak and foolish to bear. In this struggle power and wealth have gained the victory; and the huge machine has become, at last, a marvellous piece of intricate joinery, whose springs are so ingeniously contrived as to throw its weight upon the inferior masses, whilst, from the implicitness of the mechanism, none but the eye of a skilful master could perceive the series of actions by which this result is obtained. Such an eye was found in the acute and accomplished political economists both of England and France, when they came forth to denounce the injustice of the ancient systems of internal government. They saw in the monopolies and exclusive privileges which belonged to every guild; in the restraints that broke up or averted all competition in labor; and in the vicious circle of secret taxation, the hateful principle which gave permanency to vested wealth, at the expense of all liberal enterprise. It was, in their view, nothing better than a contention, on the part of the rich, to increase their store, by entailing the curse of perpetual poverty on all the rest. It was, emphatically, a struggle to preserve descriptive immunities from the encroachments of the large mass of the laboring classes. And hence arose that war of opinion which has so long raged in these nations between the two orders of the state. The assailing party called theirs the cause of free trade; whilst those on the defence were denominated the advocates of restriction.

... the free trade theory at one time had acquired a degree of favor in the United States, that would probably have excited some surprise abroad, if its application here had been well understood. There were many intrinsic circumstances to give it popularity, and render this tendency of opinion natural to our citizens. The free trade advocates in England and France are identified with the leaders of popular reform; theirs is supposed to be the liberal side: they make war against ancient abuses · their principles are whig principles: they are the assailants of old and intrenched errors, with which are associated, in our minds, all that is distasteful in monarchy; and it is natural, therefore, that the citizens of the United States should find their sympathies enlisted in favor of this party. We constitutionally feel ourselves inclined to applaud the effort they are making, and thus are easily led to adopt the idea that the like system must produce the like result when exhibited in action at home. A more careful examination would show us, that whilst we partake of and encourage the same liberal concern for the interest of the industrious classes here, we are but little likely to promote their welfare by similar laws.

The best theories of political economy are those which are formed upon an experience of facts; the wider, the older, the more minute this experience, the nearer the approach to certainty. Of all nations now existing, to none is this condition so necessary as to the United States. No community has ever before grown up under the same circumstances; none was ever exposed to the influence of such contingencies; none was ever marched forward, at the same pace, through such diversified chances;—to none, therefore, is it so

The Philosophy of Manufactures

utterly unsafe to apply, without qualification, the theories which have been founded on the experience of European nations.

It is common to say that the schoolmaster is abroad; by which figure the idea is succinctly expressed, that men are more intent than formerly upon the improvement of the arts of life. The moral and physical qualities of mankind are more diffusely developed, and the light begins to fall upon the great mass. Science may not tower more high, nor genius, like the eagle in "the pride of place," hold a less solitary or sublime eminence; but the sun and the rain of useful knowledge, that make the moral world fruitful, and generate the stock of household virtues, shine and fall through a wider atmosphere: they visit the by-places and secret corners, and vivify the good seed within humbler enclosures than they were used to do of old. The world no longer creeps upon its way by slow and lazy steps, some half century apart, but, like an impatient courser, bounds towards its goal. It leaps from experiment to experiment with hot haste, as if time were too short, and eternity too near. Science is made popular and common, and all classes seem to be busy to discard the old machine for the new. A peace of unusual duration, throughout Christendom, has assisted this process, and rendered the competition universal, eager and intense.

Just at such an era has it been the fortune of the United States, with scant forty years upon their annals, to be thrown upon their own resources. The peace of Europe had robbed us of the golden egg which our neutrality had yearly hatched; and the war which terminated in 1814 had left us—as all wars are apt to leave both victor and vanquished,—with nothing but our honor. In such a plight were we, for the first time, thrown upon our own resources, and called upon to play the game of nations. The arts we followed, and the prize we aimed at, were the arts and the prize also of Great Britain,—a stupendous power, of infinite wealth and long practised dexterity. England was unavoidably our rival; and, whether we would or no, it was our destiny to enter the lists with this giant, who, like him of old, bore a spear whose staff was as a weaver's beam, and stood in panoply, ready to encounter the young champion that came simply with his pebbles from the brook.

The control which Great Britain possessed over capital, population, and skill in the arts, confessedly placed her far above us in the means of sustaining the competition. It was in this relative condition of the two parties that the doctrines of free trade were so clamorously inculcated on the other side of the Atlantic. England had her manufactories established, and was then supplying a large portion of the world: she feared no inroad upon her domestic market, and her policy was to open all foreign ones to her trade. In no event could she be a loser by the policy;—in many particulars she had much to gain. A reciprocal reduction of duties upon manufactured fabrics would be but a harmless measure to her;—it would be thorough annihilation of manufactures with us. She had every motive of self interest to impel her to urge this measure upon foreign nations. There were, indeed, some few branches of her industry, of minor importance, which she had previously attempted to build up, though evidently disqualified by climate and position to maintain them:—an enlightened restrictive system does not pretend to foster pursuits incompatible with the capacities of the nation. Of this nature was the process of throwing silk;—a process manifestly unsuitable to the geographical position of England, and therefore hurtful to the silk weavers. The thrown silk was, to a certain extent, a raw material which, for the interest of the larger manufacture of the woven fabric, it was better to import from France and Italy. The duties, therefore, were reduced upon this article, and, what is remarkable as an evidence of the supreme care of Great Britain for her domestic industry, amidst all this profession of free trade, the throwsters were compensated by the government for the destruction of their business. Such reforms were introduced with the intent, as I have before expressed it, to give the greatest attainable vigor to her home labor; to set up and corroborate her manufactures, rather than to pull them down; and to get rid of every useless restraint that bore upon the working classes. It was like the preparations of an expert seaman, making ready his ship for battle: the unserviceable lumber and dead weights were thrown overboard, and the crew rendered more efficient by lightening the bulk. Such was the practical exposition of English free trade, as we read it in the measures of the party who maintain it.

What was our condition at the commencement of this struggle? We had followed the pursuits of agriculture and commerce. For many years it so happened, owing to the influence of extraordinary causes, that our agriculture was profitable, and our commerce, therefore, prosperous. But the causes had now been removed, and both of these concerns began to decline. Foreign pôrts were better supplied from other quarters, and were now shut against us; agriculture was overburdened by redundant crops that could find no market; commerce was paralyzed by the drying up of the spring from which it derived its supply. The wealth of our soil rotted on the field where it grew: the working classes could find no wages: enterprize was disheartened: if it attempted to enter the field of mechanical labor, it had no skill, or if it had skill, it was certain to find a foreign manufacture in the market before it at a less price: it wanted encouragement from the government, and protection to insure it a recompense. It had no protection, and therefore it was afraid to venture. All through our system there was a horrid atrophy—a dull stagnation of the fluids. In our distress we looked to the example of England; changed our policy; betook ourselves to protected manufactures; and in that pursuit speedily found relief.

It is somewhat strange that, after this course of experience, the advocates of free trade should still insist that whatever appearance of prosperity there may be in the land, it is either illusory, or exists in despite of improvident legislation; and that they should now tell us all this manufacturing industry ought to be pulled down, and the country restored to a state of trade dependent singly upon the exchange of the fruits of husbandry for the fabrics of foreign nations. It is still more strange that they should call that course of labor *free*, which is constricted by foreign rivalry into one poor, unprofitable path, from which it dare not depart without utter ruin, and yet which it can only pursue through all the embarrassments of an overloaded and unrequited competition at home—a pathway clogged up with the unconsumed and unconsumable fruits of the earth. Labor never became truly free until the provident arm of the government lifted the weight from its shoulders that pressed it to the dust, and gave it space to expand and travel through all the ways and passages of an infinitely intersected nation.

I would like to engage your attention a little longer with the object and philosophy of free trade, as it is professed in this country. Its first purpose is to break up the manufacturing system. Let us suppose it successful;—it has then a dogma to the following effect:—" Although the manufacturers are broken up, labor will gain by the change, because it has now the choice of a pursuit more congenial to it—the cultivation of the earth." I will say nothing of the freedom of a choice without an alternative, but I will remark that agriculture may become overstocked and grow unprofitable; and, as this cannot be denied, I will inquire what, in that case, free trade prescribes? "You may then," reply our opponents, "go to manufactures: it is the natural impulse of labor, when one pursuit is rendered unprofitable by too much competition, to betake itself to another which offers a better reward." I will imagine that we have taken this advice, and have determined upon erecting manufactories because agriculture has ceased to make a valuable return for the labor employed upon it. First, it is obvious that we must have been reduced to a great deal of suffering before we could have brought ourselves to abandon the paternal acres: Secondly, we have to set about the education of a new generation of mechanics, and to teach them the difficult arts of handicraft, under all the disadvantages of having but few instructors, and an intricate lesson to learn; and, lastly, we have expensive establishments to build. We accordingly sell our farms at low prices, construct workshops and manufactories, import foreign artisans to teach our own, exhaust capital and credit in the undertaking, and, perhaps, in some three or four years, find ourselves ready to furnish the market with a commodity that shall be as cheap as the imported one. Just at this stage of our adventure we make an important discovery. It is this: that up to this period the imported fabric, similar to our domestic one, has heretofore been sold to the country at some fifty or hundred per cent. above its cost of production; and that the same is now offered twenty per cent. below the price that we require for our indemnification. Ascribing this fact, perhaps, to some temporary accident, we reduce wages and other expenses, and, foregoing our own profit, we diminish our price correspondently to that of the rival fabric. Straightway the foreign production comes down another twenty per cent., and another, if necessary, and we are now convinced that this abatement of price is to be regulated by the energy of our competition; until, in despair, with the horrors of bankruptcy before us, and the clamor of our disappointed artificers ringing in our ears, we are compelled to stop our work. All that we have gained by this unlucky experiment is the satisfaction of having furnished our own people a touchstone by which they may ascertain how much the foreign manufacturer has heretofore been extorting from the American consumer in the shape of large profits upon his merchandize. When matters have arrived at this ebb, we have recourse again to the advocate of free trade, that we may learn from him what remedy he proposes for the disaster we have suffered; thinking, perhaps, that he may recommend,—what we are now prepared to believe a very obvious relief,—a duty, on the part of the government, sufficient to keep the foreign article on a level with the domestic. This, however, he does not grant us—it is against his creed; but, in place of it, he gives us an apothegm,—"that that manufacture which is not able to support itself is not worth protecting by duty; and that which is, does not require it." From this, we conclude it is his opinion, that we have gone to the wrong manufacture; and as we inform him that all our neighbors are in the same predicament, it is now his counsel that we had all better go back to farming. We consequently dismiss our workmen, and send them back to the plough, and to the labor of felling the wilderness; where, at least, they will have enough to eat, and where, if they find they have but little to wear, it will be attended with the consolation that the fashion of the forest does not require much foppery, and that every man's neighbor is about as poor as himself. We have now made our circuit, ending where we began; and upon casting up our account we find that we have travelled round this conjuring zodiac,—leaving an item of wealth behind us at each sign we passed through,—and have come out at the starting point completely stripped of all worldly possessions—gainers only in our knowledge of the experimental philosophy of free trade, and that "it is better to bear the ills we have, than fly to others that we know not of."

It is a point, however, much insisted upon, in recommendation of the free trade doctrine, that it furnishes the nation with its supplies of merchandize at much less cost than the domestic manufacture; and this fact is considered of sufficient importance to justify the abandonment of the opposite policy.

I will not affirm that, in any given branch of manufacture, we can furnish the fabric cheaper than the foreign manufacturer could do,—though such an assertion would be, doubtless, sustained, in regard to a large number of our products,—but I do affirm that our manufacture renders the foreign one cheaper, and that we get what we require at a much reduced price, by reason of the presence of our own workshops. It is not necessary to my present purpose that I should stop to discuss this principle; my object is to advance another proposition of far more importance, namely—that it is worthy of but little consideration, in the estimate of the value of our domestic system, whether the necessaries of life are rendered cheaper or dearer;—the country, we say, is benefitted to an incalculable extent, by being made the theatre of manufactures.

If it were true, as it has been affirmed by the opponents of the present policy of the government, (which, nevertheless, we deny,) that the duties levied for the encouragement and protection of our own labor were an actual tax upon the people, to the full amount of the excess above the ordinary demand for revenue, still the nation at large derives such advantages from the system as to vindicate the expediency of the tax. If, of the twenty millions raised in revenue, one half, or even the whole, were a tax of protection, it is an inconsiderable burden when compared with the wealth it creates and scatters over the land.

It may be said, that of the nine millions of free inhabitants of the United States, at least four millions, including both sexes, owe their livelihood, in whole or in part, to the work of their hands, and, in some shape or other, receive wages for their labor. The presence of manufactures has had the tendency, as all men admit and as the history of the country proves, to raise the wages of labor. It is setting it down at a low estimate,—much, in my judgment, below the fact,—to compute this increase of wages upon the whole mass of labor, agricultural as well as mechanical, at an average of six cents a day. Yet this sum would give an amount of seventy-two millions of dollars a year of increased wages, distributed amongst the working people of the United States by the operation of the domestic system: thus adding immediately to their comforts, and the improvement of their condition. Far beyond this money computation is the nation benefitted in other forms: a large number of idle hands are provided with employment; new fabrics, before unknown in our list of conveniences and luxuries, are brought into existence, and introduced to common use; industrious and thrifty habits are inculcated; the morals of the people are elevated; education diffused; trade and commerce greatly extended, both by the vast accumulation of commodities for exportation, and by the capacity for a higher scale of living, and for the consumption of a greater amount of foreign productions, communicated to the laboring population: a thousand new springs of wealth are set in motion, that swell up the sum of national prosperity much beyond our powers of calculation. Value is given to the productions of the forest, the field, and the mine, on our remotest frontier: internal improvements, necessary to facilitate the carriage of these productions to market, are rapidly extended, with instant remuneration for the cost of construction. The hunter, accustomed to waste half of his life in idleness upon the sunny hill-side, or to gather a precarious and scanty support from the desultory pursuit of game, is converted into a feeder of sheep, or the proprietor of a thriving mill; our most distant population are linked together by the bonds of internal commerce; the common inheritance of American citizens is rendered more valuable, by the vast increase of towns, roads, public works, fortifications, and navies; and, dearer than all, our UNION is corroborated, cherished, and perpetuated in the affections and the interests of the people. All this is achieved by a *tax* (if our opponents will have it so,) of some ten or twenty millions of dollars. Surely never was tax so recommended and consecrated by the virtue of its purpose!

It is obviously no answer to this argument to say that the seventy-two millions given in increased wages are still a burden upon the rest of the community; for, in regard to that sum, no consumer pays it without getting an equivalent in the immediate article he purchases: and it will also be observed that the mass of consumers is made up, in great part, by the workmen themselves.

Free trade would have us relinquish all these advantages. It would drive us to the meagre resort of household manufacture, and to the labors of the field; or it would compel that industrious, vigorous, and stirring population, who now inhabit the northern, western, and middle sections of this Union, to crowd into the pursuits of southern agriculture, until that region was also overstocked with labor and suffered the same plethora of production which heretofore befel the grain-growing states. It would dry up the sources of all prosperity but these, and the scant commerce that would be employed in the only export trade it left us. In a brief space, it would no less surely destroy the profits of planting and the foreign commerce that depended upon them; until late, and after a long interval of decline, it inevitably and irresistibly ended in the adoption of a system of restriction,—the only permanent resource of this country.

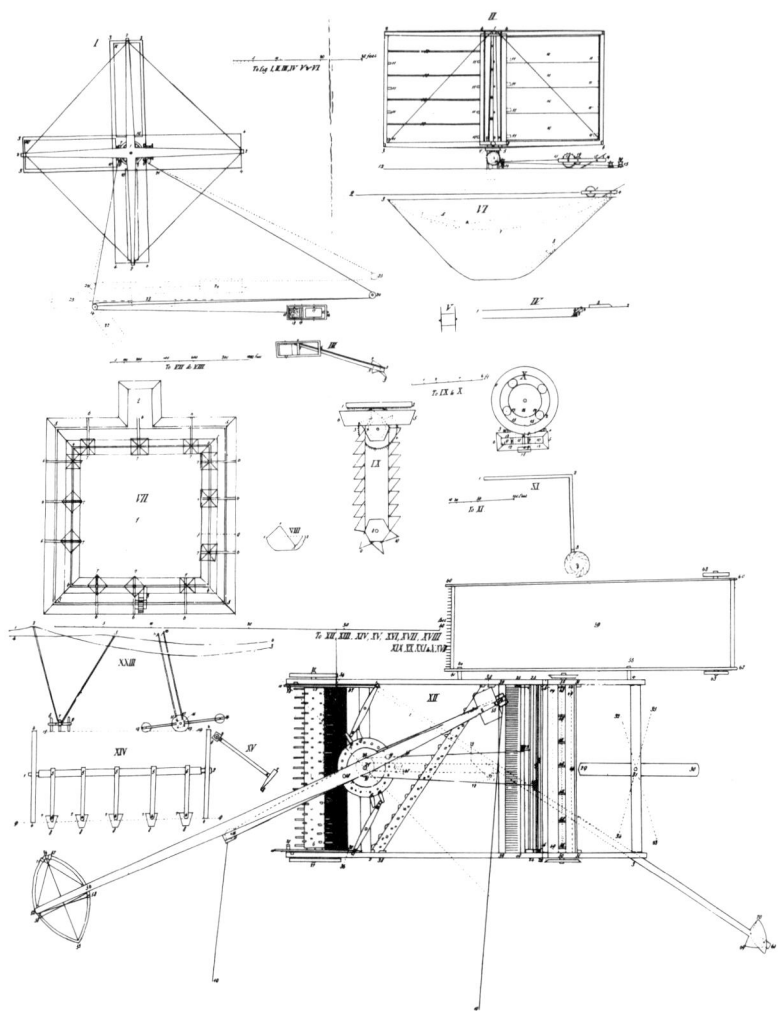

Part of Etzler's plans for agricultural machinery. From J. A. Etzler, *The New World, or Mechanical system, to perform the labours of man and beast by inanimate powers* . . . (Philadelphia, 1841).

John Adolphus Etzler

Paradise within the Reach of All Men (1833)

John Adolphus Etzler was a bird of passage to the territory of American intellect and manufactures. Although the exact years of his life are unknown, he did arrive in Pennsylvania from Germany in the 1820s in one of many moves, which took him back to Germany, to England, and eventually to Venezuela, among other places, in search of more fertile grounds for his utopian schemes and converts to people them with. His works are almost as ephemeral as his residences and would escape our attention except that they aroused the critical notice of Whittier and Thoreau and they articulate a vision of a mechanical utopia that takes to an extreme the hankering after progress through the introduction of labor-saving machinery. Etzler's schemes are especially revealing of an inherent feature of mechanization in the extent to which he blithely seeks to change the earth and the products of manufacture to suit the limitations of machinery rather than adapt machines to suit the culture and materials to which they are introduced. In addition to Etzler's apostrophe, the excerpts here include an example of the kind of power calculations his utopia rested upon and an indication of the kind of solutions he arrived at for the problems of applying that power through machinery. For Etzler, industrial technology led to pastoral ecstasy.

John Adolphus Etzler, *The Paradise Within the Reach of All Men, Without Labor, by Powers of Nature and Machinery. An Address to All Intelligent Men* (Pittsburgh, Pa., 1833); reprinted with an introduction by Joel Nydahl in *The Collected Works of John Adolphus Etzler* (Delmar, New York: Scholars' Facsimilies and Reprints, 1977). See also Patrick R. Brostowin, "John Adolphus Etzler: Scientific-Utopian During the 1830's and 1840's," Ph.D. dissertation, New York University, 1969.

FELLOW-MEN!

I PROMISE to show the means for creating a paradise within ten years, where every thing desirable for human life may be had for every man in superabundance, without labor, without pay; where the whole face of nature is changed into the most beautiful form of which it be capable; where man may live in the most magnificent palaces, in all imaginable refinements of luxury, in the most delightful gardens; where he may accomplish, without his labor, in one year, more than hitherto could be done in thousands of years; he may level mountains, sink valleys, create lakes, drain lakes and swamps, intersect everywhere the land with beautiful canals, with roads for transporting heavy loads of many thousand tons and for travelling 1000 miles in 24 hours; he may cover the ocean with floating islands moveable in any desired direction with immense power and celerity, in perfect security and in all comforts and luxury, bearing gardens, palaces, with thousands of families, provided with rivulets of sweet water; he may explore the interior of the globe, travel from pole to pole in a fortnight; he may provide himself with means, unheard of yet, for increasing his knowledge of the world; and so his intelligence; he may lead a life of continual happiness, of enjoyments unknown yet, he may free himself from almost all the evils that afflict mankind, except death, and even put death far beyond the common period of human life, and finally render it less afflicting: mankind may thus live in and enjoy a new world, far superior to our present, and raise themselves to a far higher scale of beings.

It may appear very wonderful, that none of these things, though they comprehend all the objects man may possibly desire in this world, ever existed yet, since thousands of years, and that they all should have originated from one single individual. But this wonder will greatly diminish, if not entirely cease, when it will be seen, that these great promises are founded on facts well known, that any man of common sense, if he ever had bestowed full attention upon them, would have come, ultimately, to the same or similar results as I am about to show; and when it is considered, that many contrivances of modern times have led to great comforts and advantages unknown to the ancients, though they had the same mental faculties of making them: they passed thousands of years in ignorance and errors, thinking always themselves to have reached the summit of human perfection. History teaches but too plainly, that the progress of human knowledges and intelligence was every-where most tediously slow. Individuals, who attempted sometimes to disperse new valuable truths, were not listened to, and considered insane in proportion their truths deviated from the common track of the unthinking or unreasoning multitude. Our present age is yet liable to the same great evil;—instances in proof of this are to be found in plenty; yet, as it is superior to the preceding ages, it is liable to this spiritual sloth in a less degree. After an attentive perusal of this work, after some calm reflection upon the subject, it will be found, that the promised great ends are attainable to the full extent and meaning of the words, without any wonder, without any hidden power or secret of nature, but by a few most simple contrivances.

The basis of my proposals is, that there are powers in nature at the disposal of man, million times greater than all men on earth could effect, with their united exertions, by their nerves and sinews. If I can show that such a superabundance of power is at our disposal, what should be the objections against applying them to our benefit in the best manner we can think of?—If we have the requisite power for mechanical purposes, it is then but a matter of human contrivance, to invent adapted tools or machines for application. Powers must pre-exist; they cannot be invented; they may be discovered; no mechanism can produce power: it would be as absurd to invent tools, that work without any applied power to put them in operation; machineries, of whatever contrivance they be, are nothing but tools more or less combined. I think this remark not to be superfluous, because many men, even of talents in mechanics, have erroneously cherished the idea of inventing mechanisms working of themselves without given power, and have uselessly bestowed time and expenses on the invention of a perpetual motion.—I wish my proposals not to be precipitantly confounded with such vain schemes.

The chief objects of my statements are, therefore, the powers to be applied; the application of them is but of secondary importance: this may be of an infinite variety; that of the greatest advantage is the most preferable. When you are once convinced, that there is power enough at our disposal for the great purposes in view, then you have the proof, that the attainment of the purposed ends is possible: the question is then not more, Whether the promised things are attainable, but How?

The powers are chiefly to be derived—1st, from wind; 2nd, from the tide, or the rise and fall of the ocean caused by the gravity between the moon and the ocean; and 3rd, from the sun-shine, or the heat of the sun, by which water may be transformed into steam, whose expansive power is to operate upon machineries, though by a contrivance different from that actually in use.

The waves of the ocean are also powers to be applied, but as they are caused by wind, they are included in the power of wind. Each of these powers requires no consumption of materials, but nothing but the materials for the construction of the machineries.

I shall at first endeavor to show the magnitude of each of these powers in its full extent over the whole world, beginning with well known facts; this will show the average power for any required extent of the surface of the globe. But as these powers are very irregular and subject to interruptions; the next object is to show how they may be converted into powers that operate continually and uniformly, for ever, until the machinery be worn out at length, or, in other words, into perpetual motions. After this it will be the problem, how to apply these perpetual motions of nature to the attainment of the purposes in view? I shall give a general outline of the system of machineries for effecting all promised purposes. Next to this I shall state the objects attainable by these means, and the condition of men that must result from the accomplishment of such purposes. It will then appear, from the nature of the subject, that the execution of the proposals is not qualified for single individuals; for as one machine is sufficient, under the superintendence of a few men, to supply many thousand families with all their wants, both natural and artificial ones, the consequence would be but hurtful to the laboring class, as the price of their labors would sink almost to nothing, dangers and violence would ensue, and the effects would be more destructive than beneficial, even to the undertakers themselves, until, after a series of convulsions, a different order of things should be established. It would be certainly a proper object for the government, to make the arrangement for the execution of these proposals; but as the government of our nation is the organ of the people's will, the subject must first be popular; but it cannot become popular before it is generally known and understood. Therefore the execution of the proposals is only qualified for a large body of intelligent men, who associate themselves without limiting the number, time and place, or country. I shall therefore finally propose a constitution for association. The larger this society, and the larger the means, the greater the advantages will be for every participating individual.

I shall now state:—

1st, The Power of Wind.

That there is power in wind requires no proof of me. The uses of it in navigation and wind-mills are too well known. My object is to state, how much power there is in wind. I shall state it, in the full extent, as far as it can be brought within the disposal of men over the whole surface of the globe.

*

In order to form an idea near the reality in nature, how much power of wind there may be at our disposal, we have to ascertain, by a deduction from experiences and observations, how large we may construct and expose surfaces to the effects of wind, and how close they may be brought together without intercepting the wind and diminishing its power materially. We know by experiences, that ships of the first rank carry sails 200 feet high. We may, therefore, equally, on land, oppose to the wind surfaces 200 feet high. Imagine a line of such surfaces 200 feet high, and a mile (or about 5000 feet) long; the same would then contain 1,000,000 square feet. Suppose these surfaces intersect the direction of the wind in a right angle, by some contrivances, and receive consequently the full power of the wind at all times. The average power of wind being equal to 1 horse's upon every 100 square feet, the total power these surfaces would receive would then be equal to 1,000,000 divided by 100, or 10,000 horses' power. Allowing the power of 1 horse to be equal to that of 10 men, the power of 10,000 horses is equal to 100,000 men's. But as men cannot uninterruptedly work, and want about half of the time for sleep and repose, the same power would equal to 200,000 men's. Imagine such another line of surfaces just behind or before the former at 1 mile's distance, parallel to the first and in the same circumstances. This second line would then receive the same power of wind again as the first; for the distance being 25 times greater than their height, the one line could not intercept the wind from the other in any considerable degree, both lines would receive the full power of wind, as soon as the direction of it would deviate from the horizontal more than about 2 degrees. It may be easily observed, that the wind will generally strike the ground in a steeper direction, and therefore admit a closer approach of such parallel surfaces. That the wind strikes the ground obliquely is evident on the high sea. Else whence the disturbance and rise of the waves on it?—If the wind moved parallel to the ground, the surface of the sea could not be affected by it, and would remain smooth for ever. But such is never the case. The least breeze ruffles the surface of the water. And it s too well known, to what size and powerful effects the waves may be raised by wind. Moreover, experiences in navigation teach, that vessels of the first rank sailing along a shore of about 200 feet high, trees, etc. included, at their wind-side, at a distance of 1 mile, will not suffer any considerable diminution of wind. If the supposed two lines of surfaces will receive such a power of wind as stated, that is, each equal to 200,000 men's power, a third line of the same height, at the same distance and parallel to the former, under equal circumstances, will receive the same quantity of power, so a fourth, fifth, and so on, as far as may be chosen. The length of each one may, under the supposed circumstances, be prolonged as far as we please, the power of wind will be every-where the same. Now, if we find the power of wind to be at the end of every mile equal to 200,000 men's power, and so for every mile in breadth, it follows, that every 1 square mile affords such a power.—What an immense power!—The most populous countries in the world contain in an average from 100 to 200 individuals on every square mile, of which hardly one half is able to work, or to be counted for full hands to work. But suppose even 100 full hands to work on 1 square mile, the power of wind within their places of habitation will be 2000 times greater. Yet this will not be the whole power of wind at their disposal. We are not limited to the height of 200 feet. We might extend, if required, the application of this power to the height of the clouds, by means of kites. If we extend it, for instance, to but 2000 feet high, we might increase the power 10 times as much, that is, 20,000 times greater than the inhabitants of the most populous countries could effect with their nerves and sinews. Yet we will get a more proper conception of this power, in extending this comparison over the whole globe. The surface of the globe is about 200,000,000 square miles. According to the foregoing statement of 200,000 men's power for every 1 square mile, the whole extent of the wind's power over the globe amounts to about 200,000,000 times 200,000, i. e., to 40,000,000,000,000 men's power. The number of all human individuals on, earth will not exceed 1,000,000,000, of which hardly the half may be counted for full hands to work, that is, 500,000,000; consequently, the stated power of wind is 80,000 times greater than all men on earth could effect with their nerves, when the wind is used but to the height of 200 feet.

It may now be objected, that this computation includes the surface of the ocean and uninhabitable regions of the earth, where this power could not be applied for our purposes. But you will recollect, that I have promised to show the means for rendering the ocean as inhabitable as the most fruitful dry land; and I do not even exclude the polar regions.

It may be questioned, how surfaces 200 feet high may be exposed perpendicularly to the wind for operation?—It may be done in the usual manner of wind-mills, but with greater advantage in a different way contrived by me so that every square mile may be surrounded by a continued line of surfaces or sails, to the height of 200 feet, moveable around an axis, and occupying not one tenth of the ground, with all their machineries.

What a gigantic, awful power is this! 80,000 times greater than all men on earth could effect by the united exertions of their nerves!—at the least calculation.—Suppose even one half should be lost by friction of the machineries, or more, we need not economise with such an immensity of power; let but one eighth of it be used, it would amount still to 10,000 times the power of all men on earth. If men were all and continually employed to work for useful purposes, they would effect a great deal more than we actually see, and might give to the world a far better appearance and a greater plenty of necessaries and comforts of human life. But if 10,000 times more can be done, if in 1 year, consequently, can be effected as much as hitherto in 10,000 years! —to what awful grandeur may not the human race exalt themselves?—The greatest monuments and wonders known or left us, to admire, from our progenitors, which required many millions of hands, and many centuries to be finished, are nothing but childish, insignificant trifles, in comparison to the stupendous works that may be effected by these powers. Yet it is not the only power we have at our disposal. You may startle at this idea, you may ask again and again, can it be possible, that there is such a power for our use? —like I have done. Am I perhaps grossly mistaken in my statement? Is it perhaps nothing but a fancy?—a deception of my imagination? I have taken the most common experiences of sails and wind-mills for the basis of the statement. It is now for you to judge, whether the statement of these experiences are true or materially false. It will be an easy matter to decide this question. Ask the navigator, ask the wind-miller, or observe the power of wind yourself in any way you please. The results of your enquiries or observations may vary, they may show more or less power than I have stated; but suppose even the result to be but a small portion of what I have stated, we should still have an enormity of power. However, I am confident, a close investigation will show a far greater power than I have stated. If my statement of experiences is materially true, is there perhaps some gross mistake in my conclusions and computation?—This may easily be ascertained. If you find no material mistake in my present statement, is it possible for rational men, to behold this power with indifference?—Does the subject not deserve our greatest attention and reflection?—You may ask, how is it, that no application of great extent was ever made yet?—In navigation we do make a considerable use of this power, and on land, in some places, by wind-mills. But it will occur now to your mind, that this power, on account of its irregularity, cannot always, nor any-where, be applied. Here I have to repeat, it can. There is a very material difference between the manner of application used hitherto and that which I propose. Hitherto the power of wind has been applied immediately upon the machinery for use, and they had to wait the chances of the wind's blowing; where the operation is stopped as soon as the wind ceases to blow. But the manner, which I shall state hereafter, to apply this power, is to make it operate only for collecting or storing up the power in a manner, and then to take out of this store of power, at any time, as much power for final operation upon the machineries as may be wanted for the intended purposes. The power stored up is to react, just as it may suit the purposes, and may do so long after the original power of wind has ceased. And, though the wind should cease at intervals of many months, we may have by the same power a uniform perpetual motion in a very simple way.

If you ask, perhaps, Why is this power not more used, if the statement be true? I have to ask, in return, Why is the power of steam so lately come to application? So many millions of men boiled water every day since many thousands of years; they must have frequently seen, that boiling water, in tightly-closed pots or kettles, will lift the cover or burst the vessel with great vehemence. The power of steam was, therefore, as commonly known, down to the least kitchen or wash woman, as the power of wind. But close observation and reflection was bestowed neither on the one nor the other. It is by calm reflection, by linking the elements, or first and simple observations and ideas derived therefrom, together, by little and little, that man is only capable to discover truths, which escape to immediate observations. It is thus often the case, that we arrive at truths which we never fancied or expected, beginning with the most simple truths known to every one, comprehensible even to little children, and which truths, therefore, would seem to be below the attention of mature men: man reasons from these first elements of his comprehension, he links them together into a chain, extends them further and further, applies them, and startles at last at the result: he mistrusts his judgment, suspects errors, goes back again to the most simple elements of conceptions, pursues again and again the course of his reasoning with the minutest attention, to discover errors, compares his theory with experiments, and sees finally compelled his reason to admit the discovered truth. Encouraged by the surprising result, he proceeds further with heightened curiosity. Thus mathematics took their origin, and in their consequences all sciences of certainty. Beginning with the most simple conceptions, which seem to the beginner to be the most insipid trifles, unworthy his attention, he cannot see the reason, why this minuteness of enquiry into these most simple things; he is led gradually into more complicated truths, and finally to astonishing results. He sees himself at last enabled to survey the universe without leaving his room; he discovers the size, form, and motion, of the whole earth, the distance of the sun, moon, and stars, their size, form, motions, and relations to each other; he ascertains, that they are worlds, larger even than our earth, distant many millions of miles from us and from each other; he sees a universe of many millions of large worlds—whole systems of worlds; new ideas start in his mind, he sees no end in his discoveries. But tell to the man of equal faculties, but who is unacquainted with the train of close reasoning, that led to those results,—tell him all these discoveries! talk to him about size and distances of the sun, moon, etc., where never any human being was, nor can go; tell these mathematical truths to him, whose mind is perhaps filled with erroneous notions, and prejudices, of which he cannot give any rational account, which he never thought to examine. What will he answer?—He will deride the man of these knowledges, he will take him for a fool.—But when he sees, that the same man predicts with precision eclipses of the sun, moon, etc.,—when he sees that this supposed fool makes books and astronomical tables, to show, out of his room, to the navigator the means for finding his way through the vast ocean around the world, and many other strange things, of which he has not the most distant idea,—the poor man does not know what to think of it.—Truths like these are in our days generally acknowledged; but it is not long ago when they were not. And even now the reasons of these discoveries are not generally understood; the results are but by a part of the multitude believed on authority of the learned men. Many cases might be alleged, how the multitude have lived always in the grossest errors, prejudices, and ignorance, despising and deriding all attempts of single individuals for discovering and applying usefully new truths.

I have announced to show the means for creating a paradise, a new superior world, to effect in one year more than hitherto could be done in thousands of years. People may ridicule the idea, or think the realisation of it miraculous. But where is the wonder to effect these purposes, if we have powers enough and superabundant for it? If, e. g., you have to move a weight of 1 ton, and you know 10 horses will effect it, but you have, instead of 10 horses, 100; where would be the wonder or the doubtfulness of being able to do it? Just so it is with my proposals. The removal of 1 ton by 100 horses would certainly be less easy than to effect what I have promised by a power exceeding all imaginable wants. But you may ask now, by what machineries can all the various purposes in view be effected in applying this power? Machineries are but tools. The possibility of contriving tools for any certain purpose cannot be questioned. They may be of various constructions for the same purposes. If we have sufficient power and materials for the tools to be applied, we may easily contrive and shape the tools as we please and as they suit our purpose. There is no reason to deem the making of adapted tools for certain purposes impossible. I, for one, shall resolve this problem in a very simple manner for all announced purposes. I shall speak of that hereafter. (See Etzler's Mechanical System, etc.)

*

What mechanisms, what machines, are to be applied, will be the question now, granted that there is power enough.

I shall give here a general outline of the system of machineries and establishments to be pursued.

We drudge and toil in agriculture, in architecture, in navigation, in all workshops, and in manufactories, for making many useful and many useless things for human life, for supplying many various demands of necessaries, comforts, and luxuries of life, of fancy and fashions. We little care about the real benefit the produces of our industry may afford to the buyer, provided we get pay for them, and make money by their sale. There is an endless variety of artificial productions of every kind, resulting from competition of the producers. I have promised contrivances for superseding all human labor. To imitate minutely all the infinite variety of produces of human industry by machineries, would be an endless, ungrateful, and foolish undertaking, though it might be possible. It would nearly require to invent for every little work of man a particular automaton. This is not my purpose. But the most simple contrivances I could think of, and as few as possible, for producing, not the customary articles of human industry, but all things that may either substitute or surpass the known necessaries, comforts, and luxuries of men, are my objects in view.

This problem is not so difficult as might be imagined at first. There was never any system in the productions of human labor; but they came into existence and fashion as chance directed men. Still less was there ever a thought exhibited to make a general science or system of providing for all artificial human wants. My object is to furnish, by an extremely-simple system, all what may be desirable for human life, without taking for pattern any of the existing things of industry. By abstracting from all what is in existence and fashion, I am enabled to devise means, without any artificial machinery, for producing every thing that man may want for his nourishment, dwelling, garments, furnitures, and articles of fancy and amusements.

But we have to relinquish entirely all our customary notions of human wants, and substitute them by others of a superior and more systematic order.

I shall begin with agriculture.

The first object is here, to clear the ground from all spontaneous growth and stones.

1. A machine of large size is to move along, and, while moving, to take the trees of all sizes with their roots out of the ground, to cut them in convenient pieces, to pile them up, and to take all stones out of the ground to any required depth.

2. A second machine is to follow, for taking up the piles of wood and stones and transporting the same to the places of their destination; this machine may carry thousands of tons at once.

3. The wood removed to its places for final use is then to be formed into planks, boards, beams, rails, pieces for fuel, and for any other purpose, by a simple contrivance, from whence it is to be removed to the places where it be wanted; this is done by one machine, which may also cut stones of any size.

4. The first-mentioned machine, with a little alteration, is then to level the ground perfectly, in planing it, filling the excavations or taking off the elevations of ground until all is level. If the hills or valleys are considerable, the same machine cuts terraces, winding around them up to the top in elegant shapes.

The same machine may make any excavation or elevation, cut canals, ditches, ponds of any size and shape, raise dams, artificial level roads, walls, and ramparts, with ditches around fields as enclosures, with walks on their top, from walks and paths with elevated borders.

5. The same machine, with some other little alteration, is to give to the ground its final preparation for receiving the seed; it tills the ground, in tearing the soil up to any required depth, refining or mouldering the same, sifting all small roots and stones from it, and putting the seed into the ground in any way required.

6. The same machine may take good fertile ground from one place to some other, for covering, at any required depth, poor soil with fertile soil of the best mixture.

7. The same machine, with a little addition, may reap any kind of grain or vegetable, thrash the seed out in the same time, grind it to meal, or press it to oil; it may also cut or prepare any other vegetable for final use in the kitchen or bakery.

8. Another small machine may sink wells and mines to any required depth and in any direction, and take the contents of the same up to light: it may be in earth, rocks, swamps, or water. (For the description of these machines, see Etzler's Mechanical System, page 11 to 27.)

*

Flexible Stuff.

There is yet one great desideratum: this is the making of flexible or pliable stuffs, and finishing all the articles out of them for use, such as for garments, couches, and all other commodities and ornaments, without labor. If this can be effected, without labor, then the problem of superseding all human labors is resolved completely.

This can be done, without spinning, weaving, sewing, tanning, etc., by a simple proceeding. There are cohesive substances in superabundance in nature, in the vegetable and animal kingdom, which need but be extracted: they are of various qualities: some resist water, some heat, some both, some are elastic, some soft, some hard. They all may be hardened or dissolved into fluids, just as required. They are made use of already for various purposes. In dissolving them into adapted fluids, and mixing the same with fibres of vegetables of convenient fineness, or flexible stuffs, fitly prepared, they will glutinate them together. By proper contrivance sheets of any size and form may be formed, in a similar manner as it is done with manufacturing paper. This stuff may be made as fine, and as thin or as thick, as may be desired. It may be made of any degree of stiffness or softness, which is depending from the mixture of stuffs, and from the manner of preparation and finishing. It may be calendered and polished for the sake of ornament, or for mirrors. It may be made of any color, or pattern, without any additional trouble, except admixture of dying stuff. It may then be used for any purpose, instead of any woven stuff, of leather, paper, fur, etc. It may be cast not only into sheets of any size or form, but also into any shape whatever; thus ready-made clothes of any fashion, vessels for holding dry or liquid materials, of any shape or size, any other article of commodity, may be at once cast or moulded into the appropriated form or mould.

So there will be no sewing or any kind of finishing by hands. There is no object of pliable stuff to be thought of, which could not be made completely in this way, so as to supersede all articles of that kind actually in use.

Such are all the means for the application of the immense powers of nature, for substituting all human labors, that I have in contemplation. They are simple, of a very small number in kinds, hardly three or four different contrivances. So they are by no means of such a nature as to scare the imagination from attempting the practice of them. They resolve completely the whole riddle of doing away all human labor, and make extremely easy and simple what was ever thought to be utter impossibility. But they do not only substitute all our present articles of human industry, but may create a great many things, which were never seen or thought of yet, to the greatest benefit of mankind. They are sufficient to create a paradise, surpassing in splendor, delight, enjoyments, all what human fancy ever conceived—a world altogether new for men, and this in a very short period.

Let us cast a view upon the things that immediately may be created by these means.

Any wilderness, even the most hideous and the most sterile, may be converted into the most fertile and delightful gardens. The most dismal swamps may be cleared of all their spontaneous growth, filled up and levelled, intersected with canals, ditches, and aqueducts, for draining them entirely. The soil, if required, may be meliorated, by covering or mixing it with rich soil taken from distant places. The same is to be mouldered to fine dust, levelled, sifted from all roots, weeds, and stones, and sowed and planted in the most beautiful order and symmetry, with fruit-trees and vegetables of every kind that may stand the climate. The walks and roads are to be paved with hard, vitrified, large plates, so as to be always clean from all dirt at any weather or season. They may be bordered with the most beautiful beds of flowers, fruitful vegetables, bushes, shrubs, and trees, all rising gradually in rows behind one another, arranged so as to afford almost continually delight to the organs of sight, taste, and smell. Canals and aqueducts with vitrified channels, and if required, covered, filled with the clearest water out of fountains, from the deep subterraneous recesses of water, which may spout and be led any-where. Some canals may serve for fish-ponds and for irrigating the gardens, others for draining swampy ground. Some aqueducts may be used to lead water into all parts of the garden, for irrigating the ground whenever it be required; this may be done by sprinkling the water in copious showers through moveable tubes with adapted large mouths. This water may be mixed with liquid manure, derived from all the decayed vegetables and other materials fit for manure, prepared and liquified in proper buildings to that effect. Thus the fertility of the garden will not depend from weather. When it rains too much, it may be led off from the ground in proper channels. The canals may be bordered with beautiful growth, in similar manner as the walks. The channels being of vitrified substance, the water being perfectly clear, and filtrated or distilled if required, they may afford the most beautiful sceneries imaginable, while a variety of fishes is seen clear down to the bottom playing about, and while these canals afford in the same time the chance for gliding smoothly along and between these various sceneries of art and nature, upon beautiful gondolas, and while their surface and borders may be covered with fine land and aquatic birds. The canals may end or concentrate in large beautiful ponds, where the bottom is also of vitrified substance. Thus water clear as crystal, in beds or channels like crystal, surrounded and covered by enchanting sceneries, fertilises and beautifies the gardens, and gives them the relief of a paradise. The aqueducts may be supported by the most splendid colonnades. The walks may be covered with porticos adorned with magnificent columns, statues, and sculptural works; all of vitrified substance, lasting for ever, while the beauties of nature around heighten the magnificence and deliciousness. The ponds may also be surrounded by porticos. These porticos, the fountains, the green arbors and bowers, may preserve continually freshness of air and of growth of the vegetables, and afford delightful shelter against wet weather and heat.

The crops are gathered and prepared for use without any labor. There is nothing then but enjoyment and delight. Pastures may be out of such gardens, with establishments for milking, etc., so that one or two men may do this work in less than an hour every day for thousands of consumers.

The hills and mountains may be surrounded by the most beautiful terraces imaginable, which may wind in spiral form up to their tops, affording the grandest and most beautiful prospects, as well from below in the valleys as from above down into the same.

This is what concerns agriculture, when I have supposed the most unfavorable spot to be found upon earth.

Andrew Ure

The Philosophy of Manufactures (1835)

Andrew Ure (1778–1857) was a medical doctor, a teacher of chemistry and mechanics to artisans, one of Britain's first consulting chemists, and the author of many works, including *A New System of Geology,* which attempted to reconcile biblical authority and geological evidence. His *Philosophy of Manufactures* was intended as the introductory volume to a series of books on the industries of England. Ure defended the factory system as bluntly and energetically as anyone, without blush or apology, at a time when the worst cruelties of British industrial life were subjected to widespread public scrutiny. Ure argued not only that the new factory system was necessary to the prosperity of the nation but also that it was a universal benefit to the working class. Although many of Ure's arguments were poorly supported and his book was widely ridiculed, the first chapter does contain an important early description of the new factory, emphasizing the importance of organization and discipline. Karl Marx was one who took Ure's description of industrial capitalism very seriously, and G. S. White, the biographer of Samuel Slater, adapted Ure's arguments to the American scene.

Andrew Ure, *Philosophy of Manufactures or an Exposition of the Scientific, Moral, and Commercial Economy of the Factory System of Great Britain* (London, 1835; third edition, New York: Burt Franklin, 1969), pp. 1–2, 5–9, 11–23. On Ure, see V. W. Farrar, "Andrew Ure and the Philosophy of Manufactures," *Notes and Records of the Royal Society,* volume 27 (1972–1973), pp. 299–324. See also G. S. White, *Memoir of Samuel Slater* (Philadelphia, 1836; reprint, New York: Augustus Kelley, 1966); portions of White's chapter on "The Moral Influence of Manufacturers' are reprinted in volume 2 of this series.

PHILOSOPHY OF MANUFACTURES.

BOOK THE FIRST.

GENERAL PRINCIPLES OF MANUFACTURES.

CHAPTER I.

General View of Manufacturing Industry.

MANUFACTURE is a word, which, in the vicissitude of language, has come to signify the reverse of its intrinsic meaning, for it now denotes every extensive product of art, which is made by machinery, with little or no aid of the human hand; so that the most perfect manufacture is that which dispenses entirely with manual labour. The philosophy of manufactures is therefore an exposition of the general principles, on which productive industry should be conducted by self-acting machines. The end of a manufacture is to modify the texture, form, or composition of natural objects by mechanical or chemical forces, acting either separately, combined, or in succession. Hence the automatic arts subservient to general commerce may be distinguished into Mechanical and Chemical, according as they modify the external form or the internal constitution of their subject matter. An indefinite variety of objects may be subjected to each system of action, but they may be all conveniently classified int animal, vegetable, and mineral.

A mechanical manufacture being commonly occupied with one substance, which it conducts through metamorphoses in regular succession, may be made nearly automatic; whereas a chemical manufacture depends on the play of delicate affinities between two or more substances, which it has to subject to heat and mixture under circumstances somewhat uncertain, and must therefore remain, to a corresponding extent, a manual operation. The best example of *pure* chemistry on self-acting principles which I have seen, was in a manufacture of sulphuric acid, where the sulphur being kindled and properly set in train with the nitre, atmospheric air, and water, carried on the process through a labyrinth of compartments, and supplied the requisite heat of concentration, till it brought forth a finished commercial product. The finest model of an automatic manufacture of *mixed* chemistry is the five-coloured calico machine, which continuously, and spontaneously, so to speak, prints beautiful webs of cloth with admirable precision and speed. It is in a cotton mill, however, that the perfection of automatic industry is to be seen; it is there that the elemental powers have been made to animate millions of complex organs, infusing into forms of wood, iron, and brass an intelligent agency. And as the philosophy of the fine arts, poetry, painting, and music may be best studied in their individual master-pieces, so may the philosophy of manufactures in this its noblest creation.

*

When we take into account the vastly greater proportion of young persons constantly occupied with factory labour, than of those occupied with agricultural labour, we shall then be led to conclude that at least double the amount of personal industry is engaged in the arts, manufactures, and trade, to what is engaged in agriculture. Considerably upwards of one-tenth of the population of this island is actually employed in manufactures; and probably little more than one-fifteenth in agriculture. This conclusion ought to lead our legislative landlords to treat the manufacturing interests with greater respect than they have usually been accustomed to do. If we consider, moreover, how much greater a mass of productive industry a male adult is equivalent to, in power-driven manufactures, than in agriculture, the balance in favour of the former will be greatly enhanced.

France, which has for upwards of a century and a half tried every scheme of public premium to become a great manufacturing country, has a much less proportion than one employed in trade for two employed in agriculture. M. Charles Dupin, indeed, has been led by his researches into the comparative industry of France and of the United Kingdom, to conclude that the agricultural produce of our country amounted in value to 240 millions sterling, and that of his own to 180 millions sterling, being the ratio of three to two; and that our manufacturing power is inferior to that of France in the proportion of sixty-three to seventy-two; or as seven to eight. There can be no doubt that his agricultural estimate underrates France, as much as his manufacturing estimate underrates Great Britain.

This island is pre-eminent among civilized nations for the prodigious development of its factory wealth, and has been therefore long viewed with a jealous admiration by foreign powers. This very pre-eminence, however, has been contemplated in a very different light by many influential members of our own community, and has been even denounced by them as the certain origin of innumerable evils to the people, and of revolutionary convulsions to the state. If the affairs of the kingdom be wisely administered, I believe such allegations and fears will prove to be groundless, and to proceed more from the envy of one ancient and powerful order of the commonwealth, towards another suddenly grown into political importance than from the nature of things.

In the recent discussions concerning our factories, no circumstance is so deserving of remark, as the gross ignorance evinced by our leading legislators and economists, gentlemen well informed in other respects, relative to the nature of those stupendous manufactures which have so long provided the rulers of the kingdom with the resources of war, and a great body of the people with comfortable subsistence; which have, in fact, made this island the arbiter of many nations, and the benefactor of the globe itself.* Till this ignorance be dispelled, no sound legislation need be expected on manufacturing subjects. To effect this purpose is a principal, but not the sole aim of the present volume,

* Even the eminent statesman lately selected by his Sovereign to wield the destinies of this commercial empire—Sir Robert Peel, who derives his family consequence from the cotton trade, seems to be but little conversant with its nature and condition.—See Dr. Carbutt's observations on the subject, next page.

for it is intended also to convey specific information to the classes directly concerned in the manufactures, as well as general knowledge to the community at large, and particularly to young persons about to make the choice of a profession.

The blessings which physico-mechanical science has bestowed on society, and the means it has still in store for ameliorating the lot of mankind, have been too little dwelt upon; while, on the other hand, it has been accused of lending itself to the rich capitalists as an instrument for harassing the poor and of exacting from the operative an accelerated rate of work. It has been said, for example, that the steam-engine now drives the power-looms with such velocity as to urge on their attendant weavers at the same rapid pace; but that the hand-weaver, not being subjected to this restless agent, can throw his shuttle and move his treddles at his convenience. There is, however, this difference in the two cases, that in the factory, every member of the loom is so adjusted, that the driving force leaves the attendant nearly nothing at all to do, certainly no muscular fatigue to sustain, while it procures for him good, unfailing wages, besides a healthy workshop *gratis*: whereas the non-factory weaver, having everything to execute by muscular exertion, finds the labour irksome, makes in consequence innumerable short pauses, separately of little account, but great when added together; earns therefore proportionally low wages, while he loses his health by poor diet and the dampness of his hovel. Dr. Carbutt of Manchester says, " With regard to Sir Robert Peel's assertion a few evenings ago, that the hand-loom weavers are mostly small farmers, nothing can be a greater mistake; they live, or rather they just keep life together, in the most miserable manner, in the cellars and garrets of the town, working sixteen or eighteen hours for the merest pittance."*

The constant aim and effect of scientific improvement in manufactures are philanthropic, as they tend to relieve the workmen either from niceties of adjustment which exhaust his mind and fatigue his eyes, or from painful repetition of effort which distort or wear out his frame. At every step of each manufacturing process described in this volume, the humanity of science will be manifest. New illustrations of this truth appear almost every day, of which a remarkable one has just come to my knowledge. In the woollen-cloth trade there is a process between carding and spinning the wool, called *slubbing*, which converts the spongy rolls, turned off from the cards, into a continuous length of fine porous cord. Now, though carding and spinning lie within the domain of automatic science, yet slubbing is a handicraft operation, depending on the skill of the slubber, and participating therefore in all his irregularities. If he be a steady, temperate man, he will conduct his business regularly, without needing to harass his juvenile assistants, who join together the series of card rolls, and thus feed his machine; but if he be addicted to liquor, and passionate, he has it in his power to exercise a fearful despotism over the young pieceners, in violation of the proprietor's benevolent regulations. This class of operatives, who, though inmates of factories, are not, properly speaking, factory workers, being independent of the moving power, have been the principal source of the obloquy so unsparingly

* Letter of 3rd of May, 1833, to Dr. Hawkins in his Medical Report, Factory Commission, p. 282.

cast on the cotton and other factories, in which no such capricious practices or cruelties exist. The wool slubber, when behind hand with his work, after a visit to the beer-shop, resumes his task with violence, and drives his machine at a speed beyond the power of the pieceners to accompany; and if he finds them deficient in the least point, he does not hesitate to lift up the long wooden rod from his slubbing-frame, called a billy-roller, and beat them unmercifully. I rejoice to find that science now promises to rescue this branch of the business from handicraft caprice, and to place it, like the rest, under the safeguard of automatic mechanism. The details of this recent invention will be given in describing the woollen manufacture.

* Man stands in daily want of food, fuel, clothing, and shelter; and is bound to devote the powers of body and mind, of nature and art, in the first place to provide for himself and his dependents a sufficiency of these necessaries, without which there can be no comfort, nor leisure for the cultivation of the taste and intellect. To the production of food and domestic accommodation, not many automatic inventions have been applied, or seem to be extensively applicable; though, for modifying them to the purposes of luxury, many curious contrivances have been made. Machines, more or less automatic, are embodied in the coal-mines of Great Britain; but such combinations have been mainly directed, in this as well as other countries, to the materials of clothing. These chiefly consist of flexible fibres of vegetable or animal origin, twisted into smooth, tenacious threads, which are then woven into cloth by being decussated in a loom. Of the animal kingdom, silk, wool, and hair, are the principal textile products. The vegetable tribes furnish cotton, flax, hemp, besides several other fibrous substances of inferior importance.

Wool, flax, hemp, and silk, have been very generally worked up among the nations of Europe, both in ancient and modern times; but cotton attire was, till sixty years ago, confined very much to Hindostan, and some other districts of Asia. No textile filaments however are, by their facility of production as well as their structure, so well adapted as those of cotton to furnish articles of clothing, combining comfort with beauty and convenience in an eminent degree. Hence we can understand how cotton fabrics, in their endless variety of textures and styles, plain, figured, and coloured, have within the short period of one human life, grown into an enormous manufacture, have become an object of the first desire to mankind all over the globe, and of zealous industry to the most civilized states. This business has received its great automatic development in England, though it was cultivated to a considerable extent on handicraft principles in France a century ago, and warmly encouraged by the government of that country, both as to the growth of the material and its conversion into cloth. The failure of the French however to establish a factory system prior to the English is a very remarkable fact, and proves clearly that mechanical invention, for which the former nation have long been justly celebrated, is not of itself sufficient to found a successful manufacture.

We have adverted to the mechanisms of Vaucanson. This inventive artisan directed his attention also to productive machines. He constructed one for winding silk so long ago as 1749; one for doubling and twisting it in 1751; a tapestry loom in 1758; another for winding silk in 1770; a machine for laminating stuffs in 1757, and a plan of mounting silk-mills in 1776. There can be no doubt as to the value of these inventions, as they were described with merited eulogiums in the above named years by the Academy of Paris. In 1776 he published an account of the Indian mode of weaving fine muslins in the wet state, showing that his attention had been turned likewise to the cotton trade.

The term *Factory*, in technology, designates the combined operation of many orders of work-people, adult and young, in tending with assiduous skill a system of productive machines continuously impelled by a central power. This definition includes such organizations as cotton-mills, flax-mills, silk-mills, woollen-mills, and certain engineering works; but it excludes those in which the mechanisms do not form a connected series, nor are dependent on one prime mover. Of the latter class, examples occur in iron-works, dye-works, soap-works, brass-foundries, &c. Some authors, indeed, have comprehended under the title *factory*, all extensive establishments wherein a number of people co-operate towards a common purpose of art; and would therefore rank breweries, distilleries, as well as the workshops of carpenters, turners, coopers, &c., under the factory system. But I conceive that this title, in its strictest sense, involves the idea of a vast automaton, composed of various mechanical and intellectual organs, acting in uninterrupted concert for the production of a common object, all of them being subordinated to a self-regulated moving force. If the marshalling of human beings in systematic order for the execution of any technical enterprise were allowed to constitute a factory, this term might embrace every department of civil and military engineering; a latitude of application quite inadmissible.

In its precise acceptation, the Factory system is of recent origin, and may claim England for its birthplace. The mills for throwing silk, or making organzine, which were mounted centuries ago in several of the Italian states, and furtively transferred to this country by Sir Thomas Lombe in 1718, contained indeed certain elements of a factory, and probably suggested some hints of those grander and more complex combinations of self-acting machines, which were first embodied half a century later in our cotton manufacture by Richard Arkwright, assisted by gentlemen of Derby, well acquainted with its celebrated silk establishment. But the spinning of an entangled flock of fibres into a smooth thread, which constitutes the main operation with cotton, is in silk superfluous; being already performed by the unerring instinct of a worm, which leaves to human art the simple task of doubling and twisting its regular filaments. The apparatus requisite for this purpose is more elementary, and calls for few of those gradations of machinery which are needed in the carding, drawing, roving, and spinning processes of a cotton-mill.

When the first water-frames for spinning cotton were erected at Cromford, in the romantic valley of the Derwent, about sixty years ago, mankind were little aware of the mighty revolution which the new system of labour was destined by Providence to achieve, not only in the structure of British society, but in the fortunes of the world at large. Arkwright alone had the sagacity to discern, and the boldness to predict in glowing language, how vastly productive human industry would become, when no longer proportioned in its results to muscular effort, which is by its nature fitful and capricious, but when made to consist in the task of guiding the work of mechanical fingers and arms, regularly impelled with great velocity by some indefatigable physical power. What his judgment so clearly led him to perceive, his energy of will enabled him to realize with such rapidity and success, as would have done honour to the most influential individuals, but were truly wonderful in that obscure and indigent artisan. The main difficulty did not, to my apprehension, lie so much in the invention of a proper self-acting mechanism for drawing out and twisting cotton into a continuous thread, as in the distribution of the different members of the apparatus into one co-operative body, in impelling each organ with its appropriate delicacy and speed, and above all, in training human beings to renounce their desultory habits of work, and to identify themselves with the unvarying regularity of the complex automaton. To devise and administer a successful code of factory discipline, suited to the necessities of factory diligence, was the Herculean enterprise, the noble achievement of Arkwright. Even at the present day, when the system is perfectly organized, and its labour lightened to the utmost, it is found nearly impossible to convert persons past the age of puberty, whether drawn from rural or from handicraft occupations, into useful factory hands. After struggling for a while to conquer their listless or restive habits, they either renounce the employment spontaneously, or are dismissed by the overlookers on account of inattention.

If the factory Briareus could have been created by mechanical genius alone, it should have come into being thirty years sooner; for upwards of ninety years have now elapsed since John Wyatt, of Birmingham, not only invented the series of fluted rollers, (the spinning fingers usually ascribed to Arkwright,) but obtained a patent for the invention, and erected " a spinning engine without hands" in his native town. The details of this remarkable circumstance, recently snatched from oblivion, will be given in our Treatise on the Cotton Manufactures. Wyatt was a man of good education, in a respectable walk of life, much esteemed by his superiors, and therefore favourably placed, in a mechanical point of view, for maturing his admirable scheme. But he was of a gentle and passive spirit, little qualified to cope with the hardships of a new manufacturing enterprise. It required, in fact, a man of a Napoleon nerve and ambition, to subdue the refractory tempers of work-people accustomed to irregular paroxysms of diligence, and to urge on his multifarious and intricate constructions in the face of prejudice, passion, and envy. Such was Arkwright, who, suffering nothing to stay or turn aside his progress, arrived gloriously at the goal, and has for ever affixed his name to a great era in the annals of mankind, an era which has laid open unbounded prospects of wealth and comfort to the industrious, however much they may have been occasionally clouded by ignorance and folly.

Prior to this period, manufactures were everywhere feeble and fluctuating in their development; shooting forth luxuriantly for a season, and again withering almost to the roots, like annual plants. Their perennial growth now began in England, and attracted capital in copious streams to irrigate the rich domains of industry. When this new career commenced, about the year 1770, the annual consumption of cotton in British manufactures was under four millions of pounds weight, and that of the whole of Christendom was probably not more than ten millions. Last year the consumption in Great Britain and Ireland was about two hundred and seventy millions of pounds, and that of Europe and the United States together four hundred and eighty millions. This prodigious increase is, without doubt, almost entirely due to the factory system founded and upreared by the intrepid native of Preston. If then this system be not merely an inevitable step in the social progression of the world, but the one which gives a commanding station and influence to the people who most resolutely take it, it does not become any man, far less a denizen of this favoured land, to vilify the author of a benefaction, which, wisely administered, may become the best temporal gift of Providence to the poor, a blessing destined to mitigate, and in some measure to repeal, the primeval curse pronounced on the labour of man, " in the sweat of thy face shalt thou eat bread." Arkwright well deserves to live in honoured remembrance among those ancient master-spirits, who persuaded their roaming companions to exchange the precarious toils of the chase, for the settled comforts of agriculture.

In my recent tour, continued during several months, through the manufacturing districts, I have seen tens of thousands of old, young, and middle-aged of both sexes, many of them too feeble to get their daily bread by any of the former modes of industry, earning abundant food, raiment, and domestic accommodation, without perspiring at a single pore, screened meanwhile from the summer's sun and the winter's frost, in apartments more airy and salubrious than those of the metropolis, in which our legislative and fashionable aristocracies assemble. In those spacious halls the benignant power of steam summons around him his myriads of willing menials, and assigns to each the regulated task, substituting for painful muscular effort on their part, the energies of his own gigantic arm, and demanding in return only attention and dexterity to correct such little aberrations as casually occur in his workmanship. The gentle docility of this moving force qualifies it for impelling the tiny bobbins of the lace-machine with a precision and speed inimitable by the most dexterous hands, directed by the sharpest eyes. Hence, under its auspices, and in obedience to Arkwright's polity, magnificent edifices, surpassing far in number, value, usefulness, and ingenuity of construction, the boasted monuments of Asiatic, Egyptian, and Roman despotism, have, within the short period of fifty years, risen up in this kingdom, to show to what extent, capital, industry, and science may augment the resources of a state, while they meliorate the condition of its citizens. Such is the factory system, replete with prodigies in mechanics and political economy, which promises, in its future growth, to become the great minister of civilization to the terraqueous globe, enabling this country, as its heart, to diffuse along with its commerce, the life-blood of science and religion to myriads of people still lying "in the region and shadow of death."

When Adam Smith wrote his immortal elements of economics, automatic machinery being hardly known, he was properly led to regard the division of labour as the grand principle of manufacturing improvement; and he showed, in the example of pin-making, how each handicraftsman, being thereby enabled to perfect himself by practice in one point, became a quicker and cheaper workman. In each branch of manufacture he saw that some parts were, on that principle, of easy execution, like the cutting of pin wires into uniform lengths, and some were comparatively difficult, like the formation and fixation of their heads; and therefore he concluded that to each a workman of appropriate value and cost was naturally assigned. This appropriation forms the very essence of the division of labour, and has been constantly made since the origin of society. The ploughman, with powerful hand and skilful eye, has been always hired at high wages to form the furrow, and the ploughboy at low wages, to lead the team. But what was in Dr. Smith's time a topic of useful illustration, cannot now be used without risk of misleading the public mind as to the right principle of manufacturing industry. In fact, the division, or rather adaptation of labour to the different talents of men, is little thought of in factory employment. On the contrary, wherever a process requires peculiar dexterity and steadiness of hand, it is withdrawn as soon as possible from the *cunning* workman, who is prone to irregularities of many kinds, and it is placed in charge of a peculiar mechanism, so self-regulating, that a child may superintend it. Thus,—to take an example from the spinning of cotton—the first operation in delicacy and importance, is that of laying the fibres truly parallel in the spongy slivers, and the next is that of drawing these out into slender spongy cords, called rovings, with the least possible twist; both being perfectly uniform throughout their total length. To execute either of these processes tolerably by a hand-wheel, would require a degree of skill not to be met with in one artisan out of a hundred. But fine yarn could not be made in factory-spinning except by taking these steps, nor was it ever made by machinery till Arkwright's sagacity contrived them. Moderately good yarn may be spun indeed on the *hand-wheel* without any drawings at all, and with even indifferent rovings, because the thread, under the two-fold action of twisting and extension, has a tendency to equalize itself.

The principle of the factory system then is, to substitute mechanical science for hand skill, and the partition of a process into its essential constituents, for the division or graduation of labour among artisans. On the handicraft plan, labour more or less skilled, was usually the most expensive element of production—*Materiam superabat opus;* but on the automatic plan, skilled labour gets progressively superseded, and will, eventually, be replaced by mere overlookers of machines.

By the infirmity of human nature it happens, that the more skilful the workman, the more self-willed and intractable he is apt to become, and, of course, the less fit a component of a mechanical system, in which, by occasional irregularities, he may do great damage to the whole. The grand object therefore of the modern manufacturer is, through the union of capital and science, to reduce the task of his work-people to the exercise of vigilance and dexterity,—faculties, when concentred to one process, speedily brought to perfection in the young. In the infancy of mechanical engineering, a machine-factory displayed the division of labour in manifold gradations—the file, the drill, the lathe, having each its different workmen in the order of skill: but the dexterous hands of the filer and driller are now superseded by the planing, the key-groove cutting, and the drilling-machines; and those of the iron and brass turners, by the self-acting slide-lathe. Mr. Anthony Strutt, who conducts the mechanical department of the great cotton factories of Belper and Milford, has so thoroughly departed from the old routine of the schools, that he will employ no man who has learned his craft by regular apprenticeship; but in contempt, as it were, of the division of labour principle, he sets a ploughboy to turn a shaft of perhaps several tons weight, and never has reason to repent his preference, because he infuses into the turning apparatus a precision of action, equal, if not superior, to the skill of the most experienced journeyman.

An eminent mechanician in Manchester told me, that he does not choose to make any steam-engines at present, because with his existing means, he would need to resort to the old principle of the division of labour, so fruitful of jealousies and strikes among workmen; but he intends to prosecute that branch of business whenever he has prepared suitable arrangements on the equalization of labour, or automatic plan. On the graduation system, a man must serve an apprenticeship of many years before his hand and eye become skilled enough for certain mechanical feats;

but on the system of decomposing a process into its constituents, and embodying each part in an automatic machine, a person of common care and capacity may be entrusted with any of the said elementary parts after a short probation, and may be transferred from one to another, on any emergency, at the discretion of the master. Such translations are utterly at variance with the old practice of the division of labour, which fixed one man to shaping the head of a pin, and another to sharpening its point, with most irksome and spirit-wasting uniformity, for a whole life.

It was indeed a subject of regret to observe how frequently the workman's eminence, in any craft, had to be purchased by the sacrifice of his health and comfort. To one unvaried operation, which required unremitting dexterity and diligence, his hand and eye were constantly on the strain, or if they were suffered to swerve from their task for a time, considerable loss ensued, either to the employer, or the operative, according as the work was done by the day or by the piece. But on the equalization plan of self-acting machines, the operative needs to call his faculties only into agreeable exercise; he is seldom harassed with anxiety or fatigue, and may find many leisure moments for either amusement or meditation, without detriment to his master's interests or his own. As his business consists in tending the work of a well regulated mechanism, he can learn it in a short period; and when he transfers his services from one machine to another, he varies his task, and enlarges his views, by thinking on those general combinations which result from his and his companions' labours. Thus, that cramping of the faculties, that narrowing of the mind, that stunting of the frame, which were ascribed, and not unjustly, by moral writers, to the division of labour, cannot, in common circumstances, occur under the equable distribution of industry. How superior in vigour and intelligence are the factory mechanics in Lancashire, where the latter system of labour prevails, to the handicraft artisans of London, who, to a great extent, continue slaves to the former! The one set is familiar with almost every physico-mechanical combination, while the other seldom knows anything beyond the pin-head sphere of his daily task.

It is, in fact, the constant aim and tendency of every improvement in machinery to supersede human labour altogether, or to diminish its cost, by substituting the industry of women and children for that of men; or that of ordinary labourers, for trained artisans. In most of the water-twist, or throstle cotton mills, the spinning is entirely managed by females of sixteen years and upwards. The effect of substituting the self-acting mule for the common mule, is to discharge the greater part of the men spinners, and to retain adolescents and children. The proprietor of a factory near Stockport states, in evidence to the commissioners, that by such substitution, he would save 50*l.* a week in wages, in consequence of dispensing with nearly forty male spinners, at about 25*s.* of wages each. This tendency to employ merely children with watchful eyes and nimble fingers, instead of journeymen of long experience, shows how the scholastic dogma of the division of labour into degrees of skill has been exploded by our enlightened manufacturers.

Isaac Hill

Message to Both Houses of the New Hampshire Legislature (1836)

In this excerpt from his message to the New Hampshire legislature, Governor Isaac Hill (1789–1851) expresses the views of the "conservative" wing of the state Democratic party. His strong agrarian view is balanced by only a moderate regret at some of the deleterious aspects of manufacturing. Hill himself invested in New Hampshire textile factories and came to oppose many of the anticorporation measures advocated by the "radical" Democrats in the state, like Henry Hubbard, whose attack on corporate privileges in 1842 is reprinted in this collection. Hill's resistance to accepting "funds from a superabundant national treasury" is a direct attack on Clay's "American System," under which "internal improvements" in transportation were funded by the federal government.

[Isaac Hill], Message to Both Houses of the Legislature, Journal of the Honorable Senate of the State of New Hampshire, June session (Concord, New Hampshire, 1836), pp. 40–42. For an account of the political battles over economic policies affecting manufactures in New Hampshire, see Donald B. Cole, *Jacksonian Democracy in New Hampshire* (Cambridge, Massachusetts: Harvard University Press, 1970), especially the chapter "Radical Democracy, 1836–1846," pp. 185–215.

In regard to the "promotion of agriculture, arts, siences, commerce, trades, and manufactures," enjoined in the constitution, generally speaking, it may be laid down as a rule that the several callings will be best protected by being left free from all tramels. Excessive legislation is frequently worse than no legislation: there is really more danger in giving a particular interest a legislative protection at the expense of other interests to take care of themselves. The better way to promote all is to let the government be felt in no place where governmental interference is not indispensable—to raise no revenue that is not wanted for objects useful and necessary—to furnish no dispensable object for expenditure that shall call for burdensome taxation.

It can scarcely be desirable to the people of this State that their coffers shall be filled by any indirection. The present condition of the national treasury, a temporary plethora indicating almost certain future poverty, furnishes a strong temptation to ask for distribution among the States. Desirable as is an overflowing State treasury, I cannot wish to see the time arrive when this State shall be furnished with funds from a superabundant national treasury. The public coffers must be supplied in some manner by contributions drawn from the people. If more shall be received from the customs and avails of the public domain than shall be necessary for the expenditures of the general government, how much better for the people will it be to relieve them from taxation upon their consumption, than to take the money from their pockets through the most expensive mode of collection, to be sent to the several State treasuries, not for the immediate benefit of those who contributed it, but to be scrambled for and expended in an unequal application to those points where the strongest influence will always bring it? Should the principle ever be established, in violation of the Constitution as I must think it to be, of distributing the funds of the nation among the States, I hope never to see New-Hampshire as a State commencing the work of making roads and other similar improvements; for, desirable as those works may be, when the State once begins them, there can be no end to the call for appropriations; every man who cannot have an improvement on or near his premises will have cause to complain that the State does him injustice. If the money so to be received shall be loaned to responsible companies and individuals who are willing to incur the risk of undertaking improvements, we may have assurance that the works to be prosecuted are so much wanted that they will at least reimburse the original investment: the annual income from such loans, might be substituted for the usual annual taxes laid to defray the expenses of the State Government.

There is no pursuit that tends more directly to the independence and happiness of the people than agriculture. More productive as it is more necessary than any other, it is a matter of gratification and pride that it is a calling scarcely less reputable than that requiring the highest order and severest application of intellect. The most intelligent and most meritorious citizens are of those who labor with their own hands in 'agricultural pursuits. Of such men it is safe to make not only legislators to frame our laws, but magistrates to execute them. As agriculture has risen in estimation, so have our farmers increased in wealth and all the means of independence.— In the westerly part of the State especially, of late years, the rearing of sheep and the production of wool has come in aid of other objects yielding ready money and often an unexpected profit. One new subject of enterprise succeeds another: although in a rougher soil and a severer climate, the time may arrive when wool to New-England shall be as important a staple as the wheat of the middle or cotton of the southern States.

The public attention has recently been drawn to the culture of the mulberry, the raising of silk worms and the production of silk.— That this important item of consumption and of traffic may be produced in the United States as extensively as in any country in the world, will not be disputed. The late changes of the tariff bring the bulk of articles of which silk is composed or is a component part, into the country free of duty; and from this cause it is extensively taking the place of the finer cottons and woollens. The value of silks imported into the United States for home consumption during the year 1835, according to the custom house returns, was nearly sixteen millions of dollars. The introduction of the article free of duty, instead of discouraging, seems to have given an increased impetus to preparations for planting the mulberry and hereafter extending the production and manufacture of silk. It remains to be tested whether the soil of our State shall be well adapted to the mulberry cultivation. Nothing yet appears to discourage the undertaking ; and as mulberry orchards may be planted without the investment of a large capital, it might be useful to afford legislative countenance to such towns of the state as already or may hereafter possess farms employed in support of the poor, in the cultivation of the mulberry, as would fully test the fact of the adaptation of this climate to the production of silk. As silk is deprived of what has been called protection from the general government, it might not be invidious if the State should exempt such land as is actually employed in raising the mulberry from taxation until the orchard shall become productive; or if encouragement should be given for planting the mulberry on town farms. It may, however, be laid down as a general principle, that bounties and premiums on one article of production at the expense of other articles is unwise; and that the several interests of Agriculture, as well as trades and manufactures, deserving the encouragement of legislators and magistrates, will best receive it by being left free from that kind of protection which is to be found in a high tariff.

In the rapid progress of arts and wealth, not less than in moral and intellectual improvements, the mothers and daughters of New-England have contributed their full share. Indeed to them are the most of us, as men and as social beings; indebted for whatever is commendable in conduct or in character. Many of our interior towns owe their prosperity principally to the hand of female industry. The change which within the last few years has taken a portion of our females from their homes where domestic manufactures had been carried on, to be collected in masses at the large establishments which have supplied the place of the spindle and shuttle at the domestic fireside, to many philanthropists and patriots has been a subject of regret. The silk culture generally adopted, without materially interfering with other agricultural products, might usefully give full employment to many of that class of females beneath their own parental roof who seek for a livelihood abroad—requiring the kind of labor that females and children would most naturally perform. As giving extensively that employment, the culture of the mulberry further deserves the public patronage.

Daniel Webster

Lecture before the Society for the Diffusion of Useful Knowledge (1836)

In this excerpt Webster makes a direct connection between corporate capitalism and the advance of science and technology. He attempts to correct the "great ignorance" or "great prejudice" that had led some workers to machinery-breaking riots and suggests that those who are opposed to corporations are equally hostile to the "common good." Webster, opposed to large-scale manufactures in his earlier years, had by this time become a crusader for them.

The Society for the Diffusion of Useful Knowledge was founded in Boston in 1828. It was one of many so-called Lyceums established in the 1820s and 1830s to promote the educatiion of young men by presenting series of practical and inspiring lectures. Among those who often addressed Lyceum audiences, Edward Everett and Ralph Waldo Emerson are also represented in this collection.

Daniel Webster, "Lecture before the Society for the Diffusion of Useful Knowledge," from *The Writings and Speeches of Daniel Webster,* national edition (Boston: Little, Brown, 1903), volume 13, pp. 66–77.

There has been in the course of half a century an unprecedented augmentation of general wealth. Even within a shorter period, and under the actual observation of most of us, in our own country and our own circles, vastly increased comforts have come to be enjoyed by the industrious classes, and vastly more leisure and time are found for the cultivation of the mind. It would be easy to prove this by detailed comparisons between the present and the past, showing how far the present exceeds the past, in regard to the shelter, food, clothing, and fuel enjoyed by laboring families. But this is a truth so evident and so open to common observation as matter of fact, that proof by particular enumeration of circumstances becomes unnecessary. We may safely take the fact to be, as it certainly is, that there are certain causes which have acted with peculiar energy in our generation, and which have improved the condition of the mass of society with a degree of rapidity heretofore altogether unknown.

What, then, are these causes? This is an interesting question. It seems to me the main cause is the successful application of science to art; or, in other words, the progress of scientific art.

It is the general doctrine of writers on political economy, that labor is the source of wealth. This is undoubtedly true. The materials of wealth are in the earth, and in the seas, or in their productions; and it is labor only which obtains them, works upon them, and fashions them to the uses of man. The fertility of the soil is nothing, till labor cultivates it; the iron in the mountain rock is of no value, till the strong hand of labor has drawn it forth, separated it from the neighboring earths, and melted and forged it into a manufactured article.

The great agent, therefore, that procures shelter, and food, and raiment for man, is labor; that is to say, it is an active agency, it is some moving power, it is something which has action and effort, and is capable of taking hold of the materials, with which the world supplies us, and of working them into shapes and forms such as shall administer to the wants and comforts of mankind.

The proposition of the philosophers, therefore, is true, that labor is the true source, and the only source of wealth; and it necessarily follows, that any augmentation of labor, augments, to the same degree, the productions of wealth. But when Adam Smith and his immediate followers laid down this maxim, it is evident that they had in view, chiefly, either the manual labors of agriculturists and artisans, or the active occupations of other productive classes. It was the toil of the human arm that they principally regarded. It was labor, as distinct from capital. But it seems to me that the true philosophy of the thing is, that any labor, any active agency, which can be brought to act usefully on the earth, or its materials, is the source of wealth. The labor of the ox, and the horse, as well as that of man, produces wealth. That is to say, this labor, like man's labor, extracts from the earth the means of living, and these constitute wealth in its general political sense.

Now it has been the purpose, and a purpose most successfully and triumphantly obtained, of scientific art, to increase this active agency, which, in a philosophical point of view, is, I think, to be regarded as labor, by bringing the powers of the elements into active and more efficient operation, and creating millions of automatic laborers, all diligently employed for the benefit of man. The powers are principally steam, and the weight of water. The automatic machinery are mechanisms of infinitely various kinds. Two classes: first — when a series of operations is carried through by one power, till a perfect result is had, like factories. Second — more [powers,] single or united, steamboats, cars, and printing presses. We commonly speak of mechanic inventions as labor-saving machines; but it would be more philosophical to speak of them as labor-doing machines; because they, in fact, are laborers. They are made to be active agents, to have motion, and effect, and though without intelligence, they are guided by those laws of science which are exact and perfect, and they produce results, therefore, in general, more exact and accurate than the human hand is capable of producing. When one sees Mr. Whittemore's carding machine in operation and looks at the complexity and accuracy of its operations, their rapidity, and yet their unbroken and undisturbed succession, he will see that in this machine (as well as in the little dog that turns it) man has a fellow laborer, and this fellow laborer is of immense power, of mathematical accuracy and precision, and of unwearied effort. And while he is thus a most skilful and productive laborer, he is, at the same time, a non-consumer. His earnings all go to the use of man. It is over such engines, even with more propriety than on the apiary, that the motto might be written, *vos, non vobis, laborastis.*

It is true that the machinery, in this and similar cases, is the purchase of capital. But human labor is the purchase of capital also; though the free purchase, and in communities less fortunate than ours, the human being himself, who performs the labor, is the purchase of capital. The work of machinery is certainly labor in all sense, as much as slave service, and in an enlarged sense, it is labor in [1] and regarding labor as a mere active power of production; whether that power be the hand of man, or the automatic movement of machinery, the general result is the same.

It is thus that the successful application of science to art increases the productive power and agency of the human race. It multiplies laborers without multiplying consumers, and the world is precisely as much benefited as if Providence had provided for our use millions of men, like ourselves in external appearance, who would work and labor and toil, and who yet required for their own subsistence neither shelter, nor food, nor clothing. These automata in the factories and the workshops are as much our fellow laborers, as if they were automata wrought by some Maelzel into the form of men, and made capable of walking, moving, and working, of felling the forest or cultivating the fields.

It is well known that the era of the successful application of science to arts, especially in the production of the great article of human subsistence, clothing, commenced about half a century ago. H. Arkwright predicted how productive human industry would become when no longer proportioned in its results to muscular strength. He had great sagacity, boldness, judgment, and power of arrangement. In 1770, England consumed four million pounds of cotton in manufactures — [the] United States none. The aggregate consumption of Europe and America is now five millions.[2] Arkwright deserves to be regarded as a benefactor of mankind. From the same period we may date the commencement in the general improvement in the condition of the mass of society, in regard to wealth and the means of living; and to the same period also we may assign the beginning of that spread of popular knowledge which now stamps such an imposing character on the times.

What is it, then, but this increased laboring power, what is

[1] There is a blank space here in the manuscript.
[2] Ellison's table in Mulhall's Dictionary of Statistics gives the consumption of raw cotton in Europe and the United States in 1830 as four hundred and sixty-five million pounds, and in 1840 as eight hundred and forty-two million pounds. The manuscript should doubtless have read five hundred millions.

it but these automatic allies and co-operators, who have come with such prodigious effect to man's aid in the great business of procuring the means of living, of comfort, and of wealth, out of the materials of the physical universe, which has so changed the face of society?

And this mighty agency, this automatic labor whose ability cannot be limited nor bounded, is the result of the successful application of science to art. Science has thus reached its greatest excellence, and achieved its highest attainment, in rendering itself emphatically, conspicuously, and in the highest degree useful to men of all classes and conditions. Its noblest attainment consists in conferring practical and substantial blessings on mankind.

"Practical mechanics," says a late ingenious and able writer, " is, in the pre-eminent sense, a scientific art." It is indeed true that the arts are growing every day more and more perfect, from the prosecution of scientific researches. And it ought to put to the blush all those who decry any department of science, or any field of knowledge, as barren and unproductive, that man has, as yet, learned nothing that has not been, or may not be, capable of useful application. If we look to the unclouded skies, when the moon is riding among the constellations, it might seem to us, that the distance of that luminary from any particular star could be of no possible importance, or its knowledge of no practical use to man. Yet it is precisely the knowledge of that distance, wrought out and applied by science, that enables the navigator to decide, within a few miles, his precise place on the ocean, not having seen land in many months. The high state of navigation, its safety and its despatch, are memorials of the highest character in honor of science, and the application of science to the purposes of life. The knowledge of conic sections, in like manner, may appear quite remote from practical utility, yet there are mechanical operations, of the highest importance, entirely dependent on an accurate knowledge, and a just application of the scientific rules pertaining to that subject. Therefore he who studies astronomy and explores the celestial system, a La Place or Bowditch; or he who constructs tables for finding the longitude, or works out results and proportions applicable to machinery from conic sections, or other branches of mathematics; or he who fixes by precise rule the vibration of the pendulum, or applies to use the counterbalancing and mutually adjusting centripetal and centrifugal force of bodies, are laborers for the human race of the highest character and the greatest merit.

It is unnecessary to multiply instances or examples. We are surrounded on all sides by abundant proofs of the utility of scientific research applied to the purposes of life. Every ship that swims the sea exhibits such proof; every factory exhibits it; every printing office and almost every workshop exhibits it; agriculture exhibits it; household comforts exhibit it. On all sides, wherever we turn our eyes, innumerable facts attest the great truth, that knowledge is not barren.

It is false in morals to say that good principles do not tend to produce good practice; it is equally false in matters of science and of art to declare that knowledge produces no fruit.

Perhaps the most prominent instance of the application of science to art, in the production of things necessary to man's subsistence, is the use of the elastic power of steam, applied to the operations of spinning and weaving and dressing fabrics for human wear. All this mighty discovery bears directly on the means of human subsistence and human comfort. It has greatly altered commerce, agriculture, and even the habits of life among nations. It has affected commerce by creating new objects, or vastly increasing the importance of those before hardly known; it has affected agriculture by giving new value to its products; — what would now be the comparative value of the soil of our Southern and Southwestern States if the spinning of cotton by machinery, the power loom, and the cotton gin, were struck out of existence? And it has affected habits by giving a new direction to labor and creating a multitude of new pursuits.

Bearing less on the production of the objects necessary to man's subsistence, but hardly inferior in its importance, is the application of the power of steam to transportation and conveyance by sea and by land. Who is so familiarized to the sight even now, as to look without wonder and amazement on the long train of cars, full of passengers and merchandise, drawn along our valleys, and the sides of our mountains themselves with a rapidity which holds competition [with] the winds?

This branch of the application of steam power is younger. It is not yet fully developed; but the older branch, its application to manufacturing machinery, is perhaps to be regarded as the more signal instance marking the great and glorious epoch of the application of science to the useful arts. From the time of Arkwright to our own days, and in our own country, from a period a little earlier than the commencement of the late war with England, we see the successive and astonishing effects of this principle, not new-born, indeed, in our time, but awakened, animated, and pushed forward to most stupendous results. It is difficult to estimate the amount of labor performed by machinery, compared with manual labor. It is computed that in England, on articles exported [1] [it] is thirty millions sterling per year. It would be useful, if one with competent means should estimate the products of the annual labor of Massachusetts in manufactured articles carried from the State.

If these, and other considerations may suffice to satisfy us that the application of science to art is the main cause of the sudden augmentation of wealth and comfort in modern times, a truth remains to be stated of the greatest magnitude, and the highest practical importance, and that is, that this augmentation of wealth and comfort is general and diffusive, reaching to all classes, embracing all interests, and benefiting, not a part of society, but the whole. There is no monopoly in science. There are no exclusive privileges in the workings of automatic machinery, or the powers of natural bodies. The poorest, as well as the richest man in society, has a direct interest, and generally the poor a far greater interest than the rich, in the successful operation of these arts, which make the means of living, clothing especially, abundant and cheap. The advantages conferred by knowledge in increasing our physical resources, from their very nature, cannot be enjoyed by a few only. They are all open to the many, and to be profitable, the many must enjoy it.

The products of science applied to art in mechanical inventions, are made, not to be hoarded, but to be sold. Their

[1] The words " from the " and a blank space follow the word " exported " in the manuscript.

successful operation requires a large market. It requires that the great mass of society should be able to buy and to consume. The improved condition of all classes, more ability to buy food and raiment, better modes of living, and increased comforts of every kind, are exactly what is necessary and indispensable in order that capital invested in automatic operations should be productive to the owners. Some establishments of this kind necessarily require large capital, such as the woollen and cotton factories. And in a country like ours, in which the spirit of our institutions, and all our laws, tend so much to the distribution and equalization of property, there are few individuals of sufficient wealth to build and carry on an establishment by their own means. This renders a union of capitals necessary, and this among us is conveniently effected by corporations which are but partnerships regulated by law. And this union of many to form capital for the purpose of carrying on those operations by which science is applied to art, and comes in aid of man's labor in the production of things essential to man's existence, constitutes that aggregated wealth of which complaint is sometimes heard. It would seem that nothing could be plainer than that whatever reduces the price, whether of food or of clothing, must be in the end beneficial to the laboring classes. Yet it has not unfrequently happened, that machinery has been broken and destroyed in England, by workmen, by open and lawless violence. Most persons in our country see the folly as well as the injustice and barbarism of such proceedings; but the ideas in which these violences originated are no more unfounded and scarcely more disreputable, than those which would represent capital, collected, necessarily, in large sums, in order to carry on useful processes in which science is applied to art, in the production of articles useful to all, as being hostile to the common good, or having an interest separate from that of the majority of the community. All such representations, if not springing from sinister design, must be the result of great ignorance, or great prejudice. It has been found by long experience in England, that large capitalists can produce cheaper than small ones, especially in the article of cotton. Greater savings can be made and these savings enable the proprietor to go on, when he must otherwise stop. There is no doubt that it is to her abundant capital, England is now indebted for whatever power of competition with the United States she now sustains, in producing cheap articles.

There are modes of applying wealth, useful principally to the owner, and no otherwise beneficial to the community than as they employ labor. Such are the erection of expensive houses, the embellishment of ornamental grounds, the purchase of costly furniture and equipages. These modes of expenditure, although entirely lawful and sometimes very proper, are yet not such as directly benefit the whole community. Not so with aggregate wealth employed in producing articles of general consumption. This mode of employment is, peculiarly and in an emphatic sense, an application of capital to the benefit of all. Any one who complains of it, or decries it, acts against the greatest good of the greatest number. The factories, the steamboats, the railroads, and other similar establishments, although they require capital, and aggregate capital, are yet general and popular in all the good they produce.

The unquestionable operation of all these things has been not only to increase property, but to equalize it, to diffuse it, to scatter its advantages among the many, and to give content, cheerfulness, and animation to all classes of the social system. In New England, more particularly, has this been the result. What has enabled us to be rich and prosperous, notwithstanding the barrenness of our soil and the rigor of our climate? What has diffused so much comfort, wealth, and happiness among all classes, but the diligent employment of our citizens, in these processes and mechanical operations in which science comes in aid of handicraft? Abolish the use of steam and the application of water power to machinery, and what would at this moment be the condition of New England? And yet steam and water power have been employed only, and can be employed only, by what is called aggregated wealth. Far distant be the day then, when the people of New England shall be deceived by the specious fallacy, that there are different and opposing interests in our community; that what is useful to one, is hurtful to the rest; that there is one interest for the rich, and another interest for the poor; that capital is the enemy of labor, or labor the foe of capital. And let every laboring man, on whose understanding such a fallacy is attempted to be imposed, stop the mouth of the false reasoner at once, by stating the plain and evident fact, that while aggregated wealth has for years, in Massachusetts, been most skilfully and steadily employed in the productions which result from the application of science to art; thereby reducing the cost of many of the articles most essential to human life in all conditions; labor, meantime, has been constantly rising, and is at this very moment, notwithstanding the present scarcity of money, and the constant pressure on capital, higher than it ever was before in the history of the country. These are, indeed, facts which baffle all former dogmas of political economy. In some of our most agricultural districts in the midst of our mountains, on whose tops the native forests still wave and where agricultural labor is high, in an unprecedented degree, even here automatic processes are carried on by water, and fabrics wrought out of materials which have been transported hundreds and thousands of miles, by sea and by land, and which fabrics go back again, some of them, for sale and consumption to the places where the raw material was produced. Carolina cotton is carried to the County of Berkshire, and Berkshire cotton goods are sold in Carolina. Meanwhile labor in Berkshire is not only in money price, but in comparison with the cost of the main articles of human subsistence, higher than it was ever known before. Writers on political economy may, perhaps, on facts like these see occasion to qualify their theories; meantime, it becomes every man to question and scrutinize severely, if not the motives, yet the reasoning and the logic of those who would persuade us that capital, employed in the most efficient modes of producing things useful, is hurtful to society. It would be quite as reasonable to insist, that the weaving of paper, if that be the proper term, in consequence of recent most valuable mechanical inventions, should be [suppressed], and the power press, and the hydrostatic machine taken out of the printing houses, as being all hurtful and injurious, although they may have reduced the price of books for general circulation one half.

The truth, in my opinion, rather is, that such is the enterprise of our people, such the astonishing amount of labor which they perform, and which they perform cheerfully because it is free and because it is profitable, and such the skill with which capital is used, that still more capital would be useful, and that its introduction would be advantageous, and most of all to the busy and industrious classes. And let it never be forgotten, that with us labor is free, intelligent, respecting itself, and respected by other interests; that it accumulates; that it is provident; that it lays up for itself; and that these savings become capital, and their owners in time capitalists.

I cannot omit to notice, here, another fact peculiar to this country, and which should cause us to hesitate in applying to ourselves, and our condition, European maxims respecting capital and labor. In Europe, generally speaking, the laborer is always a laborer. He is destined to no better condition on earth, ordinarily he rises no higher. We see proofs, melancholy proofs, of this truth often in the multitudes who come to our own shores from foreign countries for employment. It is not so with the people of New England. Capital and labor are much less distinctly divided with us. Few are they, on the one hand, who have need to perform no labor; few are they, on the other, who have no property or capital of their own. Or if there be those of the latter class among the industrious and the sober, they are young men who, though they are laborers to-day, will be capitalists to-morrow. A career of usefulness and enterprise is before them. If without moneyed capital, they have a capital in their intelligence, their knowledge, and their good habits. Around them are a thousand collections of automatic machinery, requiring the diligence of skilful and sober [laborers]; before them is the ocean, always inviting to deeds of hardihood and enterprise; behind them are the fertile [prairies] of the West, soliciting cultivation; and over them all is the broad banner of free institutions, of mild laws, and parental Government. Would [an] American young man of good health and good habits need say that he is without capital? Or why should he [discredit] his own understanding by listening to the absurdity, that they who have earned property, and they who have not yet lived long enough to earn it, must be enemies? The proportion of those who have not capital, such as to render them independent without personal labor, and who are yet not without some capital, is vastly larger in this community than any other. They form indeed the great mass of our society. They are its life and muscle; and long may they continue free, moral, intelligent, and prosperous as they now are.

Reproduced courtesy of Merrimack Valley Textile Museum.

Henry Hubbard

Gubernatorial Address (1842)

Henry Hubbard (1784–1857) was the leader of the "radical" Democrats in New Hampshire in the early 1840s and their successful candidate for governor. His address to the state Senate in June 1842, a portion of which is reprinted here, sets forth the "radical" position on government regulation of business corporations. Though his discussion of limited liability and eminent domain refers mainly to railroads and banks, the principles applied also to manufacturing concerns, most especially the Amoskeag mills at Great Falls on the Merrimack (Manchester), which would become the largest textile complex in the country. The ability of the Boston proprietors of New Hampshire industry to circumvent the restrictions on corporate growth and manipulate public opinion with predictions of economic disaster led to defeat of the radical Democrats in 1846 and the election of a Whig governor to replace Hubbard.

Henry Hubbard, gubernatorial address, *Journal of the Honorable Senate of the State of New Hampshire* (Concord, New Hampshire, 1842), pp. 26–32. The other most forthright spokesman for the position of the radical Democrats in New Hampshire was Edmund Burke. See his *Address Delivered before the Democratic-Republican Citizens of Lempster, N.H., on the Eighth of January, 1839* (Newport, New Hampshire, 1839).

In the course of your present session, application will probably be made to the Legislature for a renewal of those Bank Charters, which are about to expire by their own limitation, and it may be considered necessary to make further provisions in relation to the general powers and duties of private corporations, with a view to the entire security and protection of the people, in the enjoyment of their rights of property. It is undoubtedly known to every member of this assembly, that, in an answer given by me to certain interrogatories, submitted by a respectable portion of my fellow-citizens, I have already explicitly promulgated to the public my opinions upon these subjects. Although these opinions remain unchanged, although subsequent reflection has confirmed me in their correctness, yet in making this official communication I feel it to be my duty, inasmuch as the question of renewing some, and of amending other charters of private corporations in this State, may engage the consideration of the Legislature—*further*, to express my views upon a subject of such vital importance to the future liberty, interest and happiness of our people.

A disposition to multiply private corporations is one of the great evils in the legislation of the most of the States. The history of the past should admonish us of our duty. For the mere purpose of private speculation and for avoiding all personal responsibility, many of these acts of special legislation have been sought for, and too readily obtained. Could those patriotic statesmen who have gone before us, have foreseen the disastrous and demoralizing effects of creating these irresponsible bodies, they would have never lent their aid in the accomplishment of so inglorious a work, and as the friends of social order, of sound morals and of equal justice, we should, guided by the lights of experience, avoid as far as practicable, this system which has proved so debasing to the character of our community, so ruinous to the hopes and so destructive to the interests of our confiding population. I would not indiscriminately oppose every private charter; corporations of this description have been, and may be established to promote the well being of society. But they have been too frequently granted with a view to the exclusive interest of the corporators, not regarding the rights or the interests of the people. The business operations of our community may be prosecuted with much more security to the people, and with equal convenience and safety to those concerned, without, as with, the shield of such a monopoly. The exclusive privileges usually conferred by acts of incorporation make them obnoxious to the honest hearted yeomanry of the country, and they should therefore be withheld except in cases of public necessity. If the unhallowed influence of this system shall continue unrestrained and unabated, the rights of individual property, and the honest enjoyment of individual liberty will exist only in name.

The object of the Legislature should be to give protection to the interests of the people, and maintain with unyielding pertinacity their just privileges.

This is the great practical reform now demanded, and this reform is what we should be solicitous to see accomplished. It would not be a difficult work, if we are ready to apply to private corporations those principles, which necessarily govern the transactions of individuals. The stockholders of all private business corporations should be made liable for the debts and liabilities of their respective corporations. There is no good reason against this principle. In the transactions which occur between man and man, there exists a direct responsibility—and when capital is concentrated, which looks to enlarged and extensive operations, beyond the means of single individuals, this liability is continued.

It is the life spring of the concern—it inspires confidence and commands support, and it is difficult to conceive of any sufficient reason why this principle, applicable to partnerships created by voluntary association, should not extend to partnerships created by law—why individuals acting unitedly for their own interest should be made liable for their doings as natural persons, from which they should be exempted under a charter of incorporation.

This want of responsibility has been a fruitful source of the embarrassment and distress, which has pervaded and is now pervading our land. The banking corporations in this State, are unquestionably private in their character and in their purposes. Their capital is private, and their gains are all private. Created as they have been for private benefit, their stockholders, like private individuals, should be liable for the debts they contract. They only can know the exact state of their concerns. When successful, they alone share the gains, and when overtaken by reverses, there can be no justice in visiting their misfortunes upon their confiding creditors. This doctrine of the personal liability of stockholders in private corporations is founded upon the principles of immutable right. It is no new doctrine. It is the doctrine of the common law of England. It is the general principle upon which private corporations are now established in that country. The joint stock banks of Great Britain are based on this great principle. These companies are banks of issue, they emit much of the paper circulation of the country, and the redeeming feature in all these legal partnerships *is*, that the partners are liable, individually, to the full amount of their property, for the debts of their respective companies ; and as an additional security to the public, the names of all the stockholders are registered in the public archives, that the creditors of such institutions may know to whom they shall look for the payment of their liabilities. The beneficial consequences of such a provision, would be to induce that vigilance and supervision indispensably necessary to the success of the institutions, as well as to the safety of the public.

In one of the States of this Union there are but three incorporated banks. The charter of one imposes the unlimited liability of the stockholders for the debts of the corporation, while the charters of the other two are in that respect similar to the bank charters of New England. In 1837, the latter suspended specie payment, and after renewing for a short time, again suspended in 1839, and are now in a state of suspension, while the former has never suspended specie payment at all.

A principle which has worked so well elsewhere, should not be regarded as of a dangerous tendency if adopted in this State. It is true, and the fact is highly creditable to the banking corporations in New Hampshire, that with few exceptions, the people have suffered no loss through their agency—that the concerns of these institutions have been generally so well managed, as to have secured the confidence of our community. But these considerations ought not to deter us from the establishment of a *principle* important to the protection and well being of society—*a principle* which would secure the confidence of the public, *and thus advance the true interest of the corporators.*

But the operations of *some* of the banking institutions of the country, within a few years past, have been most disastrous to the interests, and most debasing to the moral sense of the community, and cannot fail to induce this patriotic assembly to do all in their power, to secure, hereafter, our common country from such disgrace, and the people from such sacrifice. It cannot be necessary for me, after having so fully expressed my views in relation to these private corporations, and the obligations which ought to rest upon the corporators, to recommend to the Legislature any particular banking system. The details of such a system are within your province, and should this Legislature, at its present session, see fit to renew the bank charters about expiring, or to establish other institutions in their stead, it is to be hoped the present system will be so improved, that such a responsibility will be imposed upon the stockholders, and such a restraint upon officers connected with these institutions, as shall effectually protect the public from all pecuniary loss.

In addition to the mere question of policy touching the renewal of the bank charters about to expire, and of introducing further provisions to afford indemnity to the people, another, graver and more important question may engage your attention at the present session—a question involving the constitutional right of the Legislature to confer upon a private corporation the power to take individual property, and appropprriate it to its own use and benefit, without the owner's consent. If such a power exists, it must be conferred upon the Legislature by the direct provisions of our Constitution—no such power can properly be implied. In the language of that instrument, "every member of the community has a right to be protected in the enjoyment of *life, liberty* and *property*." "The right of acquiring property is a natural and inherent right." "When men enter into a state of society, they surrender up some of their natural rights to that society, in order to ensure the protection of others." "No part of a man's property shall be taken from him, or applied to public uses, without his own consent or that of the representative body of the people." These are among the fundamental principles of our State constitution, and are the only provisions in reference to this subject. When taken in connection, they establish the doctrine that every man shall be protected in the acquisition and possession of property. His right of control cannot be invaded, except when the public use shall require. Then, and then only, he may be compelled to surrender such parts as may be necessary for such use. Highways, wrought at the public expense, kept in repair at the public charge, and for which the public are liable for injuries sustained through their insufficiency, comprehend that description of ways dedicated exclusively to the public use, demanded by public necessity—and which every individual who enters into and becomes a member of a political community, is bound to aid in constructing. The existence of such ways is essential to the security and well-being of society. They are the work of public corporations, and are wholly distinguishable in their character, uses and purposes, from ways wrought by individuals or by private corporations. These are constructed with private means, and for private benefit, and do not afford such a public use as is contemplated in the constitution. Of this description are the railroads which have been constructed in this State. They are the work of private corporations, and are wrought for the interest of the corporators. The public are shut out from a participation in their government and direction. Upon such a corporation power cannot be conferred to take individual property for its use, without the owner's consent. The Legislature could not confer upon a single individual for a private purpose, the right to take the property of another for his use, without the assent of its owner, nor could this power be conferred upon a voluntary association of individuals. On what better principle could it be conferred upon individuals, united under a charter of incorporation, for the accomplishment of the same purpose ? The powers which are granted by the Legislature to a private corporation, are in truth conferred upon the individual corporators by the name of the corporation. The same specific powers, described in their charters, would not be granted to the corporators as natural persons. Such grants would be regarded as transcending the constitutional right of the Legislature. When conferred upon the corporators by the name of their corporation, can the power of the Legislature be any less confined within the limits of the Constitution ? It would seem that if a grant of the specific power in the one case would be unwarranted, so would it be in the other.

The authority to establish private corporations, cannot give to the representative body of the people any new powers over the private rights of individuals. If the right exists to appropriate private property, according to the mere will and pleasure of the Legislature for the time being, it can by no means be necessary, for a constitutional exercise of the right, that there should be created by law a private corporation upon which to confer the power. No consideration of public expediency, or public policy, should swerve the representative body of the people from the plain straightforward road of constitutional duty. The plea of public use, or public benefit, is always urged upon the Legislature as a reason for the establishment of these private corporations.

There is no difference in principle whether this power shall be conferred upon a manufacturing or a railroad corporation. They will each to some extent advance the public convenience. If the Legislature, upon the ground of public use, can constitutionally confer upon private corporations the power to take for their benefit the property of individuals, without their consent, there is no principle in the way of giving the same power to all private corporations.

The principle that individual property shall not be taken except for public use is in a republic the surest guaranty of individual independence. It is in the truest sense of the word conservative, and not anarchical. Without this principle every citizen would hold his dearest rights at the shifting will and temporary caprice of Legislative assemblies. The protection of individual rights against the excitements of party and the fancied interests of those who for the time may be in power, and the determination of those rights by reference to established principles, are privileges without which a republic is but a name. It is the conviction that our rights cannot be invaded, nor our property taken from us but by the necessity that the public good should predominate over private convenience, that makes each citizen lay down his head in peace at night, trusting to the supremacy of the law alone for his protection.

The tendency of our legislation is to disregard individual rights —and unless they are sacredly preserved, we deprive the citizen of that confidence and self reliance which should characterize the freeman, and we lose one of the most substantial distinctions between a free and despotic government. Nor can the effect of this doctrine be to drive capital from our State. It would rather tend to invite investments, than to discourage them. But should the views which I have put forth in this official communication, in effect, drive capital from New Hampshire, for the reason that private corporations cannot here be established without the responsibility of the corporators, and without protection to the rights of property, it would be better to submit to that privation, than to be made to endure the greater evils resulting from the unnecessary sacrifice of individual rights.

I cannot perceive that the interests of the manufacturing classes of our community could be unfavorably affected, by imposing further restrictions upon private corporations. In England, all branches of manufacture are more extensively pursued, larger capitals invested, and a greater number of operatives employed in the same branches of business, than in this country. In a single year, the value of her cotton, woolen, linen and iron manufactures, exceeded three hundred millions of dollars, and this immense business was carried on by private individuals and by private capital, without the aid of a single corporation.

In one of the States of this Union, the manufactures of cotton and wool, the iron and coal business, are prosecuted by individual enterprise, unaided by acts of incorporation.

The same description of business may be as well done in the same manner *here* as *elsewhere*. I am not, however, for destroying, but for so far improving the charters of private corporations, that whatever may hereafter be done, should be done with a view to the security and to the protection of the public. This is the great object I have in view. This is the reform demanded by every consideration which can enter into the mind of the patriot.

This is what the moral sense of the community requires; what the just and equal rights of the people demand.

It is to be hoped that there will be no occasion for a protracted session of the Legislature at this time

There is more danger to be apprehended from too much than too little legislation.

The Philosophy of Manufactures

Reproduced courtesy of Merrimack Valley Textile Museum.

Henry David Thoreau

Paradise (to be) Regained (1843)

Thoreau's dour skepticism about the motives, methods, and consequences of industrial technology is firmly and widely established in *Walden* (1854). His celebrated pronouncement on the "factory system" in his first chapter, "Economy," is a late example of the ethical argument others had raised for half a century:

I cannot believe that our factory system is the best mode by which men may get clothing. The condition of the operatives is becoming every day more like that of the English; and it cannot be wondered at, since, as far as I have heard or observed, the principal object is, not that mankind may be well and honestly clad, but, unquestionably, that the corporations may be enriched. In the long run men hit only what they aim at. Therefore, though they should fail immediately, they had better aim at something high.

Thoreau was, however, not simply an agrarian in his opposition to manufactures. *Walden* is, if anything, a much more trenchant critique of rural economy. Thoreau's attack on the commercial farmer had the same base as his attack on the industrial corporations: the pursuit founded in private property was merely for material gain: "We know nature but as robbers."

Thoreau also was skeptical about the benefits of machinery itself as a method of production, as he demonstrated in one of his early publications, a belated review of Etzler's *Paradise within the Reach of All Men*. In this review Thoreau quoted long passages of Etzler, most of which have been deleted here. As a comparison of Etzler's opening paragraph and Thoreau's quotation of that paragraph will show, Thoreau was cavalier in his transcription of Etzler's text.

[Henry David] T[horeau], "Paradise (to be) Regained," *United States Magazine and Democratic Review*, volume 13 (November 1843), pp. 451–463.

UNITED STATES MAGAZINE,

AND

DEMOCRATIC REVIEW.

Vol. XIII.   NOVEMBER, 1843.   No. LXV.

PARADISE (TO BE) REGAINED.*

We learn that Mr. Etzler is a native of Germany, and originally published his book in Pennsylvania, ten or twelve years ago; and now a second English edition, from the original American one, is demanded by his readers across the water, owing, we suppose, to the recent spread of Fourier's doctrines. It is one of the signs of the times. We confess that we have risen from reading this book with enlarged ideas, and grander conceptions of our duties in this world. It did expand us a little. It is worth attending to, if only that it entertains large questions. Consider what Mr. Etzler proposes:

"Fellow Men! I promise to show the means of creating a paradise within ten years, where everything desirable for human life may be had by every man in superabundance, without labor, and without pay; where the whole face of nature shall be changed into the most beautiful forms, and man may live in the most magnificent palaces, in all imaginable refinements of luxury, and in the most delightful gardens; where he may accomplish, without labor, in one year, more than hitherto could be done in thousands of years; may level mountains, sink valleys, create lakes, drain lakes and swamps, and intersect the land everywhere with beautiful canals, and roads for transporting heavy loads of many thousand tons, and for travelling one thousand miles in twenty-four hours; may cover the ocean with floating islands movable in any desired direction with immense power and celerity, in perfect security, and with all comforts and luxuries, bearing gardens and palaces, with thousands of families, and provided with rivulets of sweet water; may explore the interior of the globe, and travel from pole to pole in a fortnight; provide himself with means, unheard of yet, for increasing his knowledge of the world, and so his intelligence; lead a life of continual happiness, of enjoyments yet unknown; free himself from almost all the evils that afflict mankind, except death, and even put death far beyond the common period of human life, and finally render it less afflicting. Mankind may thus live in and enjoy a new world, far superior to the present, and raise themselves far higher in the scale of being."

It would seem from this and various indications beside, that there is a transcendentalism in mechanics as well as in ethics. While the whole field of the one reformer lies beyond the boundaries of space, the other is pushing his schemes for the elevation of the race to its utmost limits. While one scours the heavens, the other sweeps the earth. One says he will reform himself, and then nature and circumstances will be right. Let us not obstruct ourselves, for that is the greatest friction. It is of little importance though a cloud obstruct the view of the astronomer compared with his own

---

* The Paradise within the Reach of all Men, without Labor, by Powers of Nature and Machinery. An Address to all intelligent Men. In two parts. By J. A. Etzler. Part First. Second English Edition. pp. 55. London, 1842.

blindness. The other will reform nature and circumstances, and then man will be right. Talk no more vaguely, says he, of reforming the world—I will reform the globe itself. What matters it whether I remove this humor out of my flesh, or the pestilent humor from the fleshy part of the globe? Nay, is not the latter the more generous course? At present the globe goes with a shattered constitution in its orbit. Has it not asthma, ague, and fever, and dropsy, and flatulence, and pleurisy, and is it not afflicted with vermin? Has it not its healthful laws counteracted, and its vital energy which will yet redeem it? No doubt the simple powers of nature properly directed by man would make it healthy and paradise; as the laws of man's own constitution but wait to be obeyed, to restore him to health and happiness. Our panaceas cure but few ails, our general hospitals are private and exclusive. We must set up another Hygeian than is now worshipped. Do not the quacks even direct small doses for children, larger for adults, and larger still for oxen and horses? Let us remember that we are to prescribe for the globe itself.

This fair homestead has fallen to us, and how little have we done to improve it, how little have we cleared and hedged and ditched! We are too inclined to go hence to a "better land," without lifting a finger, as our farmers are moving to the Ohio soil; but would it not be more heroic and faithful to till and redeem this New-England soil of the world? The still youthful energies of the globe have only to be directed in their proper channel. Every gazette brings accounts of the untutored freaks of the wind—shipwrecks and hurricanes which the mariner and planter accept as special or general providences; but they touch our consciences, they remind us of our sins. Another deluge would disgrace mankind. We confess we never had much respect for that antediluvian race. A thorough-bred business man cannot enter heartily upon the business of life without first looking into his accounts. How many things are now at loose ends. Who knows which way the wind will blow to-morrow? Let us not succumb to nature. We will marshal the clouds and restrain the tempests; we will bottle up pestilent exhalations, we will probe for earthquakes, grub them up; and give vent to the dangerous gases; we will disembowel the volcano, and extract its poison, take its seed out. We will wash water, and warm fire, and cool ice, and underprop the earth. We will teach birds to fly, and fishes to swim, and ruminants to chew the cud. It is time we had looked into these things.

And it becomes the moralist, too, to inquire what man might do to improve and beautify the system; what to make the stars shine more brightly, the sun more cheery and joyous, the moon more placid and content. Could he not heighten the tints of flowers and the melody of birds? Does he perform his duty to the inferior races? Should he not be a god to them? What is the part of magnanimity to the whale and the beaver? Should we not fear to exchange places with them for a day, lest by their behavior they should shame us? Might we not treat with magnanimity the shark and the tiger, not descend to meet them on their own level, with spears of sharks' teeth and bucklers of tiger's skin? We slander the hyæna; man is the fiercest and cruelest animal. Ah! he is of little faith; even the erring comets and meteors would thank him, and return his kindness in their kind.

How meanly and grossly do we deal with nature! Could we not have a less gross labor? What else do these fine inventions suggest,—magnetism, the daguerreotype, electricity? Can we not do more than cut and trim the forest,—can we not assist in its interior economy, in the circulation of the sap? Now we work superficially and violently. We do not suspect how much might be done to improve our relation with animated nature; what kindness and refined courtesy there might be.

*

Men having discovered the power of falling water, which after all is comparatively slight, how eagerly do they seek out and improve these *privileges?* Let a difference of but a few feet in level be discovered on some stream near a populous town, some slight occasion for gravity to act, and the whole economy of the neighborhood is changed at once. Men do indeed speculate about and with this power as if it were the only privilege. But meanwhile this aerial stream is falling from far greater heights with more constant flow, never shrunk by drought, offering mill-sites wherever the wind blows; a Niagara in the air, with no Canada side;—only the application is hard.

There are the powers too of the Tide and Waves, constantly ebbing and flowing, lapsing and relapsing, but they serve man in but few ways. They turn a few tide mills, and perform a few other insignificant and accidental services only. We all perceive the effect of the tide; how imperceptibly it creeps up into our harbors and rivers, and raises the heaviest navies as easily as the lightest ship. Everything that floats must yield to it. But man, slow to take nature's constant hint of assistance, makes slight and irregular use of this power, in careening ships and getting them afloat when aground.

The following is Mr. Etzler's calculation on this head: To form a conception of the power which the tide affords, let us imagine a surface of 100 miles square, or 10,000 square miles, where the tide rises and sinks, on an average, 10 feet; how many men would it require to empty a basin of 10,000 square miles area, and 10 feet deep, filled with sea-water, in 6¼ hours and fill it again in the same time? As one man can raise 8 cubic feet of sea-water per minute, and in 6¼ hours 3,000, it would take 1,200,000,000 men, or as they could work only half the time, 2,400,000,000, to raise 3,000-000,000,000 cubic feet, or the whole quantity required in the given time.

This power may be applied in various ways. A large body, of the heaviest materials that will float, may first be raised by it, and being attached to the end of a balance reaching from the land, or from a stationary support, fastened to the bottom, when the tide falls, the whole weight will be brought to bear upon the end of the balance. Also when the tide rises it may be made to exert a nearly equal force in the opposite direction. It can be employed whenever a *point d'appui* can be obtained.

*

The computation of the power of the waves is less satisfactory. While only the average power of the wind, and the average height of the tide, were taken before now, the extreme height of the waves is used, for they are made to rise ten feet above the level of the sea, to which, adding ten more for depression, we have twenty feet, or the extreme height of a wave. Indeed, the power of the waves, which is produced by the wind blowing obliquely and at disadvantage upon the water, is made to be, not only three thousand times greater than that of the tide, but one hundred times greater than that of the wind itself, meeting its object at right angles. Moreover, this power is measured by the area of the vessel, and not by its length mainly, and it seems to be forgotten that the motion of the waves is chiefly undulatory, and exerts a power only within the limits of a vibration, else the very continents, with their extensive coasts, would soon be set adrift.

Finally, there is the power to be derived from Sunshine, by the principle on which Archimedes contrived his burning mirrors, a multiplication of mirrors reflecting the rays of the sun upon the same spot, till the requisite degree of heat is obtained. The principal application of this power will be to the boiling of water and production of steam.

*

So much for these few and more obvious powers, already used to a trifling extent. But there are innumerable others in nature, not described nor discovered. These, however, will do for the present. This would be to make the sun and the moon equally our satellites. For, as the moon is the cause of the tides, and the sun the cause of the wind, which, in turn, is the cause of the waves, all the work of this planet would be performed by these far influences.

*

Here is power enough, one would think, to accomplish somewhat. These are the powers below. Oh ye millwrights, ye engineers, ye operatives and speculators of every class, never again complain of a want of power; it is the grossest form of infidelity. The question is not how we shall execute, but what. Let us not use in a niggardly manner what is thus generously offered.

Consider what revolutions are to be effected in agriculture. First, in the new country, a machine is to move along taking out trees and stones to any required depth, and piling them up in convenient heaps; then the same machine, "with a little alteration," is to plane the ground perfectly, till there shall be no hills nor valleys, making the requisite canals, ditches and roads, as it goes along. The same machine, "with some other little alterations," is then to sift the ground thoroughly, supply fertile soil from other places if wanted, and plant it; and finally, the same machine "with a little addition," is to reap and gather in the crop, thresh and grind it, or press it to oil, or prepare it any way for final use.

For the description of these machines we are referred to "Etzler's Mechanical System, page 11 to 27." We should be pleased to see that "Mechanical System," though we have not been able to ascertain whether it has been published, or only exists as yet in the design of the author. We have great faith in it. But we cannot stop for applications now.

*

Who knows but by accumulating the power until the end of the present century, using meanwhile only the smallest allowance, reserving all that blows, all that shines, all that ebbs and flows, all that dashes, we may have got such a reserved accumulated power as to run the earth off its track into a new orbit, some summer, and so change the tedious vicissitude of the seasons? Or, perchance, coming generations will not abide the dissolution of the globe, but, availing themselves of future inventions in aerial locomotion, and the navigation of space, the entire race may migrate from the earth, to settle some vacant and more western planet, it may be still healthy, perchance unearthly, not composed of dirt and stones, whose primary strata only are strewn, and where no weeds are sown. It took but little art, a simple application of natural laws, a canoe, a paddle, and a sail of matting, to people the isles of the Pacific, and a little more will people the shining isles of space. Do we not see in the firmament the lights carried along the shore by night, as Columbus did? Let us not despair nor mutiny.

*

Thus is Paradise to be Regained, and that old and stern decree at length reversed. Man shall no more earn his living by the sweat of his brow. All labor shall be reduced to "a short turn of some crank," and "taking the finished article away." But there is a crank, —oh, how hard to be turned! Could there not be a crank upon a crank,— an infinitely small crank?—we would fain inquire. No,—alas! not. But there is a certain divine energy in every man, but sparingly employed as yet, which may be called the crank within, —the crank after all,—the prime mover in all machinery,—quite indispensable to all work. Would that we might get our hands on its handle! In fact no work can be shirked. It may be postponed indefinitely, but not infinitely. Nor can any really important work be made easier by co-operation or machinery. Not one particle of labor now threatening any man can be routed without being performed. It cannot be hunted out of the vicinity like jackals and hyenas. It will not run. You may begin by sawing the little sticks, or you may saw the great sticks first, but sooner or later you must saw them both.

We will not be imposed upon by this vast application of forces. We believe that most things will have to be accomplished still by the application called Industry. We are rather pleased after all to consider the small private, but both constant and accumulated force, which stands behind every spade in the field. This it is that

The Philosophy of Manufactures

415

makes the valleys shine, and the deserts really bloom. Sometimes, we confess, we are so degenerate as to reflect with pleasure on the days when men were yoked like cattle, and drew a crooked stick for a plough. After all, the great interests and methods were the same.

It is a rather serious objection to Mr. Etzler's schemes, that they require time, men, and money, three very superfluous and inconvenient things for an honest and well-disposed man to deal with. "The whole world," he tells us, "might therefore be really changed into a paradise, within less than ten years, commencing from the first year of an association for the purpose of constructing and applying the machinery." We are sensible of a startling incongruity when time and money are mentioned in this connection. The ten years which are proposed would be a tedious while to wait, if every man were at his post and did his duty, but quite too short a period, if we are to take time for it. But this fault is by no means peculiar to Mr. Etzler's schemes. There is far too much hurry and bustle, and too little patience and privacy, in all our methods, as if something were to be accomplished in centuries. The true reformer does not want time, nor money, nor co-operation, nor advice. What is time but the stuff delay is made of? And depend upon it, our virtue will not live on the interest of our money. He expects no income but our outgoes; so soon as we begin to count the cost the cost begins. And as for advice, the information floating in the atmosphere of society is as evanescent and unserviceable to him as gossamer for clubs of Hercules. There is absolutely no common sense; it is common nonsense. If we are to risk a cent or a drop of our blood, who then shall advise us? For ourselves, we are too young for experience. Who is old enough? We are older by faith than by experience. In the unbending of the arm to do the deed there is experience worth all the maxims in the world.

"It will now be plainly seen that the execution of the proposals is not proper for individuals. Whether it be proper for government at this time, before the subject has become popular, is a question to be decided; all that is to be done, is to step forth, after mature reflection, to confess loudly one's conviction, and to constitute societies. Man is powerful but in union with many. Nothing great, for the improvement of his own condition, or that of his fellow men, can ever be effected by individual enterprise."

Alas! this is the crying sin of the age, this want of faith in the prevalence of a man. Nothing can be effected but by one man. He who wants help wants everything. True, this is the condition of our weakness, but it can never be the means of our recovery. We must first succeed alone, that we may enjoy our success together. We trust that the social movements which we witness indicate an aspiration not to be thus cheaply satisfied. In this matter of reforming the world, we have little faith in corporations; not thus was it first formed.

But our author is wise enough to say, that the raw materials for the accomplishment of his purposes, are "iron, copper, wood, earth chiefly, and a union of men whose eyes and understanding are not shut up by preconceptions." Aye, this last may be what we want mainly,—a company of "odd fellows" indeed.

"Small shares of twenty dollars will be sufficient,"—in all, from "200,000 to 300,000,"—"to create the first establishment for a whole community of from 3000 to 4000 individuals"—at the end of five years we shall have a principal of 200 millions of dollars, and so paradise will be wholly regained at the end of the tenth year. But, alas, the ten years have already elapsed, and there are no signs of Eden yet, for want of the requisite funds to begin the enterprise in a hopeful manner. Yet it seems a safe investment. Perchance they could be hired at a low rate, the property being mortgaged for security, and, if necessary, it could be given up in any stage of the enterprise, without loss, with the fixtures.

Mr. Etzler considers this "Address as a touchstone, to try whether our nation is in any way accessible to these great truths, for raising the human creature to a superior state of existence, in accordance with the knowledge and the spirit of the most cultivated minds of the present time." He has prepared a constitution, short and concise, consisting of twenty-one articles, so that wherever an association may spring up, it may go into operation without delay; and the editor informs us that "Communications on the subject of this book may be addressed to C. F. Stollmeyer, No. 6, Upper Charles street, Northampton square, London."

But we see two main difficulties in the way. First, the successful application of the powers by machinery, (we have not yet seen the "Mechanical System,") and, secondly, which is infinitely harder, the application of man to the work by faith. This it is, we fear, which will prolong the ten years to ten thousand at least. It will take a power more than "80,000 times greater than all the men on earth could effect with their nerves," to persuade men to use that which is already offered them. Even a greater than this physical power must be brought to bear upon that moral power. Faith, indeed, is all the reform that is needed; it is itself a reform. Doubtless, we are as slow to conceive of Paradise as of Heaven, of a perfect natural as of a perfect spiritual world. We see how past ages have loitered and erred; "Is perhaps our generation free from irrationality and error? Have we perhaps reached now the summit of human wisdom, and need no more to look out for mental or physical improvement?" Undoubtedly, we are never so visionary as to be prepared for what the next hour may bring forth.

Μέλλει τὸ θεῖον δ'ἐστι τοιοῦτον φύσει.

The Divine is about to be, and such is its nature. In our wisest moments we are secreting a matter, which, like the lime of the shell fish, incrusts us quite over, and well for us, if, like it, we cast our shells from time to time, though they be pearl and of fairest tint. Let us consider under what disadvantages science has hitherto labored before we pronounce thus confidently on her progress.

"There was never any system in the productions of human labor; but they came into existence and fashion as chance directed men." "Only a few professional men of learning occupy themselves with teaching natural philosophy, chemistry, and the other branches of the sciences of nature, to a very limited extent, for very limited purposes, with very limited means." "The science of mechanics is but in a state of infancy. It is true, improvements are made upon improvements, instigated by patents of government; but they are made accidentally or at hap-hazard. There is no general system of this science, mathematical as it is, which developes its principles in their full extent, and the outlines of the application to which they lead. There is no idea of comparison between what is explored and what is yet to be explored in this science. The ancient Greeks placed mathematics at the head of their education. But we are glad to have filled our memory with notions, without troubling ourselves much with reasoning about them."

Mr. Etzler is not one of the enlightened practical men, the pioneers of the actual, who move with the slow deliberate tread of science, conserving the world; who execute the dreams of the last century, though they have no dreams of their own; yet he deals in the very raw but still solid material of all inventions. He has more of the practical than usually belongs to so bold a schemer, so resolute a dreamer. Yet his success is in theory, and not in practice, and he feeds our faith rather than contents our understanding. His book wants order, serenity, dignity, everything,—but it does not fail to impart what only man can impart to man of much importance, his own faith. It is true his dreams are not thrilling nor bright enough, and he leaves off to dream where he who dreams just before the dawn begins. His castles in the air fall to the ground, because they are not built lofty enough; they should be secured to heaven's roof. After all, the theories and speculations of men concern us more than their puny execution. It is with a certain coldness and languor that we loiter about the actual and so called practical. How little do the most wonderful inventions of modern times detain us. They insult nature. Every machine, or particular application, seems a slight outrage against universal laws. How many fine inventions are there which do not clutter the ground? We think that those only succeed which minister to our sensible and animal wants, which bake or brew, wash or warm, or the like. But are those of no account which are patented by fancy and imagination, and succeed so admirably in our dreams that they give the tone still to our waking thoughts? Already nature is serving all those uses which science slowly derives on a much higher and grander scale to him that will be served by her. When the sunshine falls on the path of the poet, he enjoys all those pure benefits and pleasures which the arts slowly and partially realize from age to age. The winds which fan his cheek waft him the sum of that profit and happiness which their lagging inventions supply.

The chief fault of this book is, that it aims to secure the greatest degree of gross comfort and pleasure merely. It paints a Mahometan's heaven, and stops short with singular abruptness when we think it is drawing near to the precincts of the Christian's,—and we trust we have not made here a distinction without a difference. Undoubtedly if we were to reform this outward life truly and thoroughly, we should find no duty of the inner omitted. It would be employment for our whole nature; and what we should do thereafter would be as vain a question as to ask the bird what it will do when its nest is built and its brood reared. But a moral reform must take place first, and then the necessity of the other will be superseded, and we shall sail and plough by its force alone. There is a speedier way than the Mechanical System can show to fill up marshes, to drown the roar of the waves, to tame hyænas, secure agreeable environs, diversify the land, and refresh it with "rivulets of sweet water," and that is by the power of rectitude and true behavior. It is only for a little while, only occasionally, methinks, that we want a garden. Surely a good man need not be at the labor to level a hill for the sake of a prospect, or raise fruits and flowers, and construct floating islands, for the sake of a paradise. He enjoys better prospects than lie behind any hill. Where an angel travels it will be paradise all the way, but where Satan travels it will be burning marl and cinders. What says Veeshnoo Sunma? "He whose mind is at ease is possessed of all riches. Is it not the same to one whose foot is enclosed in a shoe, as if the whole surface of the earth were covered with leather?"

He who is conversant with the supernal powers will not worship these inferior deities of the wind, the waves, tide, and sunshine. But we would not disparage the importance of such calculations as we have described. They are truths in physics, because they are true in ethics. The moral powers no one would presume to calculate. Suppose we could compare the moral with the physical, and say how many horse-power the force of love, for instance, blowing on every square foot of a man's soul, would equal. No doubt we are well aware of this force; figures would not increase our respect for it; the sunshine is equal to but one ray of its heat. The light of the sun is but the shadow of love. "The souls of men loving and fearing God," says Raleigh, "receive influence from that divine light itself, whereof the sun's elasity, and that of the stars, is by Plato called but a shadow. *Lumen est umbra Dei, Deus est Lumen Luminis.* Light is the shadow of God's brightness, who is "the light of light," and, we may add, the heat of heat. Love is the wind, the tide, the waves, the sunshine. Its power is incalculable; it is many horse power. It never ceases, it never slacks; it can move the globe without a resting-place; it can warm without fire; it can feed without meat; it can clothe without garments; it can shelter without roof; it can make a paradise within which will dispense with a paradise without. But though the wisest men in all ages have labored to publish this force, and every human heart is, sooner or later, more or less, made to feel it, yet how little is actually applied to social ends. True, it is the motive power of all successful social machinery; but, as in physics, we have made the elements do only a little drudgery for us, steam to take the place of a few horses, wind of a few oars, water of a few cranks and hand-mills; as the mechanical forces have not yet been generously and largely applied to make the physical world answer to the ideal, so the power of love has been but meanly and sparingly applied, as yet. It has patented only such machines as the almshouses, the hospital, and the Bible Society, while its infinite wind is still blowing, and blowing down these very structures, too, from time to time. Still less are we accumulating its power, and preparing to act with gearter energy at a future time. Shall we not contribute our shares to this enterprise, then? T.

Drawn by J. W. Barber—Engraved by E. L. Barber, New Haven, Conn.
EAST VIEW OF LOWELL, MASS.

Reproduced courtesy of Merrimack Valley Textile Museum.

John Greenleaf Whittier

The City of a Day (1845)

John Greenleaf Whittier (1807–1892) was the only major nineteenth-century American writer who ever lived in an industrial city, and he did only for six months while he edited a reform newspaper in Lowell, Massachusetts. As fruit of his experience he published a small book of essays, *A Stranger in Lowell* (1845; later edition, 1892), in which he anticipated by half a century the literary "discovery" of urban poverty and industrial landscape. The first two paragraphs of his title essay suggest the complexities faced by a conventional literary imagination grappling with the new facts and possibilities of industrial technology.

John Greenleaf Whitter, *A Stranger in Lowell* (Lowell, Massachusetts, 1845), pp. 9–12.

THE STRANGER IN LOWELL.

I.

THE CITY OF A DAY.

THIS, then, is Lowell — a city, springing up, like the enchanted palaces of the Arabian tales, as it were in a single night — stretching far and wide its chaos of brick masonry and painted shingles, filling the angle of the confluence of the Concord and the Merrimack with the sights and sounds of trade and industry! Marvellously here has Art wrought its modern miracles. I can scarcely realize the fact, that a few years ago these rivers, now tamed and subdued to the purposes of man, and charmed into slavish subjection to the Wizard of Mechanism, rolled unchecked towards the ocean the waters of the Winnipiseogee, and the rock-rimmed springs of the White Mountains, and rippled down their falls in the wild freedom of Nature. A stranger, in view of all this wonderful change, feels himself as it were thrust forward into a new century; he seems treading on the outer circle of the millenni-

The Philosophy of Manufactures

um of steam engines and cotton mills. WORK is here the Patron Saint. Every thing bears his image and superscription. Here is no place for that respectable class of citizens called gentlemen, and their much vilified brethren, familiarly known as loafers. Over the gateways of this New World Manchester, glares the inscription, " WORK, or DIE!" Here

> " Every worm beneath the moon
> Draws different threads, and late or soon,
> Spins, toiling out his own cocoon."

The founders of this city, good Christian men, probably never dreamed of the anti-Yankee sentiment of Charles Lamb:—

> "Who first invented Work; and thereby bound
> The holiday rejoicing spirit down
> To the never-ceasing importunity
> Of business in the green fields and the town?—
> Sabbathless Satan: he who his unglad
> Task ever plies midst rotatory burnings,
> For wrath divine has made him like a wheel
> In that Red Realm from whence are no returnings!"

Rather, of course, would they adopt Carlyle's definition of "Divine labor — noble, ever fruitful — the grand, sole Miracle of Man." For this is indeed a city consecrated to the Spirit of Thrift — dedicated, every square rod of it, to the Divinity of Work. The Gospel of Industry preached daily and hourly from some thirty temples; each huger than the Milan Cathedral or the temple of Jeddo, the Mosque of St. Sophia or the Chinese Pagoda of a hundred bells; its mighty sermons uttered by steam and water power; its music the everlasting jar of Mechanism, and the organ-swell of many waters; scattering the cotton and woollen leaves of its Evangel from the wings of steamboats and rail-cars throughout the land; its thousand priests, and its thousands of priestesses, ministering around their spinning-jenny and power-loom altars, or whitening the long unshaded streets in the level light of sunset! It is truly, as Carlyle says, a miracle, neither more nor less.

As has been truly said, there is a transcendentalism in mechanics as well as ethics. A few years ago, while travelling in Pennsylvania, I encountered a small, dusky-browed German of the name of Etzler. He was possessed with the belief, that the world was to be restored to its Paradisiacal state by the sole agency of mechanics; and that he had himself discovered the means of bringing about this very desirable consummation. His whole mental atmosphere was thronged with spectral enginery — wheel within wheel — plans of hugest mechanism — Brobdignagian steam engines — Niagaras of water-power — windmills, with "sail-broad vans," like those of Satan in chaos, — by whose proper application every valley was to be exalted, and every hill laid low — old forests seized by their shaggy tops and uprooted — old morasses drained — the tropics made cool — the eternal ices melted around the poles — the ocean itself covered with artificial islands — blossoming gardens of the Blessed, rocking gently on the bosom of the deep. Give him "three hundred thousand dollars, and ten years time," and he would undertake to do the work. Wrong, pain and sin, being in his view but the results of our physical necessities, ill-gratified desires, and natural yearnings for a better state, were to vanish before the Millennium of Mechanism. "It would be," said he, "as ridiculous then to dispute and quarrel about the means of life, as it would be now about water to drink by the side of mighty rivers, or about permission to breathe the common air." To his mind the great Forces of Nature took the shape of mighty and benignant spirits, sent hitherward to be the servants of man in restoring to him his lost Paradise; waiting but for his word of command to apply their giant energies to the task, but as yet struggling blindly and aimlessly, giving ever and anon gentle hints, in the way of earthquake, fire and flood, that they are weary of idleness, and would fain be set at work. Looking down, as I now do, upon these huge, brick work-shops, I have thought of poor Etzler, and wondered whether he would admit, were he with me, that his mechanical Forces have here found their proper employment of Millennium making. Grinding on, each in his iron harness, invisible, yet shaking, by his regulated and repressed power, his huge prison-house from basement to cap-stone, is it true that the Genii of Mechanism are really at work here, raising us, by wheel and pully, steam and water power, slowly up that inclined plane, from whose top stretches the broad table-land of Promise?

1. In the 1892 edition of Whittier's prose works (*The Prose Works of John Greenleaf Whittier*, vol. 1, pp. 352–353), the last sentence of this paragraph and the first sentence of the following paragraph (echoing a phrase from Thoreau's recent review of Etzler) were deleted. A passage was then added to the end of the first paragraph, significantly sharpening Whittier's critical view of the factory city:

After all, it may well be questioned whether this gospel according to Poor Richard's Almanac, is precisely calculated for the redemption of humanity. Labor, graduated to man's simple wants, necessities, and unperverted tastes, is doubtless well; but all beyond this is weariness of flesh and spirit. Every web which falls from these restless looms has a history more or less connected with sin and suffering beginning with slavery and ending with overwork and premature death.

William Gregg

Essays on Domestic Industry (1845)

William Gregg (1800–1867), a self-educated mechanic and self-made entrepreneur, became one of the most prominent manufacturers of cotton textiles in the antebellum South. He was likewise the leading Southern spokesman for industrial interests, both in the press and in political office. He erected at Graniteville, South Carolina, the prototype of what was to become the typical Southern cotton mill village. In some striking ways Gregg's argument recapitulates the earlier debate over the role of manufacturing in American economic independence from England. The agrarian South, he said, "sent abroad" its raw materials to the industrial North and paid a premium to "import" "foreign" finished goods from the North in return. He ominously suggested that North and South were hostile nations heading for war, and the South must encourage "domestic manufactures." His essays show him very conscious of following the models of Northern manufacturers. In several important ways, however, his argument (like his situation) differed from his predecessors'. Manufacturers were, for him, not a complement to thriving agriculture, but a compensation for flagging production from the spent soils of the older slave states.

William Gregg, *Essays on Domestic Industry: An Inquiry into the Expediency of Establishing Cotton Manufactures in South Carolina* (Charleston, South Carolina, 1845). The standard biography is Broadus Mitchell, *William Gregg, Factory Master of the Old South* (1928; reprint, New York: Octagon Books, 1966). For Gregg's place in the Southern debate on manufacturing, see Herbert Collins, "The Southern Industrial Gospel Before 1860," *Journal of Southern History*, volume 12 (1946), pp. 386–402. See also Chauncy S. Boucher, "The Ante-Bellum Attitude of South Carolina Towards Manufacturing and Agriculture," *Washington University Studies*, Series 4, volume 3, whole number 14 (St. Louis, 1916); and Theodore R. Marmor, "Anti-Industrialism and the Old South; The Agrarian Perspective of J. C. Calhoun," *Comparative Studies in Society and History*, volume 9 (1967), pp. 377–406.

Industry was also an attraction for local Southern capital that threatened to move westward in search of greater returns. Whereas Northern apologists for manufactures had to reconcile factory labor with the presumed interests of the independent citizen-farmer, Gregg saw that the small farmer—the "poor white"—had been made redundant on the land by slave labor and was ready to form a new industrial work force. Gregg sheds light on the debates over industrialization by forcing the standard arguments to fit the specific conditions of the slave South.

Gregg's series of essays was written directly after a trip through the industrial regions of New England, and several of them report in detail on the economics and organization of manufactures there. The essays were first published in the Charleston *Courrier* in the autumn of 1844, then reprinted in pamphlet form the following year. The publication of these essays immediately preceded the founding of Gregg's extremely successful Graniteville factory. Reprinted here are chapters III and IV.

CHAPTER III.

However unpopular the doctrine of encouraging domestic industry, in South-Carolina, may be, I feel satisfied that there are few individuals so ultra in their notions, with regard to our being exclusively agricultural, that will not feel charmed with the idea of *domestic industry;* it carries with it the idea of an improved condition of our country—of compensated industry, and comforts around us. It is to be lamented, that our great men are not to be found in the ranks of those, who are willing to lend their aid, in promoting this good cause. Are we to commence another ten years' crusade, to prepare the minds of the people of this State for revolution ;* thus unhinging every department of industry, and paralyzing the best efforts to promote the welfare of our country. Already do we hear of persons, high in the estimation of our State, largely engaged in cotton planting, and on the side of State resistance, expressing doubts as to the permanency and safety, of any investments, that can be made in South-Carolina. Lamentable, indeed, is it to see so wise and so pure a man as Langdon Cheves, putting forth the doctrine, to South-Carolina, that manufactures should be the last resort of a country. With the greatest possible respect for the opinions of this truly great man, and the humblest pretensions on my part, I will venture the assertion, that a greater error was never committed by a statesman. No good is without its evil, and I am free to confess, that when a people become so infatuated with the spirit of manufactures, as to undertake to force large establishments into unnatural existence, at the expense of other pursuits, they are committing an error by making an evil of that, which would otherwise be a great blessing. I admit, also, that agriculture is the natural and "blessed employment of man;" but, that a country should become eminently prosperous in agriculture, without a high state of perfection in the mechanic arts, is a thing next to impossible—to be dreamed of, not realized—a picture of the imagination, not to be found in reality on the face of the globe.

What does this gentleman mean by agriculture ? Does he intend that we shall follow the footsteps of our forefathers, and still further exhaust our soil by the exclusive culture of cotton? Does he not know that this system has already literally destroyed our State, and driven from it wealth and population—that many

*Those who are disposed to agitate the State and prepare the minds of the people for resisting the laws of Congress, and particularly those who look for so direful a calamity as the dissolution of our Union, should, above all others, be most anxious so to diversify the industrial pursuits of South-Carolina, as to render her independent of all other countries; for as sure as this greatest of calamities befalls us, we shall find the same causes that produced it, making enemies of the nations which are at present, the best customers for our agricultural productions.

of its wealthiest and most enterprising citizens have left it, in search of new and more productive lands? Does he not know that money is not wealth to a nation, unless it is spent within its borders, in the improvement, mental and physical, of the condition of its inhabitants,—in the renovation of its soil,—in the construction of roads and bridges,—in the erection of fine houses, and in planting orchards, and making barns for the protection of produce and live stock. This is indeed a kind of wealth that will never be realized in South-Carolina, without domestic manufactures. And, lest I be misunderstood as to what I mean by domestic manufactures, I will here state, that I mean the erection of steam mills in Charleston, for every purpose that our mechanics may desire, to enable them to compete with foreigners in the manufacture of thousands of articles, now imported into the State—the erection of Steam Cotton manufactories to employ the poor and needy of this city, and the hundreds who seem to have little else to do, than follow our military parades through the streets,—the erection of Cotton manufactories throughout the State, to employ our poor and half starved population, whose condition could not but be improved, in working up a part of our cotton into cloth, to cover their nakedness, and to clothe our negroes and ourselves, at a cost, for the manufacture of the coarse fabrics (osnaburgs) of $2\frac{1}{4}$ cents per lb. and for the finer, such as brown and bleached shirtings, drillings, and cotton flannels, of from 3 to 8 cents per pound, instead of sending the same abroad to be returned to us, charged with 12 cts. per pound for osnaburgs, and from 20 to 65 cents, for the other articles named. I mean that, at every village and cross road in the State, we should have a tannery, a shoe-maker, a clothier, a hatter, a blacksmith, (that can make and mend our ploughshares and trace chains,) a wagon maker, and a carriage maker, with their shops stored with seasoned lumber, the best of which may be obtained in our forests. This is the kind of manufactures I speak of, as being necessary to bring forth the energies of a country, and give healthful and vigorous action to agriculture, commerce and every department of industry, and, without which, I ventured the assertion that this State can never prosper. This is the state of things that every true friend of South-Carolina ought to endeavor to bring about. If he wishes to see her worn out and desolate old fields turned into green pastures, her villages brightened up with the hand of industry, her dilapidated farm houses taken down, to be replaced by opulent mansions, her muddy and almost impassable roads graded and M'Adamized, let him use his endeavors to make the people of South-Carolina think less of their grievances and more of the peaceable means of redress—let our politicians, instead of teaching us to hate our Northern brethren, endeavor to get up a good feeling for domestic industry—let them teach our people that the true mode of resistance will be found in making more and purchasing less;—let them endeavor to satisfy our capitalists that we are not on the verge of revolution, but that there is safety in investments in South-Carolina, and no necessity of seeking, for such purposes, the stocks of others, or readily convertible ones of our own. There is no lack of capital in South-Carolina; Charleson, herself, possesses all the requisites, and it is only necessary that public attention should be properly directed to this vast field, for profitable investments, in this State, and to give assurances of political stability and safety, to bring it out, and to stop the millions which are being all the time transferred from the South to the North, and with it would be retained amongst us, the enterprising merchant, who, on his retirement from the toils of business, would forget the green fields and pleasant ways of his native land, to mingle with us in domestic industry.

Let the manufacture of cotton be commenced among us, and we shall soon see the capital that has been sent out of our State, to be invested in Georgia State, and other foreign stocks, returned to us. We shall see the hidden treasures that have been locked up, unproductive and rusting, coming forth to put machinery in motion, and to give profitable employment to the present unproductive labor of our country. To give an idea of the various sources from which capital is drawn, for such purposes, I will state how the Merrimack Company, at Lowell, is made up. It is composed of 390 Stock-holders, of whom there are, 46 merchants and traders; 68 females; 52 individuals retired from business; 80 administrators, executors, guardians and trustees; 23 lawyers; 18 physicians; 3 literary institutions; 15 farmers; 40 secretaries, clerks and students; 45 mechanics, and persons employed ed in the service of the company, who hold stock to the amount of $60,000.

Cotton manufactures have been the pioneers which have introduced and given an impetus to all other branches of mechanism in Great Britain, the continent of Europe, and this country. Taking this for granted, one would suppose, that the persons who established the extensive Iron Establishment, now in operation in the mountainous parts of our State, although, actuated by an enterprising spirit, counted without their host—it was really putting the cart before the horse. I trust, however, that, a change in our industrial pursuits is soon to take place, which will give a new aspect to things in that quarter, that those establishments are yet to thrive, proving to be inexhaustible sources of wealth to our State, and monuments to the enterprise of their projectors. If South-Carolina commence the manufacture of cotton in earnest, these works will be brought into requisition, and the iron produced by them, will no longer be sent to the Eastern States, to be turned into plough-shares for us. The endless sources of demand which will spring up for it, will cause a home consumption for it all.

The cheapness of water power, if not the chief, will at least constitute one important element of success with us. There is, probably, no State in the Union, in which water power is more abundant. Leaving out of the question, as being too tedious to enumerate, the great number of water falls on the tributary streams of the Peedee, Wateree, Broad and Saluda rivers, we will notice those only, in the immediate vicinity of our two lines of rail-road to Columbia and Hamburg, that is, within five miles of them. In the most healthy regions of the State, abounding with granite and building timber, water power may be found, sufficient to work up half the crop of South-Carolina, all of which is nearly valueless at the present time. For the information of such as are not acquainted with the manner of computing the force of falling water, I will state, that the quantity of water used by the generality of saw mills, running but one saw, with a head of 10 feet, will be sufficient to produce, if raised to a head of 15 feet, 50 horse power. From this statement, persons may easily calculate what such water-falls would be worth, if located at Lowell, or near Philadelphia. In Lowell water power is sold at $4 per spindle, which is equal to $262 for each horse power. At Manyunk, 5 miles from Philadelphia, it is sold for $100 for every square inch of under a 3 feet head, and over a 20 feet fall; this is equal to $1,016 for each horse power. It is not so valuable at places unfavorably located; but the price at which it sells in those above mentioned, accounts at once for the eagerness, with which such property is sought after, in situations remote from navigation, and even in mountainous countries.

God speed on the glorious result, that may be anticipated from so great a change, in our industrial pursuits. Were all our hopes, in this particular, consummated, South Carolina would present a delightful picture. Every son and daughter would find healthful and lucrative enployment; our roads, which are now a disgrace to us, would be improved; we would no longer be under the necessity of sending to the North for half made wagons and carriages, to break our necks; we would have, if not as handsome, at least as honestly and faithfully made ones, and mechanics always at hand to repair them. Workshops would take the place of the throngs of clothing, hat, and shoe stores, and the watch-word would be, from the seaboard to the mountains, success to domestic industry.

CHAPTER IV.

We want no laws for the protection of those that embark in the manufacture, of such cotton fabrics, as we propose to make in South-Carolina; nor does it follow, as a matter of course, that because we advocate a system which will diversify the pursuits of our people, and enable them to export a portion of one of our valuable staples, in a manufactured state, that we wish manufactures to predominate over other employments. All must admit that, to a certain extent, the system we advocate could not operate otherwise than to produce beneficial results, by regulating prices—by insuring a certain reward to labor—a profitable income to capital, and by infusing health, vigor and durability into every department of industry. It is a well established fact, that capital employed in this State, in the culture of cotton, does not, with ordinary management, yield more than 3 or 4, and in some instances, 2 per cent.; this being the only mode of employing our capital, except in the culture of rice, how can we expect to retain men of *capital* and *enterprise* among us? Those having the first, must be wholly wanting in the last—or they must possess an extraordinary attachment to the land of their nativity, to remain with us under such a state of affairs.

With this fact before us, is it surprising that South-Carolina should remain stationary in population? And let it be remembered that the same cause which has produced this result, will continue to operate hurtfully, in the same ratio, as the price of our great staple declines. In all probability, an additional outlet will soon be opened to drain us of our people and our capital. How much this is to take from us, remains to be seen. Unless we betake ourselves to some more profitable employment than the planting of cotton, what is to prevent our most enterprising planters from moving, with their negro capital, to the South-West? What is to keep our business men and moneyed capital in South-Carolina? Capital will find its way to places that afford the greatest remuneration, and in leaving our State, it will carry with it, its enterprising owner. These are truly unpleasant reflections, but they force themselves upon us. Who can look forward to the future destiny of our State, persisting, as she does, with such pertinacity, in the exclusive and exhausting system of agriculture, without dark forebodings. If we listen much longer to the *ultras* in agriculture and *croakers* against mechanical enterprise, it is feared that they will be the only class left, to stir up the indolent sleepers that are indisposed to action, and that are willing to let each day provide for itself.

Since the discovery that cotton would mature in South-Carolina, she has reaped a golden harvest; but it is feared it has proved a curse rather than a blessing, and I believe that she would at this day be in a far better condition, had the discovery never been made. Cotton has been to South-Carolina what the Mines of Mexico were to Spain, it has produced us such an abundant supply of all the luxuries and elegancies of life, with so little exertion on our part, that we have become enervated, unfitted for other and more laborious pursuits, and unprepared to meet the state of things, which sooner or later must come about. Is it out of place here to predict, that the day is not far distant, yea, is close at hand, when we shall find that we can no longer *live* by that, which has heretofore yielded us, not only a bountiful and sumptuous living, at home, but has furnished the means for carrying thousands and tens of thousands of our citizens abroad, to squander their gold in other countries—that we have wasted the fruits of a rich, virgin soil, in ease and luxury—that those who have practised sufficient industry and economy to accumulate capital, have left, or are leaving us, to populate other States.

We shall indeed soon be awakened to look about us for other pursuits, and we shall find that our soil has to be renovated—our houses and workshops have to be built—our roads and bridges have to be made, all of which ought to have been done with the rich treasures, that have been transferred to other States. Let us begin at once, before it is too late, to bring about a change in our industrial pursuits—let us set about it before the capital and enterprise of our State has entirely left us—let croakers against enterprise be silenced—let the working men of our State who have, by their industry, accumulated capital, turn out and give a practical lesson to our political leaders, that are opposed to this scheme. Even Mr. Calhoun, our great oracle—a statesman whose purity of character we all revere—whose elevation to the highest office in the gift of the people of the United States, would enlist the undivided vote of South-Carolina—even he is against us in this matter; he will tell you, that no mechanical enterprise will succeed in South-Carolina—that good mechanics will go where their talents are better rewarded—that to thrive in cotton spinning, one should go to Rhode Island—that to undertake it here, will not only lead to loss of capital, but disappointment and ruin to those who engage in it.

If we look at this subject in the abstract only, we shall very naturally come to the above conclusions; it is, however, often the case, that practical results contradict the plainest abstract propositions, and it is hoped, that in the course of these remarks, it will be proved to the satisfaction of at least, some of our men of capital and enterprise, that the spinning of cotton may be undertaken with a certainty of success, in the two Carolinas and Georgia, and that the failures which have taken place, ought not to deter others from embarking in the business, they being the result of unpardonable ignorance, and just such management on the part of those interested, as would prove ruinous in any other undertaking.

There are those who understand some things, as well as, if not better, than other people, who have taken the pains to give this subject a thorough investigation, and who could probably give, even Mr. Calhoun, a practical lesson concerning it. The known zeal with which this distinguished gentleman has always engaged in every thing relating to the interest of South-Carolina, forbids the idea that he is not a friend to domestic manufactures, fairly brought about; and, knowing, as he must know, the influence which he exerts, he should be more guarded in expressing opinions adverse to so good a cause.

Those who project new enterprises, have in all ages and countries had much to contend with, and if it were not that we have such immense advantages, in the cheapness of labor and of the raw material, we might despair of success in the manufacture of cotton in South-Carolina. But we must recollect that those who first embarked in this business in Rhode Island, had the prejudice, of the whole country against them. There were croakers then as well as now, and in addition to all the disadvantages we have to contend with, the wide ocean lay between them and the nations skilled in mechanic arts—the laws of England forbade the export of machinery, and affixed heavy penalties to prevent the emigration of artisans, and it was next to impossible to gain access to her manufacturing establishments; so that these men were completely shut out from knowledge. How is with us? We find no difficulty in obtaining the information, which money could not purchase for them, and which cost them years of toil. The New England people are anxious for us to go to spinning cotton, and they are ready and willing to give us all the requisite information. The workshops of England and America are thrown open to us, and he who has the capital at command may, by a visit to England, or to our Northern machine shops, supply himself with the best machinery that the world affords, and also the best machinists, and most skilful manufacturers to work and keep it in order. With all these advantages, what is to prevent the success of a cotton factory in South-Carolina? It may safely be asserted, that failure will be the result of nothing but the grossest mismanagement.

It will be remembered, that the wise men of the day predicted the failure of *steam navigation*, and also of our own rail road; it was said we were deficient in mechanical skill, and that we could not manage the complicated machinery of a steam engine, yet these works have succeeded—we have found men competent to manage them—they grow up amongst us, and we are not only able to keep such machines in order, but to build and fit them to steamboats, mills, locomotive carriages, &c. and the shops engaged in this sort of manufactures, do away with much of the reproach that attaches to our city—they remove many of the obstacles in erecting cotton factories, for they can furnish steam engines, water wheels, shafting, and all the running gear to put machinery in operation.

Daniel Webster

Opening of the Northern Railroad (1847)

These two brief addresses take on special significance in light of Webster's long career in political service of manufacturing interests and the special circumstances of the previous decade in New Hampshire politics. Webster was celebrating not only the triumph of technology over space and time in the interests of the human community but also the triumph of corporate capitalism over the opposition of agrarian radical Democrats with the election of a Whig governor in New Hampshire the previous year: "I rejoice most heartily that my native State has adopted a policy which has lead to these results."

Daniel Webster, *The Writings and Speeches of Daniel Webster* (Boston, 1903), volume 4, pp. 107–111, 112–117.

Opening of the Northern Railroad to Grafton, N. H.

At the opening of the Northern Railroad from Franklin to Grafton in New Hampshire, on the 28th of August, 1847, a large number of persons from all the adjacent towns were assembled at Grafton to witness the ceremonies of the occasion. Mr. Webster happened to be then at his farm in Salisbury, in the immediate neighborhood; and this fact being known to the company, he was spontaneously called upon, in the most enthusiastic manner, to address them. Mr. Webster readily complied with the unexpected summons, and made the following remarks.

I am very happy, fellow-citizens, to be here on this occasion, to meet here the Directors of the Northern Railroad, the directors of various other railroads connected with it below, and such a number of my fellow-citizens, inhabitants of this part of the State. Perhaps my pleasure and my surprise at the success of this great enterprise so far are the greater, in consequence of my early acquaintance with this region and all its localities.

But, Gentlemen, I see the rain is beginning to descend fast, and I pray you to take shelter under some of these roofs. (Cries of "Go on! go on! Never mind us!")

In my youth and early manhood I have traversed these mountains along all the roads or passes which lead through or over them. We are on Smith's River, which, while in college, I had occasion to swim. Even that could not always be done; and I have occasionally made a circuit of many rough and tedious miles to get over it. At that day, steam, as a motive power, acting on water and land, was thought of by nobody; nor were there good, practicable roads in this part of the State. At that day, one must have traversed this wilderness on horseback or on foot. So late as when I left college, there was no road from river